Gas Kinetics and Energy Transfer

Volume 4

A Specialist Periodical Report

Gas Kinetics and Energy Transfer

Volume 4

A Review of the Literature published
up to early 1980

Senior Reporters

P. G. Ashmore, *Department of Chemistry, University of Manchester
Institute of Science and Technology*
R. J. Donovan, *Department of Chemistry, University of Edinburgh*

Reporters

M. N. Ashfold, *University of Oxford*
D. L. Baulch, *University of Leeds*
I. M. Campbell, *University of Leeds*
R. A. Cox, *A.E.R.E., Harwell*
R. G. Derwent, *A.E.R.E., Harwell*
R. Grice, *University of Manchester*
G. Hancock, *University of Oxford*
A. J. McCaffery, *University of Sussex*

The Royal Society of Chemistry
Burlington House, London W1V 0BN

British Library Cataloguing in Publication Data

Gas kinetics and energy transfer. – Vol. 4. –
(Specialist periodical reports/Royal Society
of Chemistry)
 1. Chemical reaction, Rate of
 2. Gases, kinetic theory of
 3. Energy transfer
 I. Series
 541.3′9   QD501

 ISBN 0-85186-786-3
 ISSN 0309-6890

Printed in Great Britain by Spottiswoode Ballantyne Ltd.
Colchester and London

Foreword

The reports in this volume fall into three natural pairs. The first pair deals with progress in understanding the detailed dynamics of chemical reactions and energy transfer. Grice reviews research on reactive scattering in molecular beams and first provides a fascinating examination of progress in experimental techniques which have greatly extended the range of species studied. He emphasises the advantages of lasers to select specific excited states of reactant species, and the use of laser fluorescence spectroscopy for studying the vibration–rotation state distribution of product molecules. The main sections of the article contain critical surveys of the extensive studies of metal, H, halogen, and O atom reactions during the period early 1977 to late 1979. There follow reviews of work on theoretical models that have examined angular correlations, reaction dynamics, and trajectory calculations, stressing the derivation of the potential energy surfaces as the major objective.

The importance of the potential surface is again emphasised by McCaffery who reviews the reorientation of molecules, or changes in the magnetic quantum number m, during elastic and rotationally inelastic collisions, and the problems encountered in relating the theoretical treatments to the available experimental data from molecular beam, optical double resonance, and laser fluorescence studies.

The next pair of articles deals with multiphoton excitation. Ashfold and Hancock explain current models of infrared multiple photon absorption (IRMPA) and dissociation (IRMPD) processes and how information upon IRMPA can be obtained from the more common direct IRMPD investigations. They give comprehensive surveys of research up to late 1979 on the way energy is distributed in IRMP excited species and the use of thermal unimolecular theory to interpret the observations on the effect of collisions and the distribution of energy in the product fragments. The final sections review applications of i.r. lasers to provide new or more selective reaction pathways, particularly for isotopic enrichment, and to provide sources of free radicals for kinetic studies.

Donovan reviews research using ultraviolet multiphoto excitation to provide highly excited states, which may have the same parity as the ground state or involve two excited electrons. The technique provides suitable concentrations of many excited species previously inaccessible without recourse to vacuum u.v. equipment. After brief surveys of the experimental systems, some relevant aspects of multiphoton excitation theory, and the excitation and reactions of atoms and diatomic molecules, some detailed discussions ensue of recent investigations upon excited photofragments and ions derived from multiphoton absorption by polyatomics.

The final pair of articles reports on aspects of the immense activity currently devoted to atmospheric chemistry. Cox and Derwent survey recent attempts to integrate studies of those chemical and physical properties of the troposphere which involve the transport and the thermal and photo-reactions of trace gases emanating from the earth's surface. The modelling of these sources, transport, and sink processes relies on photochemical and other kinetic data on organic and nitrogen-, chlorine- and sulphur-containing compounds, and the article emphasises the evolution of techniques for assessing and using the data.

In a complementary article Baulch and Campbell review the surge of information on OH reactions, which are of major importance in atmospheric chemistry, over the period 1972 to 1979. They analyse critically the kinetic data on reactions of OH with inorganic and with organic species, after surveying advances in the preparation of OH radicals and in techniques for accurate measurement of OH concentrations. They emphasise the value of resonance fluorescence measurements and of advances in the collection and processing of transient signals from flash photolysis experiments. An interesting discussion is given of the curvature of Arrhenius plots for some reactions of OH studied over a wide range of temperatures.

This volume is published later than we planned, but our main objective in these reports has been to emphasise progress in the stronger and more lasting threads of research on gas kinetics and energy transfer, and we are confident that the critical surveys by our reporters fulfil that aim.

December 1980   P.G.A.
                                                                          R.J.D.

Contents

Chapter 1 Reactions Studied by Molecular Beam
Techniques 1
By R. Grice

1 Introduction 1
2 Experimental Techniques 2
3 Experimental Studies of Reactive Scattering 11
Alkali-metal Atom Reactions 11
Alkaline-earth Metal Atom Reactions 13
Other Metal Atom Reactions 15
Hydrogen Atom Reactions 19
Halogen Atom Reactions 24
Oxygen Atom Reactions 27
Other Non-metal Atom Reactions 31
Free-radical Reactions 33
Molecule–Molecule Reactions 34
4 Theoretical Interpretation 35
Angular Correlations 35
Dynamical Models 40
Trajectory Calculations and Potential-energy Surfaces 43
5 Conclusions 46

Chapter 2 Reorientation by Elastic and Rotationally
Inelastic Transitions 47
By A. J. McCaffery

1 Introduction 47
2 Molecular Beam Experiments 48
3 Optical Methods 52
4 Scattering Theory and Approximation Methods 67

Chapter 3 Infrared Multiple Photon Excitation and Dis-
sociation: Reaction Kinetics and Radical
Formation 73
By M. N. R. Ashfold and G. Hancock

1 Introduction 73

2 The Model for Infrared Multiple Photon Absorption (IRMPA)
   and Dissociation (IRMPD)   74
3 Investigations of the Nature of the Process   77
   Studies in Absorption   77
   Studies of Dissociation Relevant to Excitation Regions I
   and II   80
   Studies in Dissociation Relevant to Region III   86
4 Laser-induced Chemistry: Effect of Collisions   97
5 Chemical Applications of Infrared Lasers   107

Chapter 4 Ultraviolet Multiphoton Excitation: Formation
   and Kinetic Studies of Electronically Excited
   Atoms and Free Radicals   117
   By R. J. Donovan

1 Introduction   117
2 Techniques and Laser Systems   118
3 Fundamental Aspects of Multiphoton Excitation   120
   Resonant Two-photon Excitation   120
   Non-resonant Two-photon Excitation   121
4 Direct Multiphoton Excitation of Atoms and Small Molecules   122
   Atoms   122
   Molecules   123
      Hydrogen   123
      Nitric Oxide   124
5 Photofragmentation Studies   124
   Carbonyl Sulphide   124
   Mercury Halides   125
   Ammonia and Phosphine   126
   Water and Alcohols   126
   Cyanides   127
   Halogenomethanes   128
   Metal Alkyls and Carbonyls   131
   Acetylene and Ethylene   131
6 Multiphoton Ionization   133
7 Summary   136

Chapter 5 Gas Phase Reactions of Hydroxyl Radicals   137
   By D. L. Baulch and I. M. Campbell

1 Introduction   137
2 Experimental Techniques   138
   Rate Measurements   138
   Mechanistic Studies   140
   Generation of OH Radicals   140

Photolytic Methods 141
Electrical Discharge Sources 143
Thermal Sources 145
Detection Methods 146
Mass Spectrometry 147
Magnetic Resonance Methods 148
Light Absorption 149
Resonance Fluorescence 151
Fourier Transform Infrared Spectroscopy 151

3 **Reactions with Inorganic Species** 152
Bimolecular Reactions with Atoms 152
Bimolecular Reactions with Diatomic Molecules 153
Bimolecular Reactions with Inorganic Polyatomic Molecules 159
Bimolecular Reactions with Radicals 164
Three-body Combination Reactions 167

4 **Reactions with Organic Species** 170
Alkanes 170
Halogenoalkanes 174
Alkenes and Halogenoalkenes 175
Alkynes 180
Aromatic Compounds 181
Oxygen-, Nitrogen-, and Sulphur-containing Compounds 184
Organic Free Radicals 188

Chapter 6 Gas-phase Chemistry of the Minor Constituents
of the Troposphere 189
By R. A. Cox and R. G. Derwent

1 **Introduction** 189

2 **Physical and Chemical Properties of the Troposphere** 190
General Features of the Troposphere 190
Transport of Minor Trace Gases in the Troposphere 192
Gas-phase Chemistry of the Troposphere 195
Chain Carriers and their Initiation 196
Free-radical Interconversion Reactions 197
Free-radical Termination Reactions 198
The OH Steady State 198
Sources and Sinks for the Minor Constituents of the Troposphere 199
Sources of the Minor Constituents 199
Sinks of the Minor Constituents 203
Modelling the Behaviour of Minor Constituents in the Troposphere 205

3 **Photochemical and Kinetic Data**   206
   Thermal Reactions   206
   Photochemical Reactions   208
   Ozone Photochemistry and O Atom Reactions   208
   Reactions of HO_x and O_x Species   210
   Reaction Involving H_2O_2 211
   Reactions Involving NO_x 212
   CH_4 Oxidation   214
   Miscellaneous Reactions   218
     Reactions of Sulphur Compounds   218
     Reactions of Halogenated Hydrocarbons   220
     Reactions of Some Non-methane Hydrocarbon Species   220
     Reactions of Inorganic Hydrides   222

4 **The Importance of Photo-oxidation in the Life Cycles of Minor Tropospheric Constituents**   222
   Evaluation of the Tropospheric OH Distribution   222
   Life Cycles of Methane, Carbon Monoxide, and Hydrogen   224
   Atmospheric Chemistry of Sulphur Compounds   227
   Oxidation of Hydrocarbons   229
   Oxidation of Halocarbons   232

Author Index   235

1

Reactions Studied by Molecular Beam Techniques

BY R. GRICE

1 Introduction

The objective of reactive scattering experiments in molecular beams is to gain a full understanding of the dynamics of chemical reactions. Early studies[1] involving alkali-metal atom reactions required only relatively straightforward techniques by means of which many important results have been obtained. Effective models of the reaction dynamics have also been developed which rely on the simplicity of the chemical interactions in alkali-metal systems. These are frequently dominated by ionic potential-energy surfaces, which intersect the covalent surfaces and give rise to electron-jump[1] transitions. However, the most important reactions of gas-phase chemical kinetics are those of atoms and free radicals, which determine the chemistry of combustion, pyrolysis, and upper atmosphere processes. The experimental techniques required for the study of such non-alkali-metal reactions are considerably more complex than those required for alkali-metal reactions. Moreover, the chemical interactions involved are frequently much more complicated than those which govern alkali-metal reactions and hence more comprehensive experimental information is required for the development of adequate models of the reaction dynamics. Experimental techniques for the study of non-alkali-metal reactive scattering have been advanced and refined extensively over the past decade, so that it is now possible to make detailed measurements of differential reaction cross-sections for an increasing range of atom and free-radical reactions. The rapid progress of laser technology is also having a major impact on reactive scattering experiments. Reactant molecules may be promoted to excited vibrational, rotational, or electronic states and their orientations thereby selected. Similarly, the vibrational and rotational state distributions of product molecules may be determined by laser fluorescence spectroscopy. In this Report we review progress in reactive scattering experiments and examine the extent to which the measurements so far accumulated demonstrate the dependence of reaction dynamics on the electronic structure of the reaction potential-energy surface.

Progress in this field was reviewed during 1976 in *Faraday Discuss. Chem. Soc.*, 1976, **62**, on 'Potential Energy Surfaces' and the meeting on 'Energy Transfer Processes', a Report of which appeared in *Ber. Bunsenges. Phys. Chem.*, 1976, **81**.

---

[1] D. R. Herschbach, *Adv. Chem. Phys.*, 1966, **10**, 319.

1

Accordingly, this Report concentrates on work that has appeared since these meetings, from the beginning of 1977 to late 1979. Chemiluminescence in the gas phase, including molecular beam experiments, has recently been reviewed[2] in this series. Thus, discussion here of chemiluminescence measurements in molecular beams will be confined to showing their relationship to other molecular beam studies; the reader is referred to the previous review[2] for detailed discussion of spectral assignments, lifetimes, and bond energies.

2 Experimental Techniques

Early crossed-beam studies[3-5] of non-alkali-metal reactive scattering employed mass-spectrometric detection with an effusive atom or free-radical source and conventional time-of-flight velocity analysis. However, an effusive source permits only crude control of the reactant translational energy, unless a mechanical velocity selector is employed[4] at a considerable cost in beam intensity. Moreover, the inefficiency of the conventional time-of-flight method of velocity analysis and the low intensity of the effusive beam source limits the determination of differential reaction cross-sections in this form of experiment to favourable reactions with fairly large total-reaction cross-sections $Q \gtrsim 1$ Å^2. In order to overcome these limitations and to extend measurements of differential reaction cross-sections to a wider range of reactions, supersonic beams of atoms and free radicals seeded in inert buffer gas are now being used in place of effusive sources and cross-correlation time-of-flight analysis in place of the conventional method.

Supersonic nozzle beam sources[6] have been used for a considerable time to produce intense beams of stable molecules with narrow velocity distributions. In a seeded supersonic expansion[7] of a dilute mixture of a heavy gas in a light buffer gas, the heavy molecules are accelerated to the same velocity as that of the light buffer gas. Consequently, the translational energy of the heavy molecule in such a seeded beam is increased in the ratio of the molecular weight of the heavy molecule to the mean molecular weight of the gas mixture. Thus the translational energy may be controlled by varying the molecular weight of the buffer gas. Reactive scattering apparatus employing seeded supersonic nozzle beams requires powerful source differential pumping (Figure 1) by an apparatus that has pumping speeds of 4600, 1500, and 5600 l s^{-1} on the source, buffer, and scattering chambers. Supersonic beams of halogen atoms and hydrogen atoms may be produced by thermal dissociation of the diatomic molecules in a high-temperature oven. Beams of fluorine atoms seeded in helium and argon buffer gases have been produced[8] from a

[2] I. M. Campbell and D. L. Blauch, in 'Gas Kinetics and Energy Transfer,' ed. P. G. Ashmore and R. J. Donovan (Specialist Periodical Reports), The Chemical Society, London, 1978, Vol. 3, p. 42; for recent reviews of dynamics of reactive collisions see M. R. Levy, *Prog. React. Kinet.*, 1979, **10**, 1; D. R. Herschbach, *Pure Appl. Chem.*, 1976, **47**, 61; R. J. Buss and Y. T. Lee, *J. Phys. Chem.*, 1979, **83**, 34.
[3] J. D. McDonald, P. R. Le Breton, Y. T. Lee, and D. R. Herschbach, *J. Chem. Phys.*, 1972, **56**, 769.
[4] Y. T. Lee, Phys. Electron. At. Collisions, Invited Pap. Progr. Rep. Int. Conf. 7th (Amsterdam), 1971, publ., 1972, p. 357.
[5] D. St. A. G. Radlein, J. C. Whitehead, and R. Grice, *Mol. Phys.*, 1975, **29**, 1813.
[6] J. B. Anderson, R. P. Andres, and J. B. Fenn, *Adv. Chem. Phys.*, 1966, **10**, 275.
[7] N. Abauf, J. B. Anderson, R. P. Andres, J. B. Fenn, and D. G. H. Marsden, *Science*, 1967, **155**, 997.
[8] J. M. Farrar and Y. T. Lee, *J. Chem. Phys.*, 1975, **63**, 3639.

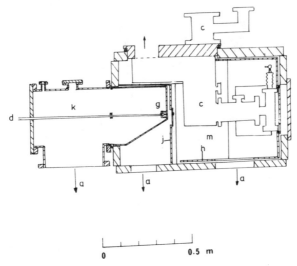

Figure 1 *Diagram of molecular beam reactive scattering apparatus: a, oil diffusion pumps; c, mass spectrometer detector; d, gas inlet; g, supersonic nozzle source; h, liquid nitrogen cooled cold shield; j, differential pumping bulkhead; k, source chamber; m, scattering chamber*

nickel oven at *ca*. 1100 K. Similarly, beams of chlorine and bromine atoms seeded in helium and argon buffer gases have been produced[9] by use of a graphite oven (Figure 2) at a higher temperature, *ca*. 2000 K. Both the nickel and the graphite ovens are heated by a direct curent $\gtrsim$450 A flowing through the oven body to obtain $\lesssim$80% dissociation of the halogen molecules at a pressure of *ca*. 10 mbar, with the inert buffer gas making up a total pressure $\gtrsim$1000 mbar. These oven materials are not corroded significantly by the halogens under these conditions. A supersonic hydrogen-atom beam seeded in undissociated hydrogen molecules has been produced[10] from a tungsten oven at *ca*. 2800 K. Owing to the high pressure of *ca*. 1500 mbar of hydrogen in the source and the high bond-strength of the hydrogen molecule only a small degree of dissociation, *ca*. 5%, can be achieved at temperatures below the point at which tungsten softens unduly. Rather higher degrees of dissociation were obtained by using helium and neon buffer gases to produce a hypothermal supersonic hydrogen-atom beam. Beams of alkali-metal atoms seeded in helium, argon, and hydrogen buffer gases may more readily be produced[11] by maintaining an appropriate alkali-metal vapour pressure in a stainless-steel oven. However, the thermal dissociation method is limited by the necessity of finding oven materials that can reach the temperatures required to attain significant dissociation and resist corrosion by the reactive species thus produced. An alternative method of dissociation involves the use of a high-pressure

[9] J. J. Valentini, M. J. Coggiola, and Y. T. Lee, *Rev. Sci. Instrum.*, 1977, **48**, 58.
[10] J. W. Hepburn, D. Klimek, K. Liu, J. C. Polanyi, and S. C. Wallace, *J. Chem. Phys.*, 1978, **69**, 4311.
[11] R. A. Larsen, S. K. Neoh, and D. R. Herschbach, *Rev. Sci. Instrum.*, 1974, **45**, 1511; A. Lübbert, G. Rotzoll, R. Viard, and K. Schügerl, *ibid.*, 1975, **46**, 1656.

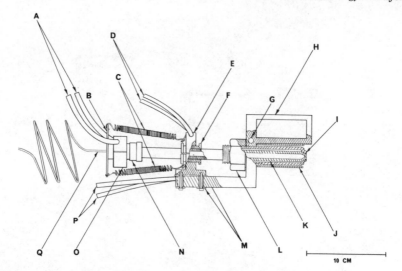

Figure 2 *Supersonic graphite source:* A, D, P, *water-cooling tubes;* B, C, E, N, O, *sprung support assembly;* F, L, *graphite sleeve and nut;* H, *copper mounting block;* I, K, *outer and inner graphite tubes;* M, *mica insulation;* Q, *gas inlet*
(Reproduced by permission from *Rev. Sci. Instrum.*, 1977, **48**, 58)

discharge source. The production of a supersonic oxygen-atom beam seeded in helium buffer gas was first achieved by Miller and Patch[12] using radio-frequency excitation. More recently, a microwave discharge source has been used[13] to produce supersonic beams of oxygen, hydrogen, and chlorine atoms seeded in helium and neon buffer gases. The source (Figure 3) operates in a vacuum to create a discharge in the quartz tube, which is foreshortened compared with that reported in ref. 13. Gas issuing from the nozzle at the end of the quartz tube is sampled by the skimmer to yield a supersonic beam with $\gtrsim 70\%$ dissociation of oxygen, hydrogen, or chlorine molecules. Microwave excitation is generally preferable to r.f. excitation owing to its stronger coupling to the discharge plasma. However, the radio-frequency method has been extended by Lee and co-workers[14] to produce supersonic beams of oxygen atoms seeded in argon and helium. In this source the plasma extends through the nozzle and the beam contains ions and metastable electronically excited $O(^1D)$ in addition to ground-state $O(^3P)$ oxygen atoms when using helium buffer gas. The possible production of metastable electronically-excited species must always be considered when using a discharge source rather than thermal dissociation. Even under the more controlled conditions of microwave excitation, where the plasma does not extend through the nozzle, the metastable electronically-excited $O_2(^1\Delta_g)$ molecule is expected[13] to be present in the beam. Fortunately, $O_2(^1\Delta_g)$ molecules are generally much less reactive than ground-state

[12] D. R. Miller and D. F. Patch, *Rev. Sci. Instrum.*, 1969, **40**, 1566.
[13] P. A. Gorry and R. Grice, *J. Phys. E*, 1979, **12**, 857.
[14] S. J. Sibener, R. J. Buss, and Y. T. Lee, XIth Int. Symp. Rarefied Gas Dynamics, 1978; S. J. Sibener, R. J. Buss, C. Y. Ng, and Y. T. Lee, *Rev. Sci. Instrum.*, 1980, **51**, 167.

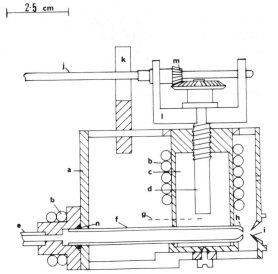

Figure 3 *Microwave discharge source* a, *framework;* b, *water-cooling tube;* c, *microwave cavity;* d, *tuning electrode;* e, *gas inlet tube;* f, *quartz discharge tube;* g, *axis of coupling stub;* h, *nozzle;* i, *skimmer;* j, *driveshaft;* k, *shaft guide;* l, *gear housing;* m, *bevel gears;* n, *O ring seal*

$O(^3P)$ atoms, but the chemistry of each reaction must be checked to ensure that $O_2(^1\Delta_g)$ makes no significant contribution to the observed reactive scattering. Direct-current discharges have been used as sources of supersonic hydrogen[15] and nitrogen[16] atom beams with very high, *ca.* 6000 K, effective source temperatures. However, this very hot plasma also extends through the nozzle of these sources and the production of metastable electronically-excited species presents a very acute problem. This may inhibit the use of these sources in the study of reactive scattering by ground-state atoms particularly for the nitrogen case. On the other hand, discharge sources may be used to study reactions of metastable electronically-excited atoms that are endoergic for ground-state atoms, provided that only a single electronically-excited state contributes to the reactive scattering. It has proved possible to use the r.f. source[14] in this way to measure reactive scattering by metastable $O(^1D)$ atoms. A supersonic beam of mercury atoms in the metastable $Hg(^3P_0)$ electronically-excited state seeded in nitrogen buffer gas has been produced[17] by illuminating a quartz nozzle tube with resonance radiation from low-pressure mercury lamps.

The conventional method[3-5] of time-of-flight velocity analysis, whereby a single pulse of molecules is transmitted by a chopper disc and the time-of-flight distribution measured in computer-generated time channels before a further pulse of

[15] K. R. Way, S. C. Yang, and W. C. Stwalley, *Rev. Sci. Instrum.*, 1976, **47**, 1049.
[16] R. W. Bickes, K. R. Newton, J. M. Hermann, and R. B. Bernstein, *J. Chem. Phys.*, 1976, **64**, 3648.
[17] S. Hayashi, T. M. Mayer, and R. B. Bernstein, *Chem. Phys. Lett.*, 1978, **53**, 419.

molecules is transmitted, suffers from a very poor duty cycle, *ca.* 5 %. This may be overcome if the single slot of the conventional disc is replaced by a pseudorandom sequence[18] of slots and teeth, which constitute a Hadamard sequence of $2^n - 1$ elements, where n is an integer and the duty cycle is raised to *ca.* 50%. The arrival times of molecules at the detector resulting from the pseudorandom chopper disc are measured in computer-generated time channels and the results cross-correlated with the appropriate complementary sequence. Hirschy and Aldridge[19] first applied this method in molecular beam experiments to measure the velocity distribution of an argon nozzle beam. A recent implementation[20] of the cross-correlation method for

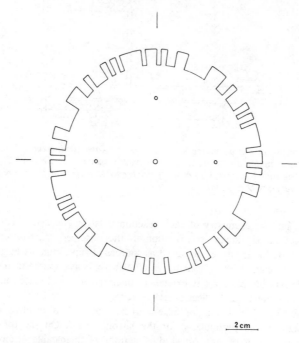

Figure 4 *Chopper disc for cross-correlation time-of-flight analysis with four sequences of 31 elements*
(Reproduced by permission from *J. Phys. E.*, 1979, **12**, 515)

measuring velocity distributions of reactive scattering is illustrated by the chopper disc shown in Figure 4 and the computer interface shown in Figure 5. In this implementation of the cross-correlation method, the rotation of the chopper disc and the advance of the channel address register in the interface are permanently synchronized. The atom or free radical beam is also modulated by a tuning-fork chopper to ensure the unambiguous measurement of scattering from the beam-intersection zone. After cross-correlation with the complementary sequence,

[18] G. Wilhelmi and F. Gompf, *Nucl. Instrum. Methods*, 1970, **81**, 36.
[19] V. L. Hirschy and J. P. Aldridge, *Rev. Sci. Instrum.*, 1971, **42**, 381.
[20] C. V. Nowikow and R. Grice, *J. Phys. E*, 1979, **12**, 515.

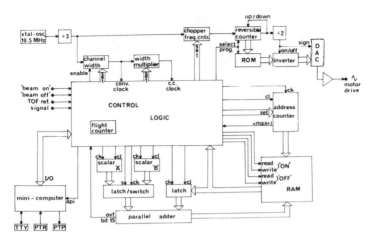

Figure 5 *Time-of-flight interface:* ck, *clock;* cl, *clear;* RAM, *random access memory;* ROM, *read-only memory;* DAC, *digital-to-analogue converter;* S, *select input word;* set, *send data word;* api, *automatic priority interrupt;* ovf, *overflow flag;* TTY, *teletype;* PTR, *papertape reader;* PTP, *papertape punch;* I/O, *input/output data highways*
(Reproduced by permission from *J. Phys. E.*, 1979, **12**, 515)

the time-of-flight data is deconvolved and transformed to velocity space using the same algorithms as in the conventional method. Consequently, the time-of-flight system may use the cross-correlation or conventional methods with equal facility. While the cross-correlation method is preferred for measuring the low intensities of reactive scattering, the conventional method is more convenient for measuring velocity distributions of energetic seeded beams where intensities must be reduced by restricting the detector aperture and the poor efficiency of the conventional method is not detrimental.

The application of lasers to reactive scattering experiments continues to be determined by the availability of certain types of high-powered lasers. The hydrogen halide chemical lasers are particularly effective at producing vibrational excitation of reactant hydrogen halide beams, as the laser transition coincides with the absorbing transition of the beam molecules. This method has been used[21-23] to excite hydrogen chloride and fluoride molecules from the vibrational ground state to specific rotational levels $J = 1$—4 of the first vibrationally-excited state. Electronic excitation of reactant molecules usually depends upon chance coincidences between an available intense laser line and an absorption transition of a reactant molecule. The excitation[24] of the $v' = 43, J' = 13$ vibration–rotational level of the $I_2[B^3 \Pi(0_u^+)]$ state by an argon ion laser has proved a useful example. Indeed, excitation of a diatomic molecule by plane or circularly polarized light yields excited molecules

[21] H. H. Dispert, M. W. Gies, and P. R. Brooks, *J. Chem. Phys.*, 1979, **70**, 5317.
[22] Z. Karny and R. N. Zare, *J. Chem. Phys.*, 1978, **68**, 3360.
[23] Z. Karny, R. C. Estler, and R. N. Zare, *J. Chem. Phys.*, 1978, **69**, 5199.
[24] F. Engelke, J. C. Whitehead, and R. N. Zare, *Faraday Discuss. Chem. Soc.*, 1977, **62**, 222.

with selected orientations and this excitation has recently been used to study[25] the orientation dependence of the reaction of $I_2[B^3\Pi(0_u^+)]$ with indium and thallium atoms. Similarly, rotation–vibrational excitation of hydrogen fluoride[23] has been used to study the orientation dependence of its reaction with strontium atoms. Atoms and molecules with low-lying excited electronic states may be excited selectively by a tunable dye laser. This method has been used to excite beams of sodium atoms[26,27] to the ^2P electronically-excited state, strontium and barium atoms[28] to the ^3P electronically-excited state, and sodium dimers[29] to the $^1\Pi_u$ electronically-excited state. In each of these cases of excitation to electronically-excited states the laser must illuminate the beam-intersection zone, since the excited atoms and molecules travel only a short distance, $\gtrsim 0.1$ mm, during their radiative lifetimes. However, excitation of a diatomic molecule prior to the beam-intersection zone allows fluorescent decay to vibration–rotation states of the ground electronic state different from the vibration–rotation state from which it was excited. Indeed excitation of a beam of sodium dimers[30] to the $Na_2(^1\Pi_u)$ electronically-excited state by a tunable dye laser followed by fluorescent decay has been used to modulate specific-rotational-state populations of the ground vibrational and electronic state.

The laser-induced fluorescence method[31,32] of determining product internal-state distributions has been improved by the introduction[33] of optical-fibre techniques in the construction of the rotatable detector (Figure 6). This permits measurement of angular distributions of product scattering while maintaining the beam sources stationary as required for the use of seeded nozzle beam sources. When the laser line used in laser-induced fluorescence detection is sufficiently narrow the Doppler shift due to the component of molecular velocity along the direction of the laser beam may be observed[26] and information on the product velocity distribution thereby obtained. Indeed it has been suggested[34] that full contour maps of reactive scattering of specific product internal states as a function of laboratory scattering angle and velocity may be determined by Fourier transformation of Doppler profiles measured by laser-induced fluorescence. In this method the laser beam passes through the beam intersection region and Doppler profiles are measured for different angles of incidence of the laser beam with respect to the molecular beams. Thus components of the laboratory product velocity distribution along the laser beam are measured by the Doppler profile at each angle of incidence and the full set of Doppler profiles may be inverted to obtain a contour

[25] R. C. Estler and R. N. Zare, *J. Am. Chem. Soc.*, 1978, **100**, 1323.
[26] W. D. Phillips, J. A. Serri, D. J. Ely, D. E. Pritchard, K. R. Way, and J. L. Kinsey, *Phys. Rev. Lett.*, 1978, **41**, 937.
[27] J. A. Silver, N. C. Blais, and G. H. Kwei, *J. Chem. Phys.*, 1977, **67**, 839; 1979, **71**, 3412; R. Duren, H. O. Hoppe, and H. Pauly, *Phys. Rev. Lett.*, 1976, **37**, 743; I. V. Hertel, H. H. Hofmann, and K. J. Rost, *ibid.*, 1976, **36**, 861; *J. Chem. Phys.*, 1979, **71**, 674.
[28] R. W. Solarz, S. A. Johnson, and R. K. Preston, *Chem. Phys. Lett.*, 1978, **57**, 514; R. W. Solarz and S. A. Johnson, *J. Chem. Phys.*, 1979, **70**, 3592.
[29] T. A. Brunner, R. D. Driver, N. Smith, and D. E. Pritchard, *J. Chem. Phys.*, 1979, **70**, 4155.
[30] K. Bergmann, R. Engelhardt, U. Hefter, P. Hering, and J. Witt, *Phys. Rev. Lett.*, 1978, **40**, 1446; *J. Chem. Phys.*, 1979, **71**, 2726.
[31] J. L. Kinsey, *Ann. Rev. Phys. Chem.*, 1977, **28**, 349.
[32] F. Engelke, *Ber. Bunsenges. Phys. Chem.*, 1977, **81**, 135; R. N. Zare, *Faraday Discuss. Chem. Soc.*, 1979, **67**, 7.
[33] K. Bergmann, R. Engelhardt, U. Hefter, and J. Witt, *J. Phys. E*, 1979, **12**, 507.
[34] J. L. Kinsey, *J. Chem. Phys.*, 1977, **66**, 2560.

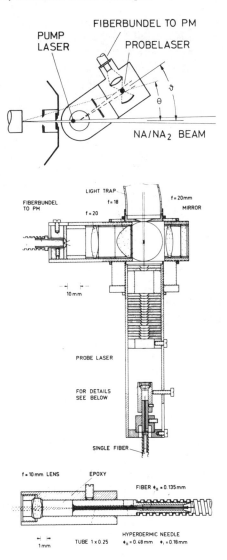

Figure 6 *Diagram of laser-induced fluorescence detector which is rotatable about the scattering centre with pump laser for excitation of the* Na/Na$_2$ *beam*
(Reproduced by permission from *J. Phys. E.*, 1979, **12**, 507)

map of the product angle–velocity distribution by standard Fourier transform techniques. Such a contour map has recently been obtained[35] for the hydroxy-radical product scattering from the reaction of an effusive hydrogen-atom beam

[35] E. J. Murphy, J. H. Brophy, G. S. Arnold, W. L. Dimpfl, and J. L. Kinsey, *J. Chem. Phys.*, 1979, **70**, 5910.

with a cross-beam of NO_2 molecules. The detection of hydroxy-radicals and also of bromine atoms[10] by the laser-induced fluorescence method illustrates the extension of tunable dye laser technology into the near-u.v., which promises to bring many more species within the scope of this detection method.

A high-power argon ion laser has been used to produce[36] nonresonant two-photon ionization of alkali-metal dimers, and tunable dye lasers to produce resonant multiphoton ionization of alkali-metal dimers,[37,38] iodine,[39] aniline, and benzene[40] molecules. Resonant multiphoton ionization is a promising method of detecting scattered reaction products since it is insensitive to scattered laser light and the detection of ions is much more efficient than the detection of photons in the laser-induced fluorescence method. Indeed resonant multiphoton ionization has been used[37] to detect BaCl product molecules from the reaction of barium atoms with hydrogen chloride molecules. However, the observed ionization spectrum[37] was not readily related to the internal-state distribution of the reaction products. This appears to present a still more severe problem in the multiphoton ionization of polyatomic molecules[40] where fragmentation may be more extensive than in electron bombardment ionization.

Finally, mention must be made of the new technique of pulsed supersonic molecular beam sources[41] whereby a nozzle source is incorporated into an electromechanical valve that can be opened for a period *ca.* 10 μs by the passage of a large current *ca.* 4×10^5 A from a charged capacitor. In this way, scattering from very intense supersonic beam pulses, *ca.* 10^{21} molecule ster^{-1} s^{-1}, may be observed[42] without the necessity of source differential pumping, since measurements are completed for each beam pulse before the later arrival of the background gas pulse. These pulsed beam sources are particularly well suited for use with pulsed lasers. The photodissociation of van der Waals dimers formed in a pulsed beam by a synchronously pulsed CO_2 laser and laser fluorescence of oxalyl fluoride have both been studied recently.[43] Such techniques may thus find application to reactive scattering experiments in which reactive species are formed by laser photodissociation. In contrast, a venerable technique of producing pulsed accelerated molecular beams from a high-speed rotor[44] has been enjoying a revival. Xenon atoms have been accelerated[45] up to translational energies $E \gtrsim 125$ kJ mol^{-1} by swatting them with the tip of a high-speed rotor moving with a velocity *ca.* 3 km s^{-1} and have been excited to the metastable $Xe(^3P_{2,0})$ electronically-excited state by

[36] E. W. Rothe, B. P. Mathur, and G. P. Reck, *Chem. Phys. Lett.*, 1978, **53**, 74; B. P. Mathur, E. W. Rothe, G. P. Reck, and A. J. Lightman, *ibid.*, 1978, **56**, 336; B. P. Mathur, E. W. Rothe, and G. P. Reck, *J. Chem. Phys.*, 1978, **68**, 2518.

[37] D. L. Feldman, R. K. Lengel, and R. N. Zare, *Chem. Phys. Lett.*, 1977, **52**, 413.

[38] A. Herrmann, S. Leutwyler, E. Schumacker, and L. Wöste, *Chem. Phys. Lett.*, 1977, **52**, 418.

[39] L. Zandee, R. B. Bernstein, and D. A. Lichtin, *J. Chem Phys.*, 1978, **69**, 3427.

[40] L. Zandee and R. B. Bernstein, *J. Chem. Phys.*, 1979, **70**, 2574; **71**, 1359; J. H. Brophy and C. T. Rettner, *Chem. Phys. Lett.*, 1979, **67**, 351.

[41] W. R. Gentry and C. F. Giese, *Rev. Sci. Instrum.*, 1978, **49**, 595.

[42] W. R. Gentry and C. F. Giese, *J. Chem. Phys.*, 1977, **67**, 5389; *Phys. Rev. Lett.*, 1977, **39**, 1259.

[43] W. R. Gentry, M. A. Hoffbauer, and C. F. Giese, VIIth Int. Symp. Molecular Beams (Riva del Garda), 1979, 273; M. G. Liverman, S. M. Beck, D. L. Monts, and R. E. Smalley, *J. Chem. Phys.*, 1979, **70**, 192.

[44] P. B. Moon, *Proc. R. Soc. London, Ser. A.*, 1978, **360**, 303; P. B. Moon, C. T. Rettner and J. P. Simons, *J. Chem. Soc., Faraday Trans.* 2, 1978, **74**, 630.

[45] M. R. Levy, C. T. Rettner, and J. P. Simon, *Chem. Phys. Lett.*, 1978, **54**, 120.

electron bombardment. The intensity of the unexcited beam *ca.* 10^{17} molecule ster^{-1} s^{-1} is somewhat less than a seeded supersonic nozzle beam but does not involve an inert buffer gas. Rotor beams are thus well suited to the study of reactive scattering of heavy atoms and molecules in metastable electronically-excited states.

3 Experimental Studies of Reactive Scattering

Alkali-metal Atom Reactions.—The dynamics of alkali-metal atom reactions have been extensively studied in molecular beams[46] and appear to be fairly well understood. However, many important features remain to be investigated and the generally tractable experimental situation has encouraged some elegant detailed studies of alkali-metal atom reactive scattering, which could not have been undertaken with more complicated systems. The dependence of the total reaction cross-section on reactant translational energy for the reactions of rubidium atoms with methyl iodide has been studied[47] by crossing a supersonic beam of CH_3I seeded in H_2 buffer gas with an effusive Rb atom beam and measuring the angular distribution of RbI reactive scattering by differential surface-ionization detection. These measurements confirm the existence of a minimum in the total reaction cross-section at a high translational energy $E \sim 87$ kJ mol^{-1}. Similar measurements[48] on the reactive scattering of K and Rb atoms with CH_3Br show a very different dependence of the total reaction cross-section on translational energy, with a low energy threshold $E_{th} \simeq 21$ kJ mol^{-1} and a monotonic increase at higher energy. The angular distributions of RbBr reactive scattering indicate a direct rebound reaction in which the product translational energy increases linearly with reactant translational energy and is always a major fraction of the total energy available to reaction products.

The dependence of the total reaction cross-section on reactant rotational state has been studied for the reactions of alkali-metal atoms with hydrogen halides. A hydrogen chloride chemical laser has been used[21] to promote HCl beam molecules to selected rotational levels $J = 1$—4 of the first vibrationally-excited state. Their reactive scattering with a K beam has been measured by differential surface ionization. The total-reaction cross-section is observed to decrease monotonically with increasing rotational quantum number. However, a much wider range of rotational states can be prepared[49] in a beam of hydrogen halide molecules formed by an appropriate pre-reaction in the beam source. The reactions of Na atoms with HF($J = 0$—14, $v = 2$—4) and HCl($J = 0$—19, $v = 1$—4) have been studied by observing[49] the depletion of infrared chemiluminescence from the hydrogen halides. In both cases the total-reaction cross-section at first decreases with increasing rotational quantum number but then increases at higher quantum number, exhibiting a minimum at $J \simeq 7$ for HF and $J \simeq 11$ for HCl. This is attributed to a bent configuration being required[50] for the reaction of alkali-metal atoms with hydrogen halides, which is most readily attained for low reactant rotation. At very

[46] For a review see R. Grice, *Adv. Chem. Phys.*, 1975, **30**, 247.
[47] S. A. Pace, H. F. Pang, and R. B. Bernstein, *J. Chem. Phys.*, 1977, **66**, 3635.
[48] H. F. Pang, K. T. Wu, and R. B. Bernstein, *J. Chem. Phys.*, 1978, **69**, 5267.
[49] B. A. Blackwell, J. C. Polanyi, and J. J. Sloan, *Chem. Phys.*, 1978, **30**, 299.
[50] G. G. Balint-Kurti and R. N. Yardley, *Faraday Discuss. Chem. Soc.*, 1977, **62**, 77.

high rates of reactant rotation the attacking alkali-metal atom responds to the orientation-averaged potential and possibly the reaction cross-section increases owing to the more ready departure of the hydrogen atom.

Low rotational states of alkali-metal halide molecules may be selectively focussed by an electrostatic quadrupole field from the thermal distribution of rotational states emerging from an effusive oven source. The effect of reactant rotational energy on the reactive scattering of K atoms with CsF[51] and RbF[52] has been investigated using this technique. These reactions proceed *via* a long-lived complex which may dissociate either to form reaction products or to reform the reactants [equation (1)]. In the case of the endoergic K + CsF reaction the fraction of collision

$$K + MF \rightleftharpoons KFM \rightarrow M + KF \qquad (1)$$
$$(M = Cs \text{ or } Rb)$$

complexes dissociating to products is higher for the thermal CsF beam $\bar{E}_R \simeq 11$ kJ mol^{-1} than for the rotationally cold beam $\bar{E}_R \simeq 0.5$ kJ mol^{-1} passed by the quadrupole state selector. The total-reaction cross-section and the reactive branching factor were measured over a range of reactant translational energies, and it was found that translational energy is more effective in promoting reaction than is CsF rotational energy. However, the reactive branching factor for the exoergic K + RbF reaction was found to be greater for the rotationally cold RbF beam than for the thermal beam and RbF rotational energy was found to be equally as effective as translational energy in promoting reaction.

Chemi-ionization has been observed[53] in the reactions of alkali-metal atoms with UF_6 at thermal energies [equation (2)]. However, chemi-ionization in the reactive

$$M + UF_6 \rightarrow M^+ + UF_6^-$$
$$(M = K \text{ or } Cs) \qquad (2)$$

scattering of alkali-metals with other metal polyfluorides MoF_6, MoF_5, WF_6, and UF_5 was shown to arise from alkali-metal dimers present in the beam. Mass analysis has now been performed[54] on the chemi-ions formed in the reactions of alkali-metal dimers[46] with halogen molecules at thermal energies [equation (3)].

$$M_2 + X_2 \rightarrow M^+ + MX + X^- \quad (a)$$
$$\rightarrow M_2X^+ + X^- \quad (b) \qquad (3)$$
$$\rightarrow M^+ + MX_2^- \quad (c)$$

Chemi-ionization is found to follow mainly reaction path (3a) when this is energetically accessible and reaction paths (3b) and (3c) predominate only in the cases when $M_2 = Li_2$ or Na_2, where reaction path (3a) is energetically inaccessible. These results are in excellent agreement with a previously proposed[55] double-electron-jump mechanism.

[51] S. Stolte, A. E. Proctor, W. M. Pope, and R. B. Bernstein, *J. Chem. Phys.*, 1977, **66**, 3468.
[52] L. Zandee and R. B. Bernstein, *J. Chem. Phys.*, 1978, **68**, 3760.
[53] B. P. Mathur, E. W. Rothe, and G. P. Reck, *J. Chem. Phys.*, 1977, **67**, 377.
[54] G. P. Reck, B. P. Mathur, and E. W. Rothe, *J. Chem. Phys.*, 1977, **66**, 3847; see also M. Roeder, W. Berneike, and H. Neuert, *Chem. Phys. Lett.*, 1979, **68**, 101.
[55] S. M. Lin, J. C. Whitehead, and R. Grice, *Mol. Phys.*, 1974, **27**, 741.

Alkaline-earth Metal Atom Reactions.—In contrast to the reactions of alkali-metal atoms, the products of alkaline-earth atom reactions are frequently free radicals with low-lying electronically-excited states, which are well suited to the laser-induced fluorescence method of detection.[31,32] Vibrational-state distributions of BaBr products have been measured[56] by laser-induced fluorescence for the reactions of Ba atoms with CH_3Br, CH_2Br_2, $CHBr_3$, and CBr_4. The proportion of the reaction exoergicity that is channelled into product vibration increases along this series of molecules in accord with the electron-jump model,[1] which has been applied[46] to the analogous alkali-metal atom reactions. The vibrational-state distribution of BaI from the reaction of Ba atoms with CF_3I was reported[57] to be bimodal, but subsequent re-analysis of the laser-induced fluorescence spectrum now indicates only a single peak at high vibrational energy. The exchange reaction of Ca atoms with NaCl has been studied[58] using laser-induced fluorescence to detect both the CaCl and Na products. The angular distributions of reactive scattering determined for $CaCl(v = 2,3)$ display the symmetric forward–backward peaking which is characteristic of a long-lived collision complex.[59] However, the $CaCl(v = 0,1)$ products in low-vibrational states exhibit angular distributions that peak more strongly in the forward than in the backward direction. Clearly, the Ca + NaCl reaction proceeds *via* an NaClCa collision complex, which yields CaCl in high-vibrational states when it persists for many rotational periods but CaCl in low-vibrational states when the lifetime of the complex is similar to its rotational period. Such a correlation between collision lifetime and product translational *versus* vibrational energy may be a more widespread phenomena in reactions proceeding *via* a long-lived collision complex mechanism than has so far been appreciated from the many studies that do not resolve individual product vibrational states.

The effect of vibrational excitation of HF molecules by an HF chemical laser on their reaction with Ca and Sr atom beams has been studied[22] by laser-induced fluorescence. The total-reaction cross-sections for both reactions are found to be higher by a factor *ca.* 10^4 for HF($v = 1$) than for HF($v = 0$) since the vibrational excitation energy exceeds the endoergicity of these reactions. The product vibrational-state distributions show that 40 and 30% of the total available energy for reaction of HF ($v = 1$) is channelled into CaF and SrF vibration. The same experimental apparatus has also been used to study[23] the reaction of Sr atoms with specific rotational states $J = 1,3$ of HF($v = 1,J$) and selected orientations with respect to the Sr atom beam. Laser-induced fluorescence detection of SrF product shows that the broadside collision orientation favours population of the highest $SrF(v' = 2)$ vibrational state compared with the collinear orientation. In addition, the higher HF($J = 3$) rotational state also favours population of $SrF(v' = 2)$ compared with HF($J = 1$). The dependence of the Ba, Sr, Ca + HF reactions on initial translational energy has also been studied[32] using a supersonic beam of HF seeded in He or H_2 driver gas. The vibrational-state distribution for BaF product

[56] M. Rommel and A. Schultz, *Ber. Bunsenges. Phys. Chem.*, 1977, **81**, 139.
[57] G. P. Smith, J. C. Whitehead, and R. N. Zare, *J. Chem. Phys.*, 1977, **67**, 4912; J. Allison, M. A. Johnson, and R. N. Zare, *Faraday Discuss. Chem. Soc.*, 1979, **67**, 124.
[58] P. J. Dagdigian, *Chem. Phys.*, 1977, **21**, 453.
[59] W. B. Miller, S. A. Safron, and D. R. Herschbach, *Discuss. Faraday Soc.*, 1967, **44**, 108.

shows only modest change indicating that most of the reactant translational energy is channelled into product translational and rotational energy. Comparison of the endoergic Sr + HF reaction with Ba + HF indicates that translational energy is much less effective than HF vibrational energy in promoting the Sr + HF reaction.

The reactive scattering of a thermal Ba atom beam by a supersonic N_2O beam has been studied[60] by mass-spectrometric detection with conventional time-of-flight analysis. The reactant translational energy is controlled by varying the temperature of the N_2O nozzle. At low reactant translational energy the BaO product recoils in the backward direction but the angular distribution broadens to smaller angles as the translational energy increases. Similar experiments[60] on the Ba + SO_2 reaction show that this follows a spin forbidden path *via* a long-lived complex at low energy, but at higher energy BaO product recoils backward. The dependence of BaO* chemiluminescence on reactant translational energy has been measured[61] over a wider energy range by seeding the N_2O in He buffer gas and is found to decrease strongly with increasing translational energy. However, the chemiluminescence cross-section increases with the temperature of the N_2O nozzle at constant total energy. From the temperature dependence of the thermal rate constant the increase in chemiluminescence is attributed to excitation of the N_2O bending mode and probably the N—O stretching mode. These observations are in good accord with an electron-jump model for the reaction [equation (4)], in which the electron affinity of

$$Ba + N_2O \rightarrow Ba^+ + N_2O^- \rightarrow BaO^* + N_2 \qquad (4)$$

N_2O increases on bending. The angular distribution of this chemiluminescence has been observed[62] in a crossed-beam experiment where the long radiative lifetime of BaO* renders chemiluminescence visible from molecules slightly displaced from the scattering zone. The angular distribution of chemiluminescence[63] from the two-body recombination reaction [equation (5)] has been observed to coincide with the

$$Ba + Cl_2 \rightarrow BaCl_2^* \qquad (5)$$

centroid distribution. This implies an excessively long lifetime, $\tau \sim 10^{-4}$ s, for dissociation of the $BaCl_2$* complex, which probably arises from stabilization of the $BaCl_2$* by vibrational energy transfer in collisions with the Cl_2 background gas. Chemiluminescence is also observed[64] in the reactions of alkaline-earth metal atoms with S_2Cl_2 owing to electronically excited $S_2(B^3\Sigma_u^-)$. This is attributed to a double-electron-jump mechanism[46] with a planar five-membered ring intermediate [equation (6)]. The reactions of alkaline-earth atoms with halogen and interhalogen

$$M + S_2Cl_2 \rightarrow M^+ + S_2Cl_2^- \rightarrow Cl^- \overset{\textstyle M^{2+}}{\underset{\textstyle S-S}{\diagdown \diagup}} Cl^- \rightarrow MCl_2 + S_2^* \qquad (6)$$

[60] T. P. Parr, A. Freedman, R. Behrens, and R. R. Herm, *J. Chem. Phys.*, 1977, **67**, 2181; A. Freedman, T. P. Parr, R. Behrens, and R. R. Herm, *J. Chem. Phys.*, 1979, **70**, 5251.
[61] D. J. Wren and M. Menzinger, *Faraday Discuss. Chem. Soc.*, 1979, **67**, 97.
[62] A. Siegel and A. Schultz, *Chem. Phys.*, 1978, **28**, 265.
[63] C. A. Mins and J. H. Brophy, *J. Chem. Phys.*, 1977, **66**, 1378.
[64] F. Engelke and R. N. Zare, *Chem. Phys.*, 1977, **19**, 327.

molecules exhibit[65] substantial cross-sections for chemi-ionization, which may again be attributed to a double-electron-jump mechanism[46] [equation (7)].

$$M + X_2 \rightarrow M^+ + X_2^- \rightarrow M^{2+} \Big\langle {}^{X^-}_{X_-} \rightarrow MX^+ + X^- \qquad (7)$$

The reactions of alkaline-earth atoms in metastable electronically-excited states have been studied[28] by using tunable dye laser excitation of the atom beam and laser-induced fluorescence detection of product vibrational state distributions. The reaction of $Sr(^3P_1)$ with HF, HCl, and $Ba(^1S, ^3D)$, $Sr(^1S, ^3P)$ with a series of halogenomethanes all yield higher vibrational excitation of the SrX, BaX products than the corresponding reactions of ground-state Sr, Ba atoms. The less-selective method of electron bombardment in a direct-current discharge has also been used to excite beams of alkaline-earth atoms. The reactions of $Ca(^3P)$ with O_2, CO_2,[66] N_2O,[67] HCl and Cl_2,[68] $Sr(^3P)$ with N_2O,[69] $Ba(^3D)$ with Br_2 and I_2,[70] and Mg^*, Ca^*, Sr^*, and Ba^* with F_2[71] all yield electronically-excited products MO^*, MX^* in addition to the ground electronic states. The absolute chemiluminescence cross-section $Q \sim 6$ Å^2 has been determined[67] for $Ca(^3P) + N_2O$ by comparison of the chemiluminescence intensity with the fluorescence arising from radiative decay of the $Ca(^3P_1)$ beam component. The increased vibrational excitation of reaction products and the widespread appearance of chemiluminescence in the reactions of electronically-excited alkaline-earth atoms compared with those of ground-state atoms are in accord with the electron-jump model[46] for these reactions. Electronic excitation decreases the ionization potential of the alkaline-earth atom so that the electron-jump transition occurs at larger internuclear distance increasing the attractive energy released in the entrance valley of the potential energy surface and thereby the vibrational excitation of reaction products. Electronic excitation of the alkaline-earth atom also alters the symmetry of the transferred electron, which renders excited electronic states of the reaction intermediate correlating with electronically-excited reaction products, accessible in the electron-jump mechanism either directly or *via* nonadiabatic transitions.

Other Metal Atom Reactions.—Molecular beam studies of metal atom reactions other than those of alkali and alkaline-earth metals have largely been confined to metals that are sufficiently volatile to yield adequate beam intensity from an effusive oven source. Reactive scattering of a Sn atom beam by a supersonic molecular cross-beam has been measured by mass-spectrometric detection and conventional time-of-flight analysis of reaction products. The reaction of Sn atoms with Cl_2 and Br_2 molecules[72] shows stripping dynamics with SnCl and SnBr product scattering in

[65] G. J. Diebold, F. Engelke, H. U. Lee, J. C. Whitehead, and R. N. Zare, *Chem. Phys.*, 1977, **30**, 265.
[66] L. Pasternack and P. J. Dagdigian, *Chem. Phys.*, 1978, **33**, 1.
[67] P. J. Dagdigian, *Chem. Phys. Lett.*, 1978, **55**, 239.
[68] U. Brinkmann and H. Telle, *J. Phys. B*, 1977, 133.
[69] B. E. Wilcomb and P. J. Dagdigian, *J. Chem. Phys.*, 1978, **69**, 1779.
[70] R. C. Estler and R. N. Zare, *Chem. Phys.*, 1978, **28**, 253.
[71] F. Engelke, *Chem. Phys.*, 1979, **39**, 279.
[72] T. P. Parr, R. Behrens, A. Freedman, and R. R. Herm, *Chem. Phys. Lett.*, 1978, **56**, 71.

the forward direction. Reaction with CH_3I, C_3H_7I molecules[73] shows rebound dynamics with SnI product scattering in the backward direction and reaction with CCl_4 molecules[73] intermediate dynamics with SnCl product scattering sideways. This sequence of reaction dynamics mirrors that displayed by alkali-metal atom reactions,[1,46] particularly those of Li atoms, and suggests the operation of an electron-jump mechanism despite the high ionization potential of Sn atoms, $I(Sn) = 709$ kJ mol^{-1}, compared with that of Li atoms, $I(Li) = 520$ kJ mol^{-1}. The reaction of Sn atoms with O_2 molecules[74] exhibits a long-lived collision complex mechanism that is very similar to the reaction of Ba atoms with O_2 molecules[75] and again indicates an electron-jump mechanism. Thus the persistence of an electron-jump mechanism in Sn atom reactions parallels the reactions of Mg atoms, which have a similarly high ionization potential, $I(Mg) = 738$ kJ mol^{-1}. However, the reactive scattering of a beam of Hg atoms seeded in H_2 buffer gas with an I_2 cross-beam has been studied[76] by mass-spectrometric detection of the angular distribution of HgI product as a function of reactant translational energy. The endoergic Hg + I_2 reaction proceeds by a long-lived collision complex mechanism in contrast to the stripping mechanism of Sn atoms[72] with halogen molecules. Thus the electron-jump mechanism is no longer applicable to metal atoms with the still higher ionization potential of Hg atoms $I(Hg) = 1006$ kJ mol^{-1}. However, the reactive scattering of a supersonic beam of metastable $Hg(^3P_0)$ atoms seeded in N_2 buffer gas with Br_2 molecules has been studied[17] by observation of chemiluminescence from the electronically-excited $HgBr(B^2\Sigma^+)$ products. The $Hg(^3P_0)$ state has a much smaller chemiluminescent cross-section $Q \sim 3$ $Å^2$ than that[77] of the $Hg(^3P_2)$ state reaction $Q \sim 90$ $Å^2$. The reactions of In and Tl atoms with electronically excited $I_2[B^3\Pi(0_u^+)]$ molecules have been studied[25] by using polarized laser excitation of the I_2 beam, which forms oriented reactant molecules. Observation of InI* and TlI* product chemiluminescence as a function of $I_2[B^3\Pi(0_u^+)]$ orientation allows the dependence of reaction on reactant orientation to be determined, as illustrated in Figure 7, where both reactions are seen to exhibit a mild preference for collinear approach of the reactants.

The reactions of Group IIIB metal atoms Sc, Y, and La with O_2, NO, and CO_2 have been studied[78] by using laser-induced fluorescence detection of the metal oxide products. The vibrational and rotational state distributions are in good agreement with the predictions of a statistical model, though a significant discrepancy appears in the rotational state distributions for the O_2 reactions. The reactions of these metal atoms with H_2O, D_2O, and alcohols[79] yield metal oxide rather than hydroxide products, with statistical vibrational energy distributions, but rotational energy distributions that lie at lower energy than the statistical prediction. The abstraction of carbonyl and carboxylic oxygen atoms by Sc, Y, and La atoms[80] also shows statistical vibrational energy distributions with rotational energy distributions that

[73] T. P. Parr, R. Behrens, A. Freedman, and R. R. Herm, *J. Chem. Phys.*, 1978, **69**, 2795.

[74] A. Freedman, R. Behrens, T. P. Parr, and R. R. Herm, *J. Chem. Phys.*, 1978, **68**, 4368.

[75] D. A. Dixon, D. D. Parrish, and D. R. Herschbach, *Faraday Discuss. Chem. Soc.* 1973, **55**, 385.

[76] T. M. Mayer, B. E. Wilcomb, and R. B. Bernstein, *J. Chem. Phys.*, 1977, **67**, 3507.

[77] H. F. Krause, S. G. Johnson, S. Datz, and F. K. Schmidt-Bleek, *Chem. Phys. Lett.*, 1975, **31**, 577.

[78] K. Liu and J. M. Parson, *J. Chem. Phys.*, 1977, **67**, 1814.

[79] K. Liu and J. M. Parson, *J. Chem. Phys.*, 1978, **68**, 1794.

[80] J. M. Parson, *J. Phys. Chem.*, 1979, **83**, 970.

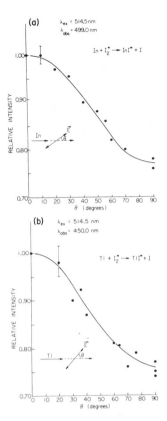

Figure 7 *Variation of chemiluminescence with reactant orientation angle θ for* a, In + I$_2^*$, *and* b, Tl + I$_2^*$
(Reproduced by permission from *J. Am. Chem. Soc.*, 1978, **100**, 1323)

are colder than those statistically predicted. Chemiluminescence from the reactions of Sc, Y, and La atoms with O$_2$ molecules have been studied over a range of reactant translational energy using beam-gas scattering[81] at thermal energy and a supersonic beam[82] of O$_2$ seeded in He at higher energy. Both the vibrational and rotational energy distributions of the electronically-excited metal oxide products conform to statistical predictions. Thus it appears that all these O atom abstraction reactions of the Group IIIB metal atoms proceed *via* a long-lived collision complex. However, interactions in the exit channel of the reaction potential-energy surface distort the rotational energy distributions for the reactions with the more complex molecules. Indeed a transition state consisting of a five-membered ring (1) leading to three product fragments has been suggested[79] for the abstraction of an O atom from ethanol, which illustrates the sort of exit channel interaction that may distort the

[81] C. L. Chalek and J. L. Gole, *Chem. Phys.*, 1977, **19**, 59; J. L. Gole, D. R. Preuss, and C. L. Chalek, *J. Chem. Phys.*, 1977, **66**, 548; D. R. Preuss and J. L. Gole, *ibid.*, pp. 2994 and 3000.
[82] D. M. Manos and J. M. Parson, *J. Chem. Phys.*, 1978, **69**, 231.

$$CH_2-CH_2$$

(1)

product rotational distribution. The reaction of Yb atoms with $CHBr_3$ molecules has also been studied[83] by using laser-induced fluorescence detection of YbBr products in a beam-gas experiment. The YbBr vibrational state distribution is found to be very similar to that measured for the Ba + $CHBr_3$ reaction,[56] indicating that the Yb + $CHBr_3$ reaction also proceeds by an electron-jump mechanism.

Angular distributions of SmO reactive scattering from an effusive Sm atom beam and a supersonic beam of O_2 seeded in He buffer gas, have been measured[83] by mass-spectrometric detection as a function of reactant translational energy. The SmO product scatters in the forward direction, showing that the reaction proceeds by a direct stripping mechanism. The dependence of chemiluminescence on reactant translational and vibrational energy has been studied[84] for the reaction of Sm atoms with a supersonic beam of N_2O seeded in He buffer gas. The chemiluminescence cross-section, which increases with translational and vibrational energy, arises from two separate transitions resulting from the multitude of SmO electronically excited states that correlate with the reactants. However the two transitions have very similar dependence on initial translational energy and it must be questioned whether the red transition arises by cascade from the blue emitter. Effusive metal-atom beams formed from relatively involatile metals in a high-temperature oven frequently populate a number of electronically-excited states of the metal atoms. The contribution of electronically-excited states of the metal atoms may be identified[81] by observing the intensity of chemiluminescence as a function of the metal oven temperature. In this way the contribution of the $Ti(^5F)$ electronically-excited state has been identified[85] in the chemiluminescence of Ti atoms from a tungsten-lined carbon crucible at *ca.* 2650 K with O_2, NO_2, and N_2O molecules in a beam-gas experiment. Chemiluminescence has also been observed[86] from B atoms produced by effusion from a graphite-lined Ta crucible at very high temperature, *ca.* 2373 K, with an O_2 cross-beam. However atom beams of very involatile materials have recently been formed by evaporation from a foil that is induced with irradiation by a high-power pulsed laser on the reverse side of the foil. The velocity of the pulsed atom beam may be varied by adjustment of the laser pulse power and the beam may contain a lower proportion of excited states in so far as it results from shock detachment rather than purely thermal evaporation of atoms. The velocity dependence of chemiluminescence has been studied in this way for the reaction of B,[87] Ho,[87,88] and Pb[89] atoms with an N_2O cross-beam, where the

[83] R. Dirscherl and H. U. Lee, VIIth Int. Symp. Molecular Beams (Riva del Garda), 1979, 104.
[84] A. Yakozeki and M. Menzinger, *Chem. Phys.*, 1977, **20**, 9.
[85] L. H. Dubois and J. L. Gole, *J. Chem. Phys.*, 1977, **66**, 779.
[86] A. Brzychey, J. Dehaven, A. T. Prengel, and P. Davidovits, *Chem. Phys. Lett.*, 1978, **60**, 102.
[87] S. P. Tang, N. G. Utterback, and J. F. Friichtenicht, *J. Chem. Phys.*, 1976, **64**, 3833.
[88] S. P. Tang, B. G. Wicke, and J. F. Friichtenicht, *J. Chem. Phys.*, 1978, **68**, 5471.
[89] B. G. Wicke, S. P. Tang, and J. F. Friichtenicht, *Chem. Phys. Lett.*, 1978, **53**, 304.

chemiluminescence cross-sections are observed to increase with translational energy over the range $E \sim 50$—400 kJ mol^{-1} in each case. High-energy beams of metal atoms may also be produced by sputtering from a metal surface caused by impact of Ar$^+$ ions at *ca.* 500 eV. Chemi-ionization reactions observed for the reactions of a wide range of metal atoms[90] with a supersonic O$_2$ molecule beam in the energy range $E \sim 10$—4000 kJ mol^{-1} show predominantly rearrangement and associative ionization [equations (8) and (9)] only when their energetic thresholds are lower

$$M + O_2 \rightarrow MO^+ + (O + e^-) \tag{8}$$

$$M + O_2 \rightarrow MO_2^+ + e^- \tag{9}$$

than that for electron transfer [equation (10)].

$$M + O_2 \rightarrow M^+ + (O_2 + e^-) \tag{10}$$

Hydrogen Atom Reactions.—The study of hydrogen atom reaction dynamics has long been dominated by the i.r. chemiluminescence method[91] of determining the vibrational and rotational distributions of hydrogen halide reaction products. This method has been used in a thorough investigation[92] of branching effects in the thermal energy reactions of hydrogen atoms from a discharge source with ICl and BrCl molecules. The rotational state distributions for the HCl product from the H + ICl reaction (Figure 8) clearly show two branches in the distribution. The branch at

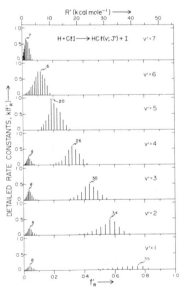

Figure 8 *Rotational state distributions for* HCl (v', J') *product from the* H + ICl *reaction* (Reproduced by permission from *Chem. Phys.*, 1977, **23**, 167)

[90] C. E. Young, R. B. Cohen, P. M. Dehmer, L. G. Pobo, and S. Wexler, *J. Chem. Phys.*, 1976, **65**, 2562.

[91] T. Carrington and J. C. Polanyi, in M.T.P. Int. Rev. Sci.: Physical Chemistry, ed. J. C. Polanyi, Butterworths, London, 1972, **9**, 135.

[92] J. C. Polanyi and W. J. Skrlac, *Chem. Phys.*, 1977, **23**, 167.

high rotational energy is most prominent for the highest vibrational states, but the branch at low rotational energy becomes predominant for the lowest vibrational states. This microscopic branching suggests the existence of two dynamical mechanisms, one yielding HCl product with high internal excitation and low translational energy, the other yielding lower internal excitation and higher translational energy. The rotational state distribution for the HCl product from the H + BrCl reaction also shows the same microscopic branching, though the two branches are no longer fully separated (Figure 9). However, the rotational state distributions for the HBr product (Figure 10) show no evidence of microscopic

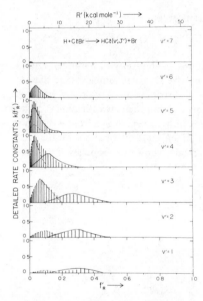

Figure 9 *Rotational state distributions for* HCl *(v′, J′) product from the* H + BrCl *reaction* (Reproduced by permission from *Chem. Phys.*, 1977, **23**, 167)

branching. The macroscopic branching of the total-reaction cross-section between the HCl and HBr products for the H + BrCl reaction $\Gamma = 0.40$ differs very greatly from the theoretical prediction $\Gamma = 3.2$ calculated from the ratio of the phase-space volumes available to each of these products. Micro- and macro-scopic branching has also been studied[93] by the i.r. chemiluminescence method for the reactions of H atoms with ClF, BrF, and IF molecules. The rotational state distributions for HF product from the H + ClF reaction[93] exhibit microscopic branching in which the branch with low internal energy is similar to the internal energy distribution for the H + F_2 reaction and the branch with higher internal excitation is similar to the internal energy distribution for the H + Cl_2 reaction. The two branches are not fully separated, as is also the case for the HCl product[92] from the H + BrCl reaction. The

[93] D. Brandt, and J. C. Polanyi, VIIth Int. Symp. Molecular Beams (Riva del Garda), 1979, 84; D. Brandt and J. C. Polanyi, *Chem. Phys.*, 1978, **35**, 23; 1980; **45**, 65.

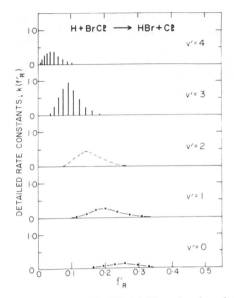

Figure 10 *Rotational state distributions for* HBr (v', J') *product from the* H + BrCl *reaction* (Reproduced by permission from *Chem. Phys.*, 1977, **23**, 167)

rotational state distribution for the HCl product[93] from the H + ClF reaction exhibits no microscopic branching, as is also the case for the HBr product[92] from the H + BrCl reaction. The rotational state distributions for the HF product[93] from the H + BrF reaction show an increased separation of the microscopic branches and an increased prominence of the branch with high internal energy as compared with the HF product from the H + ClF reaction. This parallels the changes seen in the microscopic branching for the HCl product[92] from the H + ICl, BrCl reactions. However, this trend does not extend to the HF product[93] from the H + IF reaction, where the separation of the branches is rather less complete and the branch with high internal energy rather less favoured than for the HF product from the H + BrF reaction. In all cases the yield of electronically-excited atom products[92,94] is too minor to explain the observed branching phenomena.

These elegant experimental results may be interpreted[95] in terms of the reaction dynamics which result from the attack of an H atom at each end of a heteronuclear halogen molecule XY, where the electronegativity of X is less than that of Y. The reaction path to form the HY product is more exoergic than that to form the HX product, but the activation energy for attack at the Y end of the XY molecule is higher than that for attack at the X end. Indeed the activation energies at each end of the heteronuclear XY molecule are very similar to the activation energies of the corresponding reactions of the homonuclear molecules H + X_2, Y_2. Moreover the branch of the HY product distributions with low internal energy and the HX product distributions from the H + XY reaction are both very similar to the

[94] S. H. P. Bly, D. Brandt, and J. C. Polanyi, *Chem. Phys. Lett.*, 1979, **65**, 399.
[95] J. C. Polanyi, J. L. Schreiber, and W. J. Skrlac, *Faraday Discuss. Chem. Soc.*, 1979, **67**, 66.

corresponding internal energy distributions for the homonuclear reactions. Thus these HY and HX products result from direct reaction dynamics following attack of the H atom on the Y and X ends of the XY molecule respectively. However the branch of the HY product distributions with high internal energy from the H + XY reaction results from attack of the H atom at the X end of the molecule where the activation energy is lower, followed by migration to the Y end of the molecule and formation of the more exoergic HY reaction product. The snarled trajectory executed by the H atom during this migratory encounter[95] results in insertion of the H atom into the X—Y bond after it has become considerably extended. This dissipates much of the repulsion between the X and Y atoms and results in much of the reaction exoergicity being disposed into vibrational and rotational excitation of the HY molecule. The macroscopic branching ratio between HY and HX product thus reflects the activation energies at the X and Y ends of the XY molecule and the efficiency of the migratory reaction dynamics rather than the phase-space volumes available to these products.

The vibrational and rotational state distributions for the HCl product from the H + ICl reaction have also been studied[96] at higher translational energy, $E \sim 42$ kJ mol^{-1}, thermal dissociation of H_2 in a tungsten oven at 2685 K and i.r. chemiluminescence detection being used. The HCl product distributions no longer show evidence of microscopic branching, the internal energies being confined to low values and the higher internal energy branch being undetectable. This may readily be understood[95] in terms of the preceding theory of microscopic branching, since direct formation of HY by attack of the H atom at the Y end of the XY molecule has a higher activation energy than attack at the X end, which can lead to formation of HY by migration. Consequently, the cross-section for attack at the Y end of the XY molecules increases greatly with increasing translational energy and the HY product distributions become dominated by HY formed by the direct mechanism. The reactions of H and D atoms with Br_2 molecules have been studied[10] over a range of initial translational energy E ca. 4—35 kJ mol^{-1} by using a supersonic H, D atom beam seeded in undissociated H_2, D_2 or He, Ne buffer gases expanding from a high-pressure tungsten oven and laser-induced fluorescence detection of Br atom products. Figure 11 shows that the total-reaction

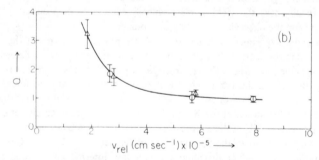

Figure 11 *Variation of the total-reaction cross-section with initial relative velocity for the H, D + Br_2 reactions shown by open circles and triangles*
(Reproduced by permission from *J. Chem. Phys.*, 1978, **69**, 4311)

[96] J. W. Hudgens and J. D. McDonald, *J. Chem. Phys.*, 1977, **67**, 3401.

cross-sections for the H and D atom reactions follow a common dependence on initial relative velocity rather than energy, with the cross-section decreasing with increasing collision velocity. The effect of reactant vibrational excitation on the reaction of D atoms with HF molecules has also been studied[97] by preparing HF molecules in vibrationally excited states $v \leqslant 6$ by a suitable pre-reaction in the beam source and observation of the depletion of the HF i.r. chemiluminescence from these states by D atoms from an effusive microwave discharge source. The endoergic abstraction reaction (11a) is observed for vibrational states $v \geqslant 3$ but the thermoneutral exchange reaction (11b) only for vibrational states $v = 5, 6$. This

$$
\begin{aligned}
D + HF\dagger &\rightarrow HD + F \quad (a) \\
&\rightarrow DF + H \quad (b)
\end{aligned}
\tag{11}
$$

corresponds to an activation energy E_a *ca.* 200 kJ mol^{-1} for the exchange reaction, which is greater than the endoergicity of the abstraction reaction.

The contour map (Figure 12) for OH reactive scattering in the $v = 0, K = 17, J = 17.5$ state from the H + NO$_2$ reaction was determined[35] by Fourier transform of the Doppler profiles of a single rotational line of the laser-induced fluorescence spectrum measured over a range of angles of incidence of the laser beam with respect to the H atom and NO$_2$ molecule beams. The Doppler profiles were approximated by Gaussian distributions and the resulting contour map of the differential reaction cross-section is correspondingly approximate. However, the results, showing OH product scattering in the forward direction, are in good agreement with reactive scattering measurements[98] by mass-spectrometric detec-

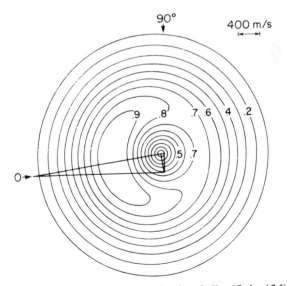

Figure 12 *Differential reaction cross-section for* OH ($v = 0, K = 17, J = 17.5$) *product from the* H + NO$_2$ *reaction*
(Reproduced by permission from *J. Chem. Phys.*, 1979, **70**, 5910)

[97] F. E. Bartoszek, D. M. Manos, and J. C. Polany, *J. Chem. Phys.*, 1978, **69**, 933.
[98] H. Haberland, P. Rohwer, and T. Schmidt, *Chem. Phys.*, 1974, **5**, 298.

tion. This is the first application of Fourier transform Doppler spectroscopy to reactive scattering measurements and is an important advance, since Doppler measurements over an angular range of 180° provide a complete map over the full range of angles and velocities of reactive scattering for a specific-product state. In some cases this will be considerably more efficient[34] than more conventional methods[33] of angular and time-of-flight velocity measurements. However, accurate determination of the contour map of reactive scattering by numerical Fourier transform methods requires very accurate experimental data for the Doppler profiles and this requirement may offset the apparent gain in efficiency in cases with small Doppler shifts requiring a very narrow laser line. Laser-induced fluorescence measurements[99] of the H + NO_2, ClO_2 reactions have shown that the $\Pi^+ \Lambda$ doublet state of the OH($v = 0,1$) products is favoured over the Π^- state, which suggests that reaction proceeds through planar HONO and HOClO collision complexes.

Scattering of a highly energetic H atom beam, E ca. 500 kJ mol^{-1}, from a plasma-arc source[15] with a cross-beam of Li atoms and Li_2 dimers results[100] in Li^+ and Li_2^+ chemi-ionization and profuse chemiluminescence. Scattering with H_2O and O_2 molecules[100] gives chemiluminescence from OH($A^2\Sigma^+$), which shows a sharp break in the rotational state distribution at $N = 23$. The difficult task of measuring H atom and HD molecule reactive scattering by mass-spectrometric detection from the exchange and abstraction reactions of D atom with hydrogen halides has been attempted.[101] However, more recent work by the same authors suggests that these measurements may not be entirely free from experimental artefacts.

Halogen Atom Reactions.—The exchange reactions of halogen atoms with halogen molecules were the first non-alkali-metal reactions to be studied[102] in molecular beams using an electron bombardment mass-spectrometer detector. Work on these reactions has now been extended by measurements[103] of BrCl reactive scattering from the Cl + Br_2 reaction by using a beam of Cl atoms seeded in He or Ar buffer gas from a high-temperature graphite oven to vary the initial translational energy over the range $E = 28$—74 kJ mol^{-1}. This reaction follows a stripping mechanism with BrCl product peaking very sharply in the forward direction $\theta \gtrsim 40°$ and approaches the spectator stripping limit at the highest initial translational energy. However, the total-reaction cross-section, Q ca. 11—14 Å^2, is found to be smaller than the hard-sphere scattering cross-section, Q ca. 30 Å^2. The observation of stripping dynamics in these circumstances indicates[102] that attractive interactions must operate between the reactants at small internuclear distance. Triatomic MO theory favours[102,104] stable tri-halogen complexes with the most electropositive

[99] R. P. Mariella and A. C. Luntz, *J. Chem. Phys.*, 1977, **67**, 5388; R. P. Mariella, B. Lantzsch, V. T. Maxson, and A. C. Luntz, *ibid*, 1978, **69**, 5411.

[100] J. B. Crooks, K. R. Way, S. C. Yang, C. Y. R. Wu, and W. C. Stwalley, *J. Chem. Phys.*, 1978, **69**, 490; K. K. Verma and W. C. Stwalley, VIIth Int. Symp. Molecular Beams (Riva del Garda), 1979, 93.

[101] W. Bauer, L. Y. Rusin, and J. P. Toennies, *J. Chem. Phys.*, 1978, **68**, 4490; J. P. Toennies, personal communication.

[102] D. Beck, F. Engelke, and H. J. Loesch, *Ber. Bunsenges. Phys. Chem.*, 1968, **72**, 1105; N. C. Blais and J. B. Cross, *J. Chem. Phys.*, 1970, **52**, 3580; Y. T. Lee, J. D. McDonald, P. R. Le Breton, and D. R. Herschbach, *J. Chem. Phys.*, 1968, **49**, 2447; 1969, **51**, 455.

[103] J. J. Valentini, Y. T. Lee, and D. J. Auerbach, *J. Chem. Phys.*, 1977, **67**, 4866.

[104] D. R. Herschbach, Proc. Conf. Potential Energy Surfaces in Chemistry, ed. W. A. Lester, I.B.M., San Jose, 1971, 44; A. D. Walsh, *J. Chem. Soc.*, 1953, 2266.

atom in the centre. The Cl—Br—Br complex with 21 valence electrons is expected[104] to prefer a bent configuration with an interbond angle a *ca.* 140°, which may impose an orientation requirement for the reaction giving a reduced total-reaction cross-section while permitting reaction at slightly larger impact parameters, which promote scattering in the forward direction.

The reactions of halogen atoms with vinyl bromide molecules[105] have also been reinvestigated[106] using F and Cl atoms seeded in Ar buffer gas. Reactive scattering of the C_2H_3Cl product from the Cl + C_2H_3Br reaction shows peaking in the forward direction and an inferred peak in the backward direction that is lower by a factor of *ca.* 2. This indicates[107] that the reaction proceeds *via* a short-lived collision complex with a lifetime of less than one rotational period. Reactive scattering of C_2H_3F product from the F + C_2H_3Br reaction was measured only in the region of the forward peak so that the lifetime of the collision complex could not be determined in this case. The differential reaction cross-section for the Cl + C_2H_3Br

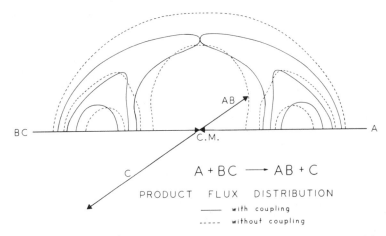

Figure 13 *Differential reaction cross-section calculated from the* RRKM–AM *model with coupling of scattering angle and product translational energy (solid curves) and without coupling (broken curves)*
(Reproduced by permission from *Faraday Discuss. Chem. Soc.*, 1979, **67**, 162)

reaction shows a mild coupling between scattering angle and product velocity with slightly higher velocity in the forward direction than at wider angles. The authors demonstrate[106] convincingly that such a coupling is indeed to be expected from the RRKM–AM model[108] of reaction *via* a long-lived collision complex (Figure 13). Collisions at larger impact parameters have high angular momentum and scatter product more sharply in the forward and backward directions, but also have a high centrifugal barrier to complex dissociation that shifts the product velocity

[105] J. T. Cheung, J. D. McDonald, P. R. Le Breton, and D. R. Herschbach, *J. Am. Chem. Soc.*, 1973, **95**, 7889.
[106] R. J. Buss, M. J. Coggiola, and Y. T. Lee, *Faraday Discuss. Chem. Soc.*, 1979, **67**, 162.
[107] G. A. Fisk, J. D. McDonald, and D. R. Herschbach, *Discuss. Faraday Soc.*, 1967, **44**, 228.
[108] S. A. Safron, N. D. Weinstein, D. R. Herschbach, and J. C. Tully, *Chem. Phys. Lett.*, 1972, **12**, 564.

distribution to higher velocities compared with collisions at smaller impact parameters, which weight sideways scattering more heavily with lower product velocities. The abstraction of I atoms from CH_3I molecules by Br atoms gives[109] IBr product recoiling backward.

The i.r. chemiluminescence method has been used[110] to measure vibration-rotational state distributions for the HF product from the F + HBr reaction at thermal energy and to compare with the distributions[111] from the F + HCl reaction. The F + HBr reaction forms HF only in low vibrational states and is closer to the statistical expectation than the higher vibrational excitation of HF from F + HCl. However, the HF rotational state distributions from F + HCl are closer to statistical values than those from F + HBr. The F + HBr reaction yeilds *ca.* 7% of the Br atom products in the $Br(^2P_{\frac{1}{2}})$ electronically-excited state; a fraction which is intriguingly similar to the fraction of electronically excited $F(^2P_{\frac{1}{2}})$, which is predicted to be present in the F atom beam on the basis of the thermal equilibrium at 300 K. The i.r. chemiluminescence method has also been used[112] to determine vibration–rotational state distributions for HF and DF products from the H, D atom abstraction reactions of F atoms with HCO_2H, DCO_2H, and H_2CO. Abstraction of a formyl hydrogen proceeds by a direct mechanism yielding high vibrational excitation $f'_V = 0.55$ of HF, DF products, while abstraction of a carboxylic hydrogen proceeds *via* a long-lived collision complex yielding a statistical distribution of HF product vibrational states corresponding to a temperature of 4300 K. The possibility of a migratory mechanism has been suggested for the abstraction of a carboxylic hydrogen with the F atom initially adding to the carbonyl group followed by migration of either the F or H atom. Measurements[113] on H atom abstraction from H_2CO and HFCO confirm the high vibrational excitation of the HF product and indicate that the HCO and FCO radical products are also vibrationally excited in the asymmetric stretch and bending modes. Product HF, DF vibrational states also show[113] nonstatistical distributions for the reaction of F atoms with $HCOND_2$, CH_3ND_2, and CH_3OD, which differ according to whether an CH, ND, or OD bond has been ruptured. Clearly the abstraction of amine and hydroxy-hydrogen atoms also proceeds by a direct mechanism. The product distributions differ due to the differing degrees of attraction experienced by the reactant F atom in each of these direct reactions. However abstraction of an H atom from HN_3 appears to proceed[114] by a long-lived collision complex mechanism.

The i.r. chemiluminescence method has also been used in an elegant first study[115] of the reaction of an atom with an unstable free radical. The pre-reaction of H

[109] E. Vietzke, M. Erdweg, J. Henschkel, L. Matus, and G. Stöcklin, VIIth Int. Symp. Molecular Beams (Riva del Garda), 1979, 102.
[110] D. Brandt, L. W. Dickson, L. N. Y. Kwan, and J. C. Polanyi, *Chem. Phys.*, 1979, **39**, 189.
[111] A. M. G. Ding, L. J. Kirsch, D. S. Perry, J. C. Polanyi, and J. L. Schreiber, *Faraday Discuss. Chem. Soc.*, 1973, **55**, 252.
[112] R. G. McDonald and J. J. Sloan, *Chem. Phys.*, 1978, **31**, 165; *Faraday Discuss. Chem. Soc.*, 1979, **67**, 128.
[113] R. G. McDonald and J. J. Sloan, *Chem. Phys.*, 1979, **40**, 321; R. G. McDonald, J. J. Sloan, and P. T. Wassell, *Chem. Phys.*, 1979, **41**, 201.
[114] J. J. Sloan, D. G. Watson, and J. S. Wright, *Chem. Phys.*, 1979, **43**, 1.
[115] B. A. Blackwell, J. C. Polanyi, and J. J. Sloan, *Chem. Phys.*, 1977, **24**, 25.

atoms with NO_2 or O_3 molecules was used in the beam source to form vibrationally excited OH radicals and their reaction with Cl atoms followed by observation of both the depletion of OH and the production of HCl i.r. chemiluminescence. The substantial vibrational excitation of the OH radicals, *ca.* 86 kJ mol^{-1} from H + NO_2 and *ca.* 300 kJ mol^{-1} from H + O_3, is channelled very efficiently into vibrational excitation of the HCl product with only a small increase in HCl rotational energy in the more energetic case. The HCl product vibrational energy distribution has a similar width to that of the OH reactants. Hence the Cl + OH reaction does not proceed *via* a long-lived collision complex mechanism as would be the case if the HOCl molecule were involved as a reaction intermediate.

Oxygen Atom Reactions.—Full contour maps of the differential reaction cross-section have been determined[116] as a function of translational energy for a range of oxygen atom reactions using a supersonic beam of oxygen atoms seeded in inert buffer gas from a microwave discharge source[13] and mass-spectrometric detection with cross-correlation[20] time-of-flight analysis. The reaction of O atoms seeded in He buffer gas with CS_2 molecules[117, 118] at an initial translational energy E = 38 kJ mol^{-1} gives OS product scattering in the forward direction as illustrated by a contour map of the differential reaction cross-section (Figure 14). The differential reaction cross-section for O atoms seeded in Ne with CS_2 molecules[117, 118] at an initial translational energy E = 13 kJ mol^{-1} is found to be identical, to within

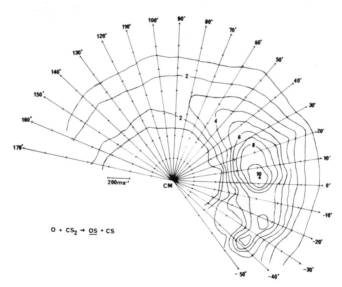

Figure 14 *Differential reaction cross-section for OS product from the O + CS_2 reaction at an initial translational energy E = 38 kJ mol^{-1}*
(Reproduced by permission from *Mol. Phys.*, 1979, **37**, 329)

[116] R. Grice, *Faraday Discuss. Chem. Soc.*, 1979, **67**, 16.
[117] P. A. Gorry, C. V. Nowikow, and R. Grice, *Chem. Phys. Lett.*, 1977, **49**, 116; 1978, **55**, 19.
[118] P. A. Gorry, C. V. Nowikow, and R. Grice, *Mol. Phys.*, 1979, **37**, 329.

experimental uncertainty, with that obtained at the higher initial translational energy. Thus the O + CS₂ reaction follows a stripping mechanism independent of initial translational energy in this energy range despite the small total reaction cross-section $Q \simeq 3$ Å². A minor fraction of the total energy available to reaction products $f_T' \simeq 0.31$ is disposed into product translation which, when taken with the fraction $f_V' \simeq 0.26$ disposed into vibration determined by flash photolysis experiments,[119] indicates a substantial fraction disposed into product rotation $f_R' \simeq 0.4$. These stripping dynamics may be attributed either to reaction at small impact parameters with short-range attractive interactions or reaction at large impact parameters with a strong orientation requirement for reaction or a combination of both effects. Qualitative MO theory offers a possible rationalization[118] for the high rotational excitation of reaction products by proposing a planar bent *cis*-configuration of the OSCS transition state, which causes bending of the SCS molecule followed immediately by scission of the S—C bond so that bending goes over to rotation of the OS and CS fragments in opposed senses. Direct observation of the rotational state distributions of the OS and CS fragments by laser-induced fluorescence detection should be forthcoming in the near future as a more quantitative test of this qualitative mechanism.

Reactive scattering of O atoms seeded in He buffer gas with Cl₂ molecules[117,120] at an initial translational energy $E = 31$ kJ mol⁻¹ also gives OCl product scattering in the forward direction as illustrated by a contour map of the differential reaction cross-section (Figure 15). However more limited angular distribution data suggest

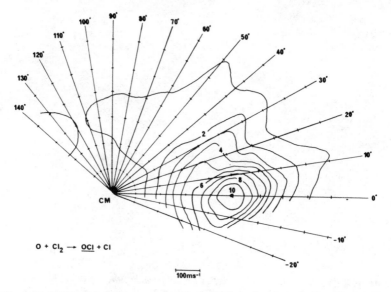

Figure 15 *Differential reaction cross-section for* OCl *product from the* O + Cl₂ *reaction at an initial translational energy* $E = 31$ kJ mol⁻¹
(Reproduced by permission from *Mol. Phys.*, 1979, **37**, 347)

[119] I. W. M. Smith, *Discuss. Faraday Soc.*, 1967, **44**, 194.
[120] P. A. Gorry, C. V. Nowikow, and R. Grice, *Mol. Phys.*, 1979, **37**, 347.

the presence of a subsidiary backward peak of height 0.3 ± 0.1 and measurements with O atoms seeded in Ne buffer gas at a lower initial translational energy $E = 13$ kJ mol^{-1} indicate that the backward peak increases to 0.55 ± 0.15. Thus the O + Cl$_2$ reaction proceeds *via* a short-lived collision complex with an OClCl configuration similar to that suggested for the O + Br$_2$ and I$_2$ reactions.[121,5] However, Figure 15 shows substantially higher OCl product velocity for scattering in the forward direction than for scattering at wide angles. Thus the osculating complex model[107] for reaction *via* a short-lived collision complex, which assumes that the product recoil velocity distribution is essentially independent of scattering angle, is not applicable to the O + Cl$_2$ reaction. It seems more appropriate to regard the OCl reactive scattering as consisting of two components, forward scattering arising from large impact parameter collisions and wide-angle scattering arising from smaller impact parameters. The small total-reaction cross-section, Q *ca.* 1.5 Å^2, implies an orientation requirement for the O + Cl$_2$ reaction that restricts the directions of approach of reactive collisions to a total effective solid angle of *ca.* $\Pi/10$.

Supersonic beams of O atoms seeded in He and Ne buffer gas have also been used[116,122] to measure differential reaction cross-sections of OI scattering from the O + CF$_3$I, C$_2$F$_5$I, C$_3$F$_7$I reactions. Reactive scattering of O atoms seeded in He buffer gas with CF$_3$I molecules at an initial translational energy $E = 32$ kJ mol^{-1} gives an isotropic distribution, but O atoms seeded in Ne buffer gas with CF$_3$I molecules at lower initial translational energy $E = 14$ kJ mol^{-1} favour OI scattering into the backward hemisphere. Thus the I atom abstraction reaction of O atoms with CF$_3$I molecules occurs in collisions at small impact parameters at low energy with hard-sphere repulsive interactions at small internuclear distances. The maximum impact parameter for reaction increases slightly with initial translational energy sufficient to yield isotropic scattering in accord with a hard-sphere scattering model[123] originally developed to explain the rebound reactions of alkali-metal atoms with alkyl iodides. However, the product translational energy distributions for the thermoneutral O + CF$_3$I reaction are strongly skewed with respect to the initial translational energy distributions, indicating that significant energy exchange also occurs with internal modes of the CF$_3$ radical. The importance of energy transfer to the departing fluoro-alkyl radical is confirmed[116] by the product translational energy distributions for the O + C$_2$F$_5$I, C$_3$F$_7$I reactions,[124] which show decreasing energy in product translation as the complexity of the radical increases along the series CF$_3$, C$_2$F$_5$, C$_3$F$_7$, despite the increasing reaction exoergicities. However, the energy disposed into product translation is in all cases greater than would occur for reactions proceeding *via* a long-lived collision complex with full energy equilibration between all the modes of the complex.

More recently, reactive scattering measurements[116,124] of O atoms seeded in He buffer gas with C$_2$F$_4$ molecules at an initial translational energy $E = 36$ kJ mol^{-1} have determined the angular distribution of F$_2$CO product and the product

[121] D. D. Parrish and D. R. Herschbach, *J. Am. Chem. Soc.*, 1973, **95**, 6133.
[122] P. A. Gorry, C. V. Nowikow, and R. Grice, *Chem. Phys. Lett.*, 1978, **55**, 24; *Mol. Phys.*, 1979, **38**, 1485.
[123] J. L. Kinsey, G. H. Kwei, and D. R. Herschbach, *J. Chem. Phys.*, 1976, **64**, 1914.
[124] R. J. Browett, J. H. Hobson, P. A. Gorry, and R. Grice, to be published.

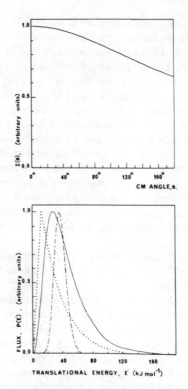

Figure 16 *Product angular and translational energy distributions for the* $O + C_2F_4$ *reaction at an initial translational energy* $E = 36$ kJ mol^{-1} *(solid curves), together with the product translational energy distribution predicted by a long-lived complex model (dotted curve), and the initial translational energy distribution (dashed curve)*

translational energy distribution (Figure 16). This reaction (12) is of particular

$$O + C_2F_4 \rightarrow F_2CO + CF_2 \qquad (12)$$

interest since the O atom cleaves a C=C double bond. The F_2CO product scattering favours the forward hemisphere and disposes only a small fraction $f' = 0.10$ of the total energy available to reaction products into product translation. However, more energy is disposed into product translation than is predicted by the long-lived collision complex model. Thus the observed scattering is in accord with the Cvetanovic mechanism[125] whereby the O atom bonds to the Π orbital of the C_2F_4 molecule forming a short-lived diradical intermediate with a lifetime of approximately one rotational period before dissociating to reaction products. The product translational energy distribution would be compatible with the formation of either ground-state $CF_2(^1A_1)$ or electronically excited $CF_2(^3B_1)$. Reactive scattering of O atoms seeded in Ar buffer gas with ICl molecules has also been measured

[125] R. J. Cvetanovic, *J. Phys. Chem.*, 1970, **74**, 2730; R. E. Huie and J. T. Herron, *Prog. React. Kinet.*, 1975, **8**, 1.

recently[14] and confirms previous observations[5,121] that reaction proceeds *via* a long-lived O—I—Cl collision complex.

The abstraction of H atoms from saturated hydrocarbons has been studied[126] by laser-induced fluorescence detection of OH radical products using an O atom beam from a low-pressure discharge source and a supersonic beam of hydrocarbon molecules seeded in H_2 or He buffer gas. Abstraction of primary H atoms yields OH only in the $v = 0$ state, secondary H atoms OH equally in the $v = 0$ and 1 states, and tertiary H atoms OH predominantly in the $v = 1$ state. The OH products populate only low rotational states in all cases, while variation of the translational energy shows that the activation energies decrease along the sequence of primary, secondary, and tertiary H atoms. This indicates that reaction occurs only for collinear approach of the O atom to the H—C bond, with the barrier to reaction on the potential energy surface moving from the exit valley to midway along the reaction co-ordinate to the entrance valley for the abstraction of primary, secondary, and tertiary H atoms.

Angular distributions of OS and CS reactive scattering from a thermal O atom beam from a low-pressure discharge source and an effusive CS_2 beam have been measured[127] using photoionization mass spectroscopy and confirm that the O + CS_2 reaction proceeds by a stripping mechanism. The use of photoionization in the detection of reactive scattering is important because it permits the unambiguous identification of free radical products even in the presence of their parent polyatomic molecules. The angular distribution of OI scattering from an effusive O atom beam formed in an Ir oven at 2200 K with an effusive I_2 molecule beam has been measured[128] by using electron bombardment mass-spectrometric detection and suggests that scattering is peaked in the backward direction. Chemiluminescence has also been reported[129] from the O + NO reaction, but the high pressures employed in the experiment make the attribution of the results to this reaction somewhat uncertain. Reactive scattering of $O(^1D_2)$ atoms from a r.f. discharge source with CH_4 molecules has recently[14] identified the products as CH_3O + H.

Other Non-metal Atom Reactions.—A preliminary study of the angular distribution of N_2O reactive scattering of an effusive N atom beam from a low-pressure microwave discharge with an NO_2 cross-beam has recently been reported.[130] The results indicate that N_2O product is scattered in the forward direction with most of the exoergicity of the N + NO_2 reaction being disposed into N_2O internal excitation. However, the preliminary nature of these results illustrates the great difficulty of gaining significant dissociation of the strongly bound N_2 molecule in a discharge source without also producing N atoms in metastable electronically-excited states. A high degree of dissociation may be obtained from a direct-current arc source[16] with an extremely high effective source temperature, *ca.* 7000—15000 K. Reactive scattering of a supersonic N atom beam from such a

[126] A. C. Luntz and P. Andresen, VIIth Int. Symp. Molecular Beams (Riva del Garda), 1979, 96.
[127] R. E. Graham and D. Gutman, Int. Mass Spectrometry Symp. (*Salford*), 1977; *Dyn. Mass Spectrom.*, 1978, **5**, 156.
[128] P. N. Clough, G. M. O'Neill, and J. Geddes, *J. Chem. Phys.*, 1978, **69**, 3128.
[129] T. Kasai, T. Masui, H. Nakane, I. Hanazaki, and K. Kuwata, *Chem. Phys. Lett.*, 1978, **56**, 84.
[130] R. A. R. Porter, G. R. Brown, and A. E. Grosser, *Chem. Phys. Lett.*, 1979, **61**, 313.

source with a halogen molecule cross-beam has been reported,[131] but subsequent analysis by the authors indicates that the observed reactive scattering arises from metastable electronically-excited states of the N atoms rather than from ground-state $N(^4S)$ atoms. The associative electron detachment of $N(^2D)$ atoms with $O(^3P)$ atoms[132] has been studied in a merged beam experiment, where both atom beams were formed by charge exchange neutralization of beams of the corresponding atomic ions [equation (13)]. The total-reaction cross-section rises

$$N(^2D) + O(^3P) \rightarrow NO^+ + e^- \qquad (13)$$

from a threshold at an initial translational energy, $E = 37$ kJ mol^{-1}, to a maximum $Q = 0.6$ Å^2 at $E = 1000$ kJ mol^{-1} before declining to zero at $E = 2000$ kJ mol^{-1}.

The dependence of chemiluminescence on initial translational energy over the range 25—125 kJ mol^{-1} has been measured[133,134] for the reaction of metastable electronically excited $Xe(^3P_{2,0})$ atoms with Br_2, ICl, and CCl_4 molecules by electron bombardment excitation of a rotor accelerated Xe beam. The cross-section for XeBr* chemiluminescence, Q *ca.* 200 Å^2, at thermal energy is much greater than that for XeCl* from CCl_4, Q *ca.* 25 Å^2, but both decrease in a very similar manner with increasing initial translational energy. The reaction with ICl yields predominantly XeCl* rather than XeI*. Polarization of the XeBr* chemiluminescence[134,135] is also greater than that for XeCl* and Figure 17 shows that the degree of polarization for both the Br_2 and CCl_4 reactions increases with initial translational energy. Only the XeBr* chemiluminescence approaches the limiting

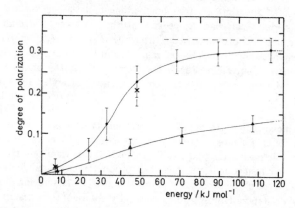

Figure 17 *Polarization of* XeBr* *and* XeCl* *chemiluminescence from the* $Xe(^3P_{2,0})$ + Br_2, CCl_4 *reactions as a function of initial translational energy, shown by solid circles and triangles*
(Reproduced by permission from *Faraday Discuss. Chem. Soc.*, 1979, **67**, 329)

[131] R. L. Love, J. M. Herrmann, R. W. Bickes, and R. B. Bernstein, *J. Am. Chem. Soc.*, 1977, **99**, 8316; R. B. Bernstein, personal communication.
[132] G. Ringer and W. R. Gentry, *J. Chem. Phys.*, 1979, **71**, 1902.
[133] M. R. Levy, C. T. Rettner, and J. P. Simons, *Chem. Phys. Lett.*, 1978, **54**, 120.
[134] C. T. Rettner and J. P. Simons, *Faraday Discuss. Chem. Soc.*, 1979, **67**, 329; J. P. Simons, Y. Ono, and R. J. Hennessy, *ibid.*, p. 358.
[135] C. T. Rettner and J. P. Simons, *Chem. Phys. Lett.*, 1978, **59**, 178.

degree of polarization, $p = 1/3$, within the energy range of the experiments. The breadth and width of the chemiluminescence spectra indicate high vibrational excitation of both the XeBr* and XeCl* products. All of these features mirror the reactions of alkali-metal atoms with Br_2, ICl, and CCl_4 molecules indicating that the reaction of $Xe(^3P_{2,0})$ also follows an electron-jump mechanism.[1] This is to be expected, since the ionization potential of $Xe(^3P_{2,0})$ (369 kJ mol^{-1}) is similar to that of Cs (376 kJ mol^{-1}). However, the symmetry and spin–orbit interaction of the electron transfer from $Xe(^3P_{2,0})$ allows access to a wider manifold of XeBr* and XeCl* electronic states than the electron transfer from an alkali-metal atom, which correlates[136] only with alkali-metal halide product in the ground electronic state in collisions at low energy.

Collisions of Ar, Kr, and Xe atoms with CsCl molecules have been studied[137] up to very high initial translational energies, E ca. 1000 kJ mol^{-1}, by seeding the inert gas atoms in a supersonic expansion of H_2 driver gas from a W nozzle heated to 1500 K and a high source pressure of ca. 8 bar. Collision-induced ionization is the predominant reaction path observed in all cases with cross-sections $Q \gtrsim 4$ Å^2, which increase with translational energy up to the highest energies explored [equation (14)]. However, a step onset for rearrangement ionization is observed at

$$A + CsCl \rightarrow A + Cs^+ + Cl^- \qquad (14)$$
$$(A = Ar, Kr, or Xe)$$

the threshold for collision-induced ionization, which reaches only a very small cross-section, Q ca. 0.04 Å^2, before declining at higher energy. Collisions of Xe with

$$A + CsCl \rightarrow CsA^+ + Cl^- \qquad (15)$$
$$(A = Ar, Kr, or Xe)$$

CsI molecules [equation (16)] also exhibit[138] the alternative channel for rearrangement ionization that is not observed for CsCl.

$$Xe + CsI \rightarrow Cs^+ + XeI^- \qquad (16)$$

This indicates that the outcome of these high-energy collisions is determined by hard-sphere interactions where the masses of the individual atoms have a dominant effect. Rearrangement ionization arises from collinear collisions in which the central ion of the collision complex can depart with the inert gas atom only when it has a mass similar to those of the other atoms of the complex. Collisional dissociation does not require a stringently collinear collision complex and consequently has a much larger total cross-section.

Collisional ionization has also been observed[137,138] for alkali-metal halide dimers, but rearrangement ionization was not detected.

Free-radical Reactions.—In contrast to the extensive range of atom reactions detailed earlier, there have been very few studies of free-radical reactions in

[136] J. L. Magee, *J. Chem. Phys.*, 1940, **8**, 687; R. Grice and D R. Herschbach, *Mol. Phys.*, 1974, **27**, 159.
[137] S. H. Sheen, G. Dimoplon, E. K. Parks, and S. Wexler, *J. Chem. Phys.*, 1978, **68**, 4950.
[138] S. Wexler and E. K. Parks, VIIth Int. Symp. Molecular Beams (Riva del Garda), 1979, 107.

molecular beams. Early studies[139] involved the production of a CH_3 radical beam by pyrolysis of appropriate precursors in a Ta tube at *ca.* 1800 K and the measurement of CH_3 reactive scattering with cross-beams of halogen molecules by use of mass spectrometric detection and conventional time-of-flight analysis. More recently OH radicals have been produced[115] by the reaction of H atoms from a low-pressure microwave discharge with NO_2 or O_3 molecules in a beam source consisting of two concentric glass tubes and have been used in an i.r. chemiluminescence study of the Cl + OH reaction that was discussed earlier. A development of this form of OH radical source whereby a jet of NO_2 molecules passes through a low-pressure flow of H atoms in a glass beam source, has been used to measure[140] reactive scattering of OH radicals with a Br_2 cross-beam by mass-spectrometric detection of HOBr product with cross-correlation time-of-flight analysis. The contour map of HOBr reactive scattering was found to be in good agreement with the RRKM–AM model[108] for scattering *via* a long-lived collision complex. Indeed the reactive scattering is very similar to that observed for the reactions[121,5] of O atoms with Br_2 and I_2 molecules, which are believed to proceed *via* O—Br—Br and O—I—I collision complexes. Qualitative MO theory[140] suggests that the HO—Br—Br collision complex may gain stability by adopting a bent staggered configuration.

Chemiluminescence from electronically excited $CN(B^2\Sigma^+)$ has been observed[141] by crossing a carbon beam from a graphite sublimation source with an NO molecule beam. The time-delay distribution of $CN(B^2\Sigma^+)$ chemiluminescence measured by chopping the carbon beam was used to identify the reactive species as the C_2 molecule from the three principal species C, C_2, and C_3 present in the carbon beam [equation (17)]. The chemiluminescence spectrum indicates a $CN(B^2\Sigma^+)$

$$C_2 + NO \rightarrow CN(B^2\Sigma^+) + CO \tag{17}$$

vibrational state distribution that is approximately statistical. Electronic-state correlation diagrams have been constructed to suggest that the lowest triplet state $C_2(X'^3\Pi_u)$ may be the reactant leading to chemiluminescence rather than the ground singlet state $C_2(X^1\Sigma_g^+)$.

Molecule–Molecule Reactions.—Chemiluminescence from the reaction of NO molecules with O_3 molecules has been thoroughly investigated[142] by scattering a supersonic beam of NO seeded in H_2 and He driver gases with O_3 gas contained in a scattering cell. Both visible chemiluminescence from electronically-excited NO_2^* and i.r. chemiluminescence from vibrationally excited $NO_2^\dagger$ in the electronic ground-state are observed to increase rapidly with initial translational energy but at different rates. Hence it appears that the two sources of chemiluminescence come from different reaction paths and that the i.r. chemiluminescence does not arise by

[139] J. A. Logan, C. A. Mims, G. W. Stewart, and J. Ross, *J. Chem. Phys.*, 1976, **64**, 1804; L. C. Brown, J. C. Whitehead and R. Grice, *Mol. Phys.*, 1976, **31**, 1069.

[140] I. Veltman, A. Durkin, D. J. Smith, and R. Grice, *Faraday Discuss. Chem. Soc.*, 1979, **67**, 248; *Mol. Phys.*, 1980, **40**, 213.

[141] H. F. Krause, *J. Chem. Phys.*, 1979, **70**, 3871.

[142] A. E. Redpath, M. Menzinger, and T. Carrington, *Chem. Phys.*, 1978, **27**, 409; see also D. van den Ende and S. Stolte, *ibid.*, 1980, **45**, 55.

conversion from the electronically-excited state that yields visible chemi-luminescence. Indeed, heating the nozzle of the seeded NO beam enhances the visible chemiluminescence suggesting that the $NO(^2\Pi_{3/2})$ fine-structure component forms NO_2^* while the ground-state component $NO(^2\Pi_{1/2})$ forms $NO_2^\dagger$.

Chemiluminescence has also been observed[143] by crossing a beam of metastable electronically excited $O_2(^1\Delta_g)$ molecules, from a low-pressure O_2–He discharge in a tube coated with HgO to remove O atoms, with supersonic beams of olefin molecules seeded in H_2 driver gas. The chemiluminescence cross-section for *NN*-dimethylisobutenylamine was independent of initial translational energy but the cross-sections for methyl vinyl ether and 1,1-diethoxyethylene increased rapidly with initial translational energy. The same apparatus has also been used[144] with a supersonic beam of SbF_5 molecules seeded in H_2 driver gas crossing a similar beam of various organic halide molecules, RX. Chemi-ionization is observed from the abstraction of a halogen anion by SbF_5 [equation (18)] with cross-sections which

$$SbF_5 + RX \rightarrow SbF_5X^- + R^+ \qquad (18)$$

are insensitive to initial translational energy but increase rapidly with the vibrational energy of the organic halide molecules.

4 Theoretical Interpretation

This section will be restricted to reviewing theoretical models that have a direct application to the interpretation of experimental data.

Angular Correlations.—A significant extension in the means of interpretation has been formulated by Herschbach in a series of papers[145–149] dealing with correlations between the vectors involved in a reactive collision. These vectors may be taken to be the observable initial and final relative velocities **k**, **k′** and reactant and product angular momenta **j**, **j′** together with the conserved total angular momentum **J**, as illustrated in Figure 18 for the reaction of an atom with a diatomic molecule where **l**, **l′** denote the initial and final orbital angular momenta. The correlation[145] of the four observable vectors, denoted by $W(\hat{\mathbf{k}},\hat{\mathbf{j}},\hat{\mathbf{k}}',\hat{\mathbf{j}}')$ where the caret diadem indicates

$$\mathbf{l} + \mathbf{j} = \mathbf{J} = \mathbf{l}' + \mathbf{j}' \qquad (19)$$

unit vectors, may in principle be measured in an experiment with laser excitation of reactants and laser fluorescence detection of products. However, the most complete experimental data that has been measured so far[146] is the orientation of CsI angular momentum as a function of scattering angle for the Cs + CH_3I reaction by using an electrostatic deflecting field. This represents the triple vector correlation [equation (20)] obtained by averaging over the reactant angular momentum **j**. More familiar

$$W(\hat{\mathbf{k}}, \hat{\mathbf{k}}', \hat{\mathbf{j}}') = \langle W(\hat{\mathbf{k}}, \hat{\mathbf{j}}, \hat{\mathbf{k}}', \hat{\mathbf{j}}') \rangle_j \qquad (20)$$

[143] K. T. Alben, A. Auerbach, W. M. Ollison, J. Weiner, and R. J. Cross, *J. Am. Chem. Soc.*, 1978, **100**, 3274.

[144] A. Auerbach, R. J. Cross, and M. Saunders, *J. Am. Chem. Soc.*, 1978, **100**, 4908.

[145] G. M. McClelland and D. R. Herschbach, *J. Phys. Chem.*, 1979, **83**, 1445.

[146] D. A. Case and D. R. Herschbach, *Mol. Phys.*, 1975, **30**, 1537.

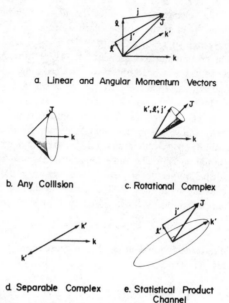

a. Linear and Angular Momentum Vectors

b. Any Collision

c. Rotational Complex

d. Separable Complex

e. Statistical Product Channel

Figure 18 *Vector symmetries for reactions proceeding* via *a long-lived collision complex. a, linear and angular momentum vectors; b, any collision; c, rotational complex; d, separable complex; and e, statistical product channel*
(Reproduced by permission from *J. Phys. Chem.*, 1979, **83**, 1445)

experiments measure the angular distribution $I(\theta)$ of reactive scattering, which is the double vector correlation $W(\hat{\mathbf{k}}, \hat{\mathbf{k}}')$ or measure the polarization of product angular momentum averaged over scattering angles $I(\chi)$, which is $W(\hat{\mathbf{k}}, \hat{\mathbf{j}}')$. Double vector correlations require the specification of one angle, triple vector correlations three angles, and the quadruple vector correlation five angles. In all cases the correlation functions may be expected to depend on the magnitude of the vectors involved. Thus the full set of correlation functions[147] represent a complete classification of the information which may be obtained from molecular beam scattering experiments.

The special case of a long-lived collision complex imposes specific symmetries on the distributions of vectors and this may be used to classify the type of complex (Figure 19). A scattering experiment with unpolarized reactants will have an azimuthally symmetric distribution of the vectors **j**, **l**, **J**, **j'**, **l'**, and **k'** about **k** but a *rotational* complex with a lifetime distribution both longer and broader than the rotational period will also have an azimuthally symmetric distribution of the vectors **k'**, **l'** and **j'** about **J**. However, the product angular distribution for a rotational complex does not necessarily exhibit forward–backward symmetry contrary[145] to earlier conclusions. The azimuthal symmetry of **k'** about **J** and the forward–backward symmetry of **J** imposed by isotropic coupling of **l** and **j** are

[147] D. A. Case and D. R. Herschbach, *J. Chem. Phys.*, 1976, **64**, 4212.

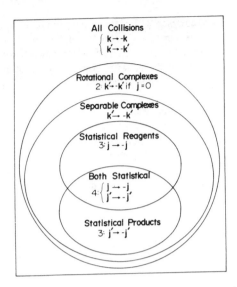

Figure 19 *Venn diagram showing the classification of long-lived collision complex mechanisms according to the symmetry properties of their vector correlations* (Reproduced by permission from *J. Phys. Chem.*, 1979, **83**, 1445)

insufficient to guarantee forward–backward symmetry for k' unless J is perpendicular to k or k'. Hence forward–backward symmetry will only be observed in the product angular distribution for a rotational complex in experiments with reactants produced with low rotational energy from a supersonic nozzle expansion or experiments using laser excitation or laser-induced fluorescence detection to observe molecules in low j or j' states. However a *separable* complex has a long rotational lifetime and a dynamical motion which is sufficiently random that it gives equal probability to each final relative velocity k' and its inverse $-k'$. This imposes forward–backward symmetry on the product angular distribution. Thus the observation of forward–backward symmetry in the product angular distribution in experiments with higher values of j and j' requires dynamical randomness rather than just a long rotational lifetime for the complex. The dynamical randomness required for a *separable* complex may be achieved under suitable conditions without rotational–vibrational energy transfer only by a complex possessing a plane of symmetry but rotational–vibrational energy transfer is always required for complexes lacking a plane of symmetry. The classification of *rotational* and *separable* complexes concerns correlations between reagent and product trajectories with the directional properties being uncorrelated in the *separable* case except for the conservation of total angular momentum. A *statistical* complex has uncorrelated vectors within either a reactant or product channel or both subject only to the conservation of angular momentum. This requires for a *statistical* product channel only that J, l', and j' lie in a plane (since $J = l' + j'$) and that k' be perpendicular to l'. Hence k' must be azimuthally symmetric about l' as shown in Figure 18. This implies that the triple vector correlation $W(\hat{k}, \hat{k}', \hat{j}')$ is invariant to inversion of $\hat{\jmath}' \to -\hat{\jmath}'$ for a complex with a *statistical* product channel and similarly $W(\hat{k}, \hat{\jmath}, \hat{k}')$ is invariant to inversion of $\hat{\jmath} \to -\hat{\jmath}$ for a complex with a *statistical* reactant channel. Such invariances may be detected in experiments with laser-induced fluorescence detection of products or with laser pumping of reactants with a circularly polarised laser beam, by the independence of the observed triple vector correlation of the sign of circular polarisation. Clearly a *statistical* channel requires that there be no orientation requirement for passage of trajectories over the transition state in that channel for the formation or dissociation of the collision complex. If a complex has both *statistical* reactant and product channels, the quadruple vector correlation $W(\hat{k}, \hat{\jmath}, \hat{k}', \hat{\jmath}')$ is invariant to simultaneous inversion of $\hat{\jmath}$ and $\hat{\jmath}'$ and this may be detected in experiments using laser excitation of reactants and laser-induced fluorescence detection of products at a selected scattering angle. Thus the classification of collision complexes into the series *rotational, separable,* and *statistical* of increasing randomness may be performed on the basis of the symmetry properties of their vector correlations as indicated by the Venn diagram of Figure 19. This elegant classification identifies the dynamical properties of the complex and specifies appropriate experimental tests. It should therefore provide the plan of campaign for all further work on reactions proceeding *via* long-lived collision complexes.

The double vector correlation of two vectors $\hat{a}, \hat{b}$ may be expanded[145-147] in a Legendre series [equation (21)] where the coefficients are determined by averaging

$$W(\hat{a}, \hat{b}) = \sum_n A_n P_n(\hat{a} \cdot \hat{b}) \tag{21}$$

with the angular correlation as the weight function [equation (22)]. Similarly, triple

$$A_n = (2n + 1) \langle P_n(\hat{a} \cdot \hat{b}) \rangle \tag{22}$$

vector correlations may be expanded in a series of Biedenharn polynomials, which are the appropriate orthogonal, rotationally invariant functions[145-147] formed from products of spherical harmonics

$$W(\hat{a}, \hat{b}, \hat{c}) = \sum_{v\lambda\mu} B_{v\lambda\mu} P_{\mu\lambda v}(\hat{a}, \hat{b}, \hat{c}) \tag{23}$$

The angular distribution of reaction products may thus be written

$$I(\theta) = W(\hat{k}, \hat{k}') = \sum_n A_n P_n(\hat{k} \cdot \hat{k}') = \sum_n A_n P_n(\cos \theta) \tag{24}$$

where θ is the scattering angle between the initial and final relative velocities and the coefficients are given by

$$A_n = (2n + 1) \langle P_n(\hat{k} \cdot \hat{k}') \rangle = (2n + 1) \int_0^\Pi P_n(\cos \theta) I(\theta) \sin \theta \, d\theta \tag{25}$$

If the vectors are considered in the classical approximation they may be related to the total angular momentum vector $\mathbf{J}$ *via* the addition theorem of Legendre polynomials. For a separable complex the polynomial of equation (25) may be factored[147] into a product of reactant and product channel polynomials

$$\langle P_n(\hat{k} \cdot \hat{k}') \rangle = \langle P_n(\hat{k} \cdot \hat{\jmath}) P_n(\hat{\jmath} \cdot \hat{k}') \rangle \tag{26}$$

In the case $j = j' = 0$ the total angular momentum is perpendicular to $\mathbf{k}$ and $\mathbf{k}'$, and hence

$$\langle P_n(\hat{\mathbf{k}} \cdot \hat{\mathbf{k}}') \rangle = [P_n(0)]^2 \tag{27}$$

in which case the summation of equation (24) reduces to

$$I(\theta) \propto 1/\sin \theta \tag{28}$$

However when $j \neq 0$, $j' \neq 0$, the fact that $\mathbf{j}$ and $\mathbf{j}'$ are isotropically coupled to $\mathbf{l}$ and $\mathbf{l}'$ while $\mathbf{k}$ is perpendicular to $\mathbf{l}$ and $\mathbf{k}'$ perpendicular to $\mathbf{l}'$, reduces[147] equation (26) to

$$\langle P_n(\hat{\mathbf{k}} \cdot \hat{\mathbf{k}}') \rangle = [P_n(0)]^2 \langle P_n(\hat{\mathbf{l}} \cdot \hat{\mathbf{J}}) \, P_n(\hat{\mathbf{J}} \cdot \hat{\mathbf{l}}') \rangle \tag{29}$$

Similarly, the polarization of the product angular momentum $I(\chi)$ may be expanded in a Legendre series analogous to equation (24) where χ denotes the angle between the initial relative velocity and the product angular momentum vector and the coefficients of the expansion involve

$$\langle P_n(\hat{\mathbf{j}}' \cdot \hat{\mathbf{k}}) \rangle = P_n(0) \langle P_n(\hat{\mathbf{l}} \cdot \hat{\mathbf{J}}) \, P_n(\hat{\mathbf{J}} \cdot \hat{\mathbf{j}}') \rangle \tag{30}$$

It is remarkable[147] that the averages over the Legendre moments for the reactant channel $\langle P_n(\hat{\mathbf{l}} \cdot \hat{\mathbf{J}}) \rangle$ and the product channel $\langle P_n(\hat{\mathbf{J}} \cdot \hat{\mathbf{k}}') \rangle$, $\langle P_n(\hat{\mathbf{J}} \cdot \hat{\mathbf{j}}') \rangle$ prove to depend only on the parameters $\Lambda = \langle 1/(1 + j) \rangle$ and $\Lambda' = \langle l'/(l' + j') \rangle$ for a wide range of reactions, at least for the lower $n = 2$, 4 moments. This dependence has been summarized[147] in the empirical equation

$$\langle P_n(\hat{\mathbf{l}} \cdot \hat{\mathbf{J}}) \rangle = \exp[-q(1 - \Lambda)^p] \tag{31}$$

where $q = 6.7$, $p = 2.6$ for $n = 2$, and $q = 10.5$, $p = 2.0$ for $n = 4$. The equation also applies to the product channel $\langle P_n(\hat{\mathbf{J}} \cdot \hat{\mathbf{l}}') \rangle$ with Λ', and to $\langle P_n(\hat{\mathbf{J}} \cdot \hat{\mathbf{j}}') \rangle$ with the mirror image of equation (31) reflected through $\Lambda' = \frac{1}{2}$. The averages over the products of Legendre moments in equations (29) and (30) can be approximately replaced by the product of the averages

$$\langle P_n(\hat{\mathbf{k}} \cdot \hat{\mathbf{k}}') \rangle \simeq [P_n(0)]^2 \langle P_n(\hat{\mathbf{l}} \cdot \hat{\mathbf{J}}) \rangle \langle P_n(\hat{\mathbf{J}} \cdot \hat{\mathbf{l}}') \rangle \tag{32}$$

Hence the empirical equation (31) provides a simple approximate method of determining double vector correlations for *separable* complexes. The results of this method have been compared with a previous model[59] for reaction *via* a long-lived collision complex which assumes the transition state for dissociation of the complex to be a symmetric top. The models are in excellent agreement in predicting that long-lived collision complexes have product angular distributions lying between sharp forward–backward peaking ($\Lambda \rightarrow 1$, $\Lambda' \rightarrow 1$) and isotropic ($\Lambda \rightarrow 1$, $\Lambda' \rightarrow 0$). However, the formal possibility of having strong sideways peaking for complexes dissociating *via* an oblate transition state has rarely been observed experimentally. Sideways peaking requires $\mathbf{l}$ or $\mathbf{l}'$ to be perpendicular to $\mathbf{J}$ on average, which implies $j > J$ or $j' > J$. In this case the average over orientations should vary $\hat{\mathbf{l}} \cdot \hat{\mathbf{J}}$ or $\hat{\mathbf{J}} \cdot \hat{\mathbf{l}}'$ over a wide range and smooth out the effect of $\langle P_2(\hat{\mathbf{l}} \cdot \hat{\mathbf{J}}) \rangle$ or $\langle P_2(\hat{\mathbf{J}} \cdot \hat{\mathbf{l}}') \rangle$. The polarization of the product angular momentum is isotropic when the angular distribution is sharply forward–backward peaked and is sharply peaked at $\chi = 90°$ when the angular distribution is isotropic. In experiments with supersonic nozzle beams, where rotational cooling of the reactant molecules gives $\Lambda \rightarrow 1$, a simple sum rule applies for the distributions of product scattering and polarization of product angular momentum

$$0.3 \gtrsim \langle \cos^2 \theta \rangle - \tfrac{1}{2} \langle \cos^2 \chi \rangle \gtrsim 1/3 \qquad (33)$$

where the right-hand equality holds for $\Lambda' \to 0, 1$ and the left-hand equality for $\Lambda' \to \tfrac{1}{2}$. Thus measurement of the polarization of product angular momentum in experiments with supersonic nozzle beams gives information on the product angular distribution *via* the sum rule of equation (33). This may be of particular value in cases where the collision kinematics are unfavourable to the direct observation of the product angular distribution.

The significance of angular correlations is not confined to reactions proceeding *via* a long-lived collision complex. Information theory has been used[148] to show that measurement of the triple vector correlation $W(\hat{k}, \hat{k}', \hat{j}')$ for an impulsive model of direct reaction dynamics gives a greater surprisal than the double vector correlations and product energy distributions together. While this result suggests that measurements of triple vector correlations give much information on the dynamics of direct reactions, it is unable to assess the quality of this information. The classical and quantum theory[149] for laser-induced fluorescence detection of the polarization of product angular momentum shows that plane polarized laser light allows the determination of eight moments of the distribution symmetric with respect to the plane defined by the initial and final relative velocity vectors. Circularly polarized laser light allows the determination of four moments antisymmetric with respect to this plane.

Dynamical Models.—Recent work on models of reaction dynamics has involved the extension and elaboration of previous concepts rather than the introduction of completely new models. This is well exemplifed by an eclectic model[150] for the reactions of alkali-metal atoms with CCl_4 molecules that proceed *via* an electron-jump transition forming $M^+ + CCl_4^-$ with strong repulsion between the $MCl + CCl_3$ reaction products arising from the repulsive CCl_4^- anion state. These reactions exhibit sideways scattering with strong coupling between the product angular and velocity distributions that is reminiscent of the rainbow effect observed in elastic scattering. The eclectic model combines an optical model whereby the reactant and product trajectories are governed by two-body central force potentials, with (a) attraction between reactants arising from covalent–ionic configuration interaction, (b) product repulsion estimated by analogy[151] with photodissociation, and (c) energy transfer to the CCl_3 radical during this repulsion estimated from an impulsive model similar to that of Landau and Teller. The abrupt switch between reactant and product trajectories is assumed to occur only at a single representative geometry of the collision complex which was found to favour a bent $M-Cl-C$ configuration. This model produces a local minimum in the derivative of the scattering angle *vs.* impact parameter and thereby reactive rainbow scattering, due to the angular dependence of the energy transfer efficiency. Hence the product

[148] D. A. Case and D. R. Herschbach, *J. Chem. Phys.*, 1978, **69**, 150.
[149] D. A. Case, G. M. McClelland, and D. R. Herschbach, *Mol. Phys.*, 1978, **35**, 541.
[150] S. J. Riley, P. E. Siska, and D. R. Herschbach, *Faraday Discuss. Chem. Soc.*, 1979, **67**, 27 and 144.
[151] D. R. Herschbach, *Faraday Discuss. Chem. Soc.*, 1973, **55**, 233.

velocity is coupled with the product scattering angle. The scattering predicted by the model is strongly determined by the range of impact parameters that lead to reaction. It gives rebound dynamics when reaction is confined to small impact parameters and stripping dynamics when reaction occurs in collisions at large impact parameters. Thus the eclectic model provides a comprehensive rationalization for the full range of direct reaction dynamics observed in the reactions of alkali-metal atoms with halogen-containing molecules. The application of this product repulsion model to the rebound reactions of alkali-metal atoms with CH_3Br and CH_3I molecules has been modified[152] using information theory to estimate the dependence of product translational energy on reactant translational energy. However a very simple model[153] for the $H + H_2$ reaction, whereby hard-sphere scattering for the reactant and product trajectories connected by specular refection is used to relate the probability of reaction as a function of impact parameter $P(b,E)$ to the angular distribution of reactive scattering $I(\theta,E)$, has been found to give good agreement with quantum scattering calculations. A similar model[154] has been used previously to analyse the results of classical trajectory calculations. Opacity functions $P(b,E)$ have also been calculated [155] using a quadratic form for $\theta(b)$ with several adjustable parameters to analyse alkali-metal atom reactive scattering.

The problem[156] of intramolecular energy exchange and product translational energy distributions for reactions proceeding *via* a long-lived collision complex continues to attract theoretical attention. A statistically adiabatic model[157] for product translational energy distributions from long-lived collision complexes dissociating *via* a tight transition state takes cognisance of the correlation of bending modes of the transition state with rotation of the reaction products. When the quantum spacing of the bending mode is greater than that of the corresponding product rotation, adiabatic correlation of their quantum states will ensure that the excess of energy is disposed into product translation as the transition state dissociates to products. Addition reactions of F atoms with unsaturated hydrocarbons[158] yield product translational energy distributions in good agreement with statistical-phase space theory when a Cl atom[159] but not when a CH_3 radical[160] or an H atom[161] is displaced. This discrepancy in the case of CH_3 radical or H atom products was attributed[158-161] to incomplete randomization of energy over the vibrational modes of the long-lived complex. However, the agreement obtained in the case of Cl product indicates that energy is randomized for these collision complexes which dissociate *via* a loose transition state. Moreover, addition of CH_3 radicals to unsaturated hydrocarbons is typically associated with a steric factor, *ca.* 10^{-3}, which indicates that complexes dissociate to form CH_3 radical products *via* a

[152] E. Pollak and R. B. Bernstein, *J. Chem. Phys.*, 1979, **70**, 3995; see also A. González Ureña, V. J. Herrero, and F. J. Aoiz, *Chem. Phys.*, 1969, **44**, 81.
[153] G. H. Kwei and D. R. Herschbach, *J. Phys. Chem.*, 1979, **83**, 1550.
[154] R. Grice, *Mol. Phys.*, 1970, **19**, 501.
[155] S. M. McPhail and R. G. Gilbert, *Chem. Phys.*, 1978, **34**, 319.
[156] R. A. Marcus, *Ber. Bunsenges. Phys. Chem.*, 1977, **81**, 190.
[157] G. Worry and R. A. Marcus, *J. Chem. Phys.*, 1977, **67**, 1636.
[158] J. M. Parson, K. Shobatake, Y. T. Lee, and S. A. Rice, *Faraday Discuss. Chem. Soc.*, 1973, **55**, 344.
[159] K. Shobatake, Y. T. Lee, and S. A. Rice, *J. Chem. Phys.*, 1973, **59**, 1435.
[160] J. M. Parson, K. Shobatake, Y. T. Lee, and S. A. Rice, *J. Chem. Phys.*, 1973, **59**, 1402.
[161] K. Shobatake, J. M. Parson, Y. T. Lee, and S. A. Rice, *J. Chem. Phys.*, 1973, **59**, pp. 1416, 1427.

tight transition state. Thus the discrepancy in the case of CH_3 products may be explained by energy transfer from bending modes of a tight transition state without invoking any lack of energy randomization. The case of H atom products has been studied more closely in recent experiments[162] by use of a supersonic F atom beam seeded in He and Ar buffer gases to measure the energy dependence of reactive scattering with C_2H_4 molecules [equation (34)]. The product translational energy

$$F + C_2H_4 \rightarrow C_2H_4F \rightarrow C_2H_3F + H \qquad (34)$$

distributions for this reaction continue to show discrepancies with statistical-phase space theory up to the highest initial translational energy $E = 50$ kJ mol^{-1}, and the product angular distribution becomes essentially isotropic. The authors argue that this indicates an incomplete randomization of energy in the C_2H_4F complex since the effect of the potential barrier in the exit valley of the potential-energy surface on the product translational energy should decrease as the energy of the complex increases. However it has been pointed out[163] that the light H atom product of reaction (34) is unable to carry away much of the orbital angular momentum l of the collision. Consequently the total angular momentum of the C_2H_4F complex J is disposed almost entirely into rotational angular momentum j' of the C_2H_3F product. In this limit, the angular correlation theory detailed above predicts zero correlation between J and l'. Hence the distribution of product relative velocity k' is expected to be isotropic even if the C_2H_4F complex does not persist for many rotational periods. Accordingly, the lack of energy randomization observed at high initial translational energy for reaction (34) may be attributable to the lifetime of the C_2H_4F complex decreasing and becoming comparable to its rotational period. Thus we must conclude that unequivocal evidence for incomplete energy randomization in reactions proceeding *via* a long-lived collision complex has not yet been established. However, theoretical studies of polyatomic oscillators[156] often find quasiperiodic motion at low energies with stochastic behaviour dominating only at higher vibrational energies. Reactive collisions will usually sample such a wide range of initial conditions[164] that the observed energy distributions are considerably broadened on this account. Thus, experiments in which photon absorption prepares excited molecules in better defined initial states should prove more fertile ground for measuring the rates of intramolecular energy transfer. Indeed, slow redistribution of vibrational energy is indicated[165] by the laser fluorescence spectrum of laser-excited glyoxal molecules in a recent beam experiment.

An improvement has recently been suggested by Holmlid and Rynefors[166] in the formulation of the RRKM–AM model[108] for the product translational energy distribution from a long-lived collision complex. In the revised model the total particle flux is explicitly conserved in calculating the product translational energy

[162] J. M. Farrar and Y. T. Lee, *J. Chem. Phys.*, 1976, **65**, 1414.
[163] P. E. Siska, *Faraday Discuss. Chem. Soc.*, 1979, **67**, 225; G. M. McClelland and D. R. Herschbach, *ibid.*, p. 251.
[164] K. K. Freed, *Faraday Discuss. Chem. Soc.*, 1979, **67**, 231.
[165] D. M. Lubman, R. Naaman, and R. N. Zare, *Faraday Discuss. Chem. Soc.*, 1979, **67**, 238; see also J. E. Kenny, D. V. Brunsbaugh, and D. H. Levy, *J. Chem. Phys.*, 1979, **71**, 4757.
[166] L. Holmlid and K. Rynefors, *Chem. Phys.*, 1977, **19**, 261; *Faraday Discuss. Chem. Soc.*, 1979, **67**, 228.

distribution $p(E')$ for a complex with specific reactant and product barriers E_B, E'_B, by dividing the flux f' at a specific product translational energy E' by the total outward flux in the reactant and product channels

$$p(E') = f'/(\int_{E_B}^{E_t} f \, dE + \int_{E'_B}^{E'_t} f \, dE') \qquad (35)$$

where E_t and E'_t denote the total energies available to reactants and products. In the RRKM–AM model[108] the dependence of the denominator of equation (35) on E_B and E'_B was ignored in favour of gaining a mathematically handy final expression. Substituting the classical level densities for the fluxes f and f' in the reactant and product channels, equation (35) becomes

$$p(E') = (s-1)(E'_t - E')^{s-2}/[(E'_t - E'_B)^{s-1} + (E_t - E_B)^{s-1}] \qquad (36)$$

where the number of oscillators s is assumed to be the same for both transition states. When the reactant and product barriers are identified with the centrifugal barriers and $\mathbf{l} \simeq \mathbf{J} \simeq \mathbf{l'}$, they are related[108] by

$$E_B = (\mu'/\mu)^{3/2} \cdot (C'/C)^{1/2} \cdot E'_B = DE'_B \qquad (37)$$

where μ, μ' and C, C' denote the reactant and product reduced masses and van der Waals coefficients. In this case the product translational energy distribution may be obtained by integrating over the range of product barriers

$$P(E') \propto (s-1)(E'_t - E')^{s-2} \int_0^{min} [(E'_t - E'_B)^{s-1} + (E_t - DE'_B)^{s-1}]^{-1}$$

$$\times E'^{-1/3}_B \cdot dE'_B \qquad (38)$$

where $min = min[(E'_{BM}/E'_t)^{1/3}, (E'/E'_t)^{1/3}]$ and $E'_t \geqslant E'_B$, otherwise $E'_t - E'_B$ is replaced by zero. When the product barrier is low compared with E'_t equation (38) approximates closely to the RRKM–AM expression[108] but significant differences may occur in other cases.

Trajectory Calculations and Potential-energy Surfaces.—Monte Carlo calculations of classical trajectories have long provided the principal means of relating experimentally observed reaction dynamics to the underlying structure of the reaction potential-energy surface. The effectiveness of reactant vibration in overcoming barriers in the exit valley of the potential energy surface is often associated with a further requirement for a smaller amount of reactant translational energy. This has now been investigated[167] in trajectory calculations with gradual and sudden late-energy barriers, which show that the gradual barrier extends into the entrance valley and translational energy is required to overcome this part of the barrier. Sudden barriers do not extend into the entrance valley and translational energy is unnecessary. Trajectory studies on the $O + CS_2$ reaction[168] show that excitation of the symmetric stretch vibration of the CS_2 molecule is more effective in promoting reaction than excitation of the asymmetric stretch vibration. This arises because the symmetric stretch mode lies at lower frequency than the asymmetric

[167] J. C. Polanyi and N. Sathyamurthy, *Chem. Phys.*, 1978, **33**, 287.
[168] G. C. Schatz, *J. Chem. Phys.*, 1979, **71**, 542.

stretch and correlates with motion along the reaction co-ordinate in the OSCS transition state. Trajectory calculations on the motion of a collinear collision complex[169] in a deep potential-energy well identify two types of complex trajectory: one reflects into the product valley and becomes direct with increasing initial translational energy while the other reflects toward the plateau and remains complex with increasing translational energy. Statistical behaviour has been related[169] to exponential divergence of adjacent trajectories in calculations on the K + NaCl and H + ICl reactions. The branching factor between HF and DF products from the F + HD reaction determined[170] by trajectory calculations demonstrates a dependence on the HD rotational state that is not in agreement with the predictions of information theory. The activation energies differ for the two reaction paths and this is not accommodated in the information theory. However, information theory has been extended[171] to cover the relationship between energy distributions over two different product degrees of freedom. This demonstrates that translational energy distributions measured in reactive scattering experiments and internal energy distributions measured by laser-induced fluorescence or i.r. chemiluminescence experiments are not redundant. Polarization of the HBr product angular momentum as a function of scattering angle has been studied[172] in trajectory calculations on the H + Br_2 reaction.

The diatomics-in-molecules method of calculating potential energy surfaces[173] from the potential energy curves of the ground and electronically-excited states of the constituent diatomic molecules has been reformulated by use of a projection operator technique and applied to the F + H_2 potential energy surface. The valence bond method[174] has been used to calculate a potential energy surface for the H + Br_2 reaction. Each method has the advantage of incorporating the correct topology[175] for the surface, provided that an adequate number of diatomic states or basis functions are included in the calculation. However, such semi-empirical and *ab initio* methods are usually confined to light systems, and for heavier systems with a greater number of electrons more empirical methods become necessary. An empirical potential-energy surface has been constructed[176] for the Hg + I_2 reaction that agrees with the main features observed in reactive scattering experiments[76] on this system. Semi-empirical surfaces have been calculated[177] for the reactions Li + HF, F_2, which involved an electron-jump transition, and the $^1A'$, $^3A''$, $^1A''$, B^1A'

[169] D. E. Fitz and P. Brumer, *J. Chem. Phys.*, 1979, **70**, 5527; J. W. Duff and P. Brumer, *J. Chem. Phys.*, 1979, **71**, pp. 2693, 3895.

[170] J. C. Polanyi and J. L. Schreiber, *Chem. Phys.*, 1978, **31**, 113.

[171] G. S. Arnold and J. L. Kinsey, *J. Chem. Phys.*, 1977, **67**, 3530.

[172] N. C. Blais and D. G. Truhlar, *J. Chem. Phys.*, 1977, **67**, 1540.

[173] M. B. Faist and J. T. Muckerman, *J. Chem. Phys.*, 1979, **71**, pp. 225, 233; for reviews of semi-empirical calculations of potential energy surfaces, see P. J. Kuntz, in 'Atom-Molecule Collision Theory,' ed. R. B. Bernstein, Plenum Press, New York, 1979, p. 79; J. C. Tully, in 'Semi-empirical Methods of Electronic Structure Calculation,' ed. G. A. Segal, Plenum Press, New York, 1977, p. 173.

[174] P. Baybutt, F. W. Babrowicz, L. R. Kahn, and D. G. Truhlar, *J. Chem. Phys.*, 1978, **68**, 4809.

[175] D. J. Mascord, P. A. Gorry, and R. Grice, *Faraday Discuss. Chem. Soc.*, 1977, **62**, 255; P. A. Gorry and R. Grice, *ibid.*, pp. 318, 320; W. H. Gerber and E. Schumacher, *J. Chem. Phys.*, 1978, **69**, 1692; J. Kendrick, and I. H. Hillier, *Mol. Phys.*, 1977, **33**, 635; E. R. Davidson, *J. Am. Chem. Soc.*, 1977, **99**, 397.

[176] T. M. Mayer, J. T. Muckerman, B. E. Wilcomb, and R. B. Bernstein, *J. Chem. Phys.*, 1977, **67**, 3522.

[177] Y. Zeiri and M. Shapiro, *Chem. Phys.*, 1978, **31**, 217; *J. Chem. Phys.*, 1979, **70**, 5264.

surfaces for the $O(^3P, ^1D) + H_2$ system have been studied in configuration interaction calculations.[178] Walsh triatomic MO theory, which has long served[3, 179] as a qualitative method of rationalizing reactive scattering measurements, has now been extended[118, 140] to the tetratomic HAAB and BAAB cases. While data on tetratomic molecules with which to test the theory are rather sparse, the comparisons that can be made are reassuring in cases where the electronegativity of B is much greater than that of A (Figure 20). Thus tetratomic MO theory should be of assistance in rationalizing the dynamics of four-atom reactions where the range

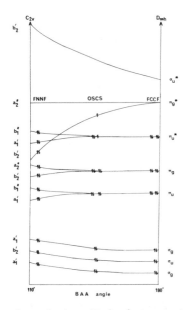

Figure 20 *Walsh diagram for molecular orbitals of tetra-atomic molecules of the form BAAB, where the electronegativity of A is less than that of B, predicts a planar bent configuration for the transition state of the O + CS₂ reaction*
(Reproduced by permission from *Mol. Phys.*, 1979, **37**, 329)

of geometrical configurations which the system may adopt is much more extensive than in three-atom reactions.

Finally, mention must be made of calculations[180] on the effect of intense non-resonant laser irradiation on the dynamics of chemical reactions. It is natural to hope that direct spectroscopic measurements on reaction transition states might be achieved in this way, but calculations[180] on the $F + H_2$ reaction irradiated by Nd-glass laser are discouraging in that terawatts per cm² of laser power were found to be necessary to produce observable effects. The aim of performing spectroscopy

[178] R. E. Howard, A. D. McLean, and W. A. Lester, *J. Chem. Phys.*, 1979, **71**, 2412.
[179] R. Grice, M. R. Cosandey, and D. R. Herschbach, *Ber. Bunsenges. Phys. Chem.*, 1968, **72**, 975; C. F. Carter, M. R. Levy, and R. Grice, *Faraday Discuss. Chem. Soc.*, 1973, **55**, 357.
[180] P. L. De Vries, T. F. George, and J. M. Yuan, *Faraday Discuss. Chem. Soc.*, 1979, **67**, 90; A. E. Orel and W. H. Miller, *Chem. Phys. Lett.*, 1978, **57**, 362.

on reaction transition states might more readily be approached[181] by observing the shift of atomic emission lines from electronically-excited atomic reaction products due to the small fraction of emission that occurs while the atom is still interacting with the reaction transition state. Indeed the absorption analogue of this type of experiment might be performed[181] in a beam experiment with more modest laser power using the high sensitivity of laser resonance fluorescence detection or two-photon ionization.

5 Conclusions

This review demonstrates the rapidly increasing maturity of the field of molecular beam studies of reaction dynamics. Experiments are becoming steadily more accurate while exploring a wider range of reactant and product variables and an increasing range of chemical species. However, many aspects are only just beginning to be explored: the analysis of reactant orientations,[182] the polarization of product angular momenta, and the study of free-radical reactions. The increasing detail of experimental measurements and complexity of the reactions that are now accessible to study is beginning to reveal deficiencies in the earlier models of reaction dynamics. In particular, no simple theory for non-alkali-metal reactions has the effectiveness of the electron-jump model for alkali-metal reactions. The diatomics-in-molecules method seems the most promising means of exploring non-alkali-metal potential-energy surfaces for three- and four-atom systems. Polyatomic systems have been investigated in detail only when reactions proceed *via* a long-lived collision complex. Thus the extensive range of techniques now available for the study of reaction dynamics in molecular beams promises to extend our knowledge into the dynamics of polyatomic systems proceeding by direct mechanisms while deepening our understanding of the subtleties of the reactions of an atom with a diatomic molecule:

> 'Till out of chaos comes in sight
> Clear fragments of a Whole;
> Man, learning Nature's ways aright,
> Obeying, can control.'
> W.C. Dampier (1929)

---

[181] J. C. Polanyi, *Faraday Discuss. Chem. Soc.*, 1979, **67**, 129.

[182] For a very recent study of K atoms with sideways oriented CF_3I, see P. R. Brooks, J. S. McKillop, and H. G. Pippin, *Chem. Phys. Lett.*, 1979, **66**, 144.

2

Reorientation by Elastic and Rotationally Inelastic Transitions

BY A. J. McCAFFERY

1 Introduction

Numerous experimental methods have been devised over the years to study energy transfer in collisions between molecules and the information on intermolecular interactions that may be extracted from a particular investigation is a function of the degree to which the experimenter is able to specify the variables of the collision; the internal states of the molecules, their relative velocities for example. Quite recently it has become feasible to select internal molecular states very precisely prior to collision and to achieve equally high resolution of the quantum states of product molecules. Different types of experiment have been used to achieve this degree of resolution and in the special case of atom–diatomic molecule collisions there have been many interesting theoretical studies of the collision dynamics of fully state resolved molecule–atom encounters. The main theme in this Review will be the subject of *reorientation* by elastic and inelastic collisions. In quantal terms this concerns the fate of the magnetic quantum number, m, in molecular collisions. In more physical language, reorientation is about the torques exerted on a rotating molecule to turn it about an axis perpendicular to the rotational angular momentum vector. In either vocabulary it is clear that reorientation is a function of the anisotropy of the intermolecular potential and is therefore of considerable interest since it is difficult to obtain information on this part of the potential energy surface.

Much of this survey will concern atom–diatomic molecule collisions. The reason for this is that a wide range of both experiment and theory can be brought within the discussion: molecular beam studies, laser fluorescence, and optical double resonance on the experimental side, while from the theoretical view-point, atom–diatom scattering is sufficiently tractable that rigorous solutions are feasible, and moreover a variety of approximation methods has been developed. It is of interest to discuss what evidence is available concerning reorientation by collisions and to see how various theoretical models cope with this. This Report will not, however, solely consider diatomics, and some important work on polyatomics and its relation to the diatomic work will also be reviewed.

Before embarking on the details of the various experimental and theoretical studies of reorientation it is worth clarifying in physical terms the nature of this process for the case of the rotating diatomic molecule. We begin by considering diatomics that have been produced in well-defined initial rotational levels, j, and which have a definite orientation with respect to some axis: the relative velocity vector or an electric field in the case of a molecular beam experiment, or the light beam or observation direction in the case of a thermal-cell laser-fluorescence experiment. If this can be achieved then the individual molecules within the beam have their *j* vectors aligned in a preferential direction with respect to this reference or quantization axis, *i.e.*, the array of molecules is *polarized*. Reorientation consists of a rotation of the plane in which the diatomic spins such that the *j* vector tilts and projects a new value onto the quantization axis. It may result from either an elastic or an inelastic collision. In terms of the quantum states involved the problem may be described more succinctly. An array of diatomics is produced having definite specified values before collision of its quantum numbers: electronic, vibrational, rotational, and magnetic. Following the encounter the flux into each of the molecule's quantum *channels* is determined. Cross-sections for changing *j*, *m*, etc., may be determined and related back to the intermolecular potential effective during the collision. In this way, maximum 'resolution' of the collision process is obtained at the microscopic level. In reality, numerous factors impede these high quantum resolution experiments and some of the problems will be discussed.

The macroscopic behaviour of real gases reflects the forces at work on the microscopic level and full understanding of bulk behaviour comes only from the most detailed experiment and theory. As a further justification for studies in this field, it will be clear from the foregoing that reorientation phenomena are an indication of *steric* forces in collisional interactions and are of considerable importance in discussing the specificity of elementary chemical processes.

In this Report we shall first discuss the very small number of fully state-selected molecular beam experiments that have been performed. A range of optical experiments is then described, and finally scattering theory calculations on fully state-selected molecules are briefly reviewed in relation to these experiments. Thus we start with the most detailed experiments and then move on to consider the more 'averaged' techniques. In principle we expect the former, which are generally difficult experiments, to yield more detailed information than the latter, though it should be emphasized that significant discoveries in this field have been made in experiments on molecules in thermal cells where the results are quite highly averaged.

2 Molecular Beam Experiments

The most detailed observations use crossed molecular beams, each with definite velocity and with well-defined internal energy states $|njm\rangle$. Determination of velocity and angular dependence of scattered products for a range of specified final internal energy states $|n' j' m'\rangle$ represents the ideal experiment and would yield cross-sections $\sigma_{njm,n'j'm'}$ as a function of these variables. Not surprisingly, this degree of resolution has rarely been achieved. Some years ago Bennewitz and co-workers[1,2] began a series of experiments on the orientation dependence of the total cross-section for state selected molecules. Beams of TlF molecules[1] were produced

[1] H. G. Bennewitz, K. H. Kramer, W. Paul, and J. P. Toennies, *Z. Phys.*, 1964, **177**, 84.
[2] H. G. Bennewitz, R. Gegenbach, R. Haerten, and G. Müller, *Z. Phys.*, 1969, **226**, 279; **227**, 399.

in the $|jm\rangle = |1, m\rangle$ rotational state by an electrostatic four-pole field, which selectively refocusses specific $|jm\rangle$ states of polar diatomics by a second-order Stark interaction. This beam was crossed with a secondary beam in the scattering region and by changing the direction of the electric field in this region it was possible to produce $|1, 0\rangle$ or $|1, 1\rangle$ states with respect to the secondary beam direction. In this way the ratio of total scattering cross-sections $\sigma(1, 1)/\sigma(1, 0)$ was measured for scattering from rare-gas partners. The ratio was found to be very close to unity for the rare gases and the results were interpreted with an assumed purely attractive r^{-6} potential of the form $A(1 + q \cos^2 \theta)$ where q is an angular anisotropy parameter. The value of $q = 0.40$ was derived from the experimental data. Later work by Bennewitz *et al.*[2-4] extended the techniques to CsF in the $|2, 2\rangle$, $|2, 0\rangle$ states scattered off rare gases; again the results were interpreted with a simple attractive potential in terms of a second-order Legendre function to yield values of the anisotropy parameter $q = 0.23$ for TlF and 0.28 for CsF.

The inversion of this experimental data[1-4] to yield potential parameters was by means of the 'high energy' or 'sudden' approximation. In this the rate of change of classical trajectory is assumed fast compared with rotational period enabling the S matrix to be written in terms of a time-dependent potential for internal co-ordinates fixed throughout the collision, which are then averaged over at the end. The potential is determined by assuming a linear trajectory and the strengths and weaknesses of this procedure (and of related approximations) are described in the review by Balint-Kurti.[5] For example, it may only be used to consider small-angle scattering.[5]

The experiments carried out by Reuss and co-workers[6] are in a similar category to those of Bennewitz in that a total cross-section is determined for the scattering of two orientations of a diatomic by the collision partner. The experimental technique is of interest since results on a homonuclear diatomic (H_2) were obtained and the inversion procedure to yield anisotropy parameters differed markedly from that adopted by Bennewitz *et al.* Orientations of the hydrogen molecules in these experiments[6] was by use of a molecular beam magnetic resonance method of the type discussed by Ramsey.[7] Selection of $|jm\rangle$ levels was by deflection in an inhomogeneous magnetic field and an r.f. field was used to monitor populations in initial $|m\rangle$ states from the change of intensity produced by the field, *i.e.*, by the alteration of the so-called 'flop-out' signal. In this manner the selectivity of the scattering region to initial states $m = 0$ and $m = 1$ was determined. The effect of initial orientation on total cross-section is very small. Reuss and co-workers[6,8] have used the distorted-wave Born approximation as the basis of their inversion procedure and have found good agreement with experiment in the H_2–Ar case. The potential used contained both attractive (r^{-6}) and repulsive (r^{-12}) forces with Legendre functions of second-order representing the angular dependent part of the

[3] H. G. Bennewitz, R. Haerten, and G. Müller, *Z. Phys.*, 1969, **226**, 139.
[4] H. G. Bennewitz and R. Haerten, *Z. Phys.*, 1969, **227**, 399.
[5] G. G. Balint-Kurti, 'Int. Rev. Sci., Phys. Chem.', ed. A. D. Buckingham and C. A. Coulson, Butterworths, Boston, 1975, p. 283.
[6] H. Moerkerken, M. Prior, and J. Reuss, *Physica*, 1970, **50**, 499.
[7] N. F. Ramsey, 'Molecular Beams,' Oxford University Press, Oxford, 1956.
[8] J. Reuss and S. Stolte, *Physica*, 1969, **42**, 111.

intermolecular potential. Two anisotropy parameters result from this procedure, namely $q_{2.12}$ and $q_{2.6}$ using an obvious form of nomenclature. As those readers with a knowledge of molecular beams experiments are aware, the velocity-dependence of total cross-sections can yield information on both the long- and short-range potentials when glories are observed. Non-glory contributions sample the long-range forces and enable $q_{2.6}$ to be evaluated, whereas the glory contribution that results from the 'well' region is a function of both $q_{2.6}$ and $q_{2.12}$.

Similar experiments on nitric oxide by Reuss and co-workers[9,10] also yielded values of the anisotropy parameters for a wide range of collision partners. For state-selected beam experiments NO has some advantages, since it has an approximately linear Stark effect, is light, and is relatively easy to handle. In the study by Stolte *et al.*,[10,11] a beam of NO molecules was focused by a six-pole field and the molecule's magnetic moment was utilized for orientation with a magnetic field parallel and perpendicular to the relative velocity vector defined in the scattering region. The electronic structure of NO is quite complex and the experiment selects $j = m = \frac{3}{2}$ levels of the $^2\Pi_{3/2}$ ground state. This necessitates higher-order Legendre polynomials in the potential, a P_4 term being the lowest order of these. Reuss has contributed a useful review of this and the work on H_2,[11] which summarizes the results in terms of anisotropy parameters for the glory and non-glory contributions. The variation with collision partner in the non-glory results is not great, the $q_{2.6}$ parameter being around 0.20–0.25 for the rare gases and for N_2. The long-range anisotropy is considerably less than 0.20 for CCl_4 as collision partner, but only somewhat smaller for SF_6, whereas CO_2 is substantially higher. CO_2 also shows unusual behaviour in one of the optical techniques for studying reorientation, and will be discussed later. It is interesting to note that CS_2 has a relatively small long-range anisotropy parameter and behaves more like a rare gas than like CO_2.

Glory results on the same system yield the ratio of the short- to the long-range anisotropy parameters $q_{2.12}/q_{2.6}$. These values are mostly quite close to unity for H_2–rare gas and NO–Ar, but increase somewhat for NO–Xe, Kr. The review by Reuss[11] discusses these data and the assumptions made in deriving them from experiment. With some assumptions it is possible to obtain from $q_{2.12}/q_{2.6}$ the ratio of the positions of minima for the angularly dependent and angularly independent potentials and, from $q_{2.6}$, the value of the former potential at roughly twice the R value of the potential minimum. In other words the experiment yields, for the systems studied, a sketch of the P_2 angular dependent part of the intermolecular potential as a function of distance. There is still, of course, the dependence of this interpretation on the model chosen to reduce the dimension of the coupled equations in the theoretical treatment, the distorted wave approximation in this instance. We shall consider the validity of this later.

Finally in this section on molecular beam work is what must be regarded as the Rolls-Royce of this type of experiment: a high-resolution measurement of the differential cross-section of state-selected LiF colliding with argon, by Wharton and

[9] S. Stolte, J. Reuss, and H. L. Schwartz, *Physica*, 1972, **57**, 254.
[10] S. Stolte, J. Reuss, and H. L. Schwartz, *Physica*, 1973, **66**, 211.
[11] J. Reuss, *Adv. Chem. Phys.*, 1975, **30**, 389.

co-workers.[12] Both high-frequency oscillations and rainbow features were resolved for polarized and unpolarized LiF and the paper by Wharton[12] describes the experiment in considerable detail. The LiF beam was rotationally and translationally cooled by expansion through a nozzle and passed through a quadrupole lens, which was in fact a three-quarters quadrupole since one pole was at earth potential. A second quadrupole 'purified' the beam, which then intersected a supersonic Ar beam at 90°. The quadrupole lenses were used as mentioned earlier to refocus selectively particular $|jm\rangle$ states that were detected by a hot tungsten wire surface-ionization detector. Wharton *et al.*[12] attributed the success of these very difficult measurements to the intense supersonic beam of LiF molecules, the unusual quadrupole selectors, and to novel stops and apertures as spatial filters, which removed all the background molecules. Also of interest was the employment of the phase space method,[13] used in the design of particle accelerators, as the basis for designing the optical system. A double modulation method was used to obtain differential cross-section data while an on-line PDP 11 controlled the angular scan, modulated the quantization field, *etc.*

This degree of experimental detail has been given here as an indication of the reliance that can be placed on the results obtained in these very careful experiments.[12] Typical results are shown in Figure 1 for polarized $|m| = 1$ and unpolarized LiF molecules and the high-frequency oscillations may be clearly seen. At high energy, rainbow features are present and occur in the same place as in non-state-selected experiments.

The method chosen to reduce the dimension of the quantal formulation, *i.e.*, the coupled radial differential equations of Arthurs and Dalgarno,[14] was the

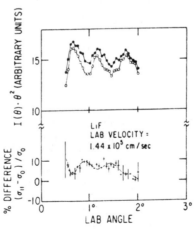

Figure 1 *Laboratory θ^2 weighted differential cross-sections at two different polarizations. Filled circles for $(J,M)(1,1)$; open circles for unpolarized $J = 1$. The % difference is shown in the bottom*
(Reproduced by permission from *J. Chem. Phys.*, 1979, **70**, 5296)

[12] L. Y. Tsou, D. J. Auerbach, and L. Wharton, *J. Chem. Phys.*, 1979, **70**, 5296.
[13] T. C. English and T. F. Gallagher, *Rev. Sci. Instrum.*, 1975, **46**, 24.
[14] A. M. Arthurs and A. Dalgarno, *Proc. R. Soc. London, Ser. A*, 1960, **256**, 540.

infinite-order sudden approximation first introduced by Pack.[15] Wharton *et al.*[12] make the point that the distorted-wave approximation used by Reuss *et al.*[11] to treat total cross-sections, as mentioned earlier, is not suited to systems having substantial long-range anisotropies at high energies, since it is based on the assumption that the anisotropic potential is a small perturbation on the isotropic part. Dalgarno *et al.*[16] have also commented on this unreliability of the distorted-wave approximation under certain circumstances. Not enough is known yet about which simplification of the coupled equations to use in a particular circumstance, though Fitz[17] has compared the IOS with the exponential Born method of Balint-Kurti and Levine[18] for the scattering of oriented CsF by Ar. He concluded that both are reasonable methods for studying m-dependence in small Δj rotational transitions in which the energy of the system is much larger than rotational separations. The IOS method is much faster for computation as several authors have noted.

The potential introduced by Wharton *et al.*[12] is more sophisticated than that used by earlier workers, the spherical part for example containing both r^{-6} and r^{-8} contributions to the attractive term. The angular part is also quite complex with r^{-6}, r^{-7}, and r^{-8} expressions. It has the form:[12]

$$V_{r\theta} = \frac{-C_6 \, q_{2.6}}{r^6} \, P_2(\cos \theta) - \frac{\varepsilon r_m^7}{r^7}[q_{1.7} \, P_1(\cos \theta) + q_{3.7} \, P_3(\cos \theta)] - \frac{C_8 \, q_{2.8}}{r^8} \qquad (1)$$

The experimental differential cross-sections could not be fitted using just the spherical model, nor the spherical part plus a $q_{2.6}$ term, and for this reason the $q_{1.7}$ and $q_{3.7}$ terms were added. Even this was insufficient and the $q_{2.8}$ term was introduced and varied to yield the fit shown in Figure 2. The anisotropic potential parameters used in the calculation are given in the caption and are of considerable interest since very little is known about the anisotropic potential in atom–diatomic molecule systems, particularly at the level of detail described by Wharton.[12] The dependence on the IOS assumptions remains a cautionary note and one to which Wharton *et al.* draw attention.[12]

3 Optical Methods

In this section various optical experiments are described in which the selection and/or detection of molecular $|jm\rangle$ states is by radiative transitions, both induced and spontaneous. It is worth emphasizing the somewhat different approach in many of the experiments discussed here, since it will become clear that the questions asked are different from those asked by molecular beam experimenters. This latter group were directly concerned with the intermolecular potential, particularly the anisotropic part, and have designed experiments to measure this as straightforwardly as possible. In the optical work, a question which is clearly discernible in many experiments is: are there 'selection rules' in atom–molecule collisions analogous with the familiar rules governing photon–molecule interactions? This

[15] T. P. Tsien, G. A. Parker, and R. T. Pack, *J. Chem. Phys*, 1973, **59**, 5373.
[16] A. Dalgarno, R. J. W. Henry, and C. S. Roberts, *Proc. Phys. Soc.*, 1966, **88**, 611.
[17] D. E. Fitz, *Chem. Phys.*, 1977, **24**, 133.
[18] G. Balint-Kurti and R. D. Levine, *Chem. Phys. Lett.*, 1970, **7**, 107.

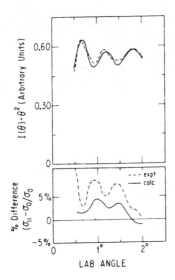

Figure 2 *Calculated differential cross-section in unpolarized condition and difference differential cross-section with anisotropic parameters $q_{1.7} = 0.32$, $q_{3.7} = 0.31$, $q_{2.6} = 0.488$, and $q_{2.8} = 0.40$. Differential cross-sections summed over final j states from j $= 0$ to $j = 3$. Experimental data are shown for comparison*
(Reproduced by permission from *J. Chem. Phys.*, 1979, **70**, 5296)

essentially spectroscopic viewpoint may provide information on the intermolecular potential since the observation of selection or propensity rules is an indication of restrictions on the form of the interaction. There are of course considerable differences in the atom–molecule and photon–molecule interactions. The (almost) zero linear momentum and the well-defined values of angular momentum and parity of the photon, which determine its interaction with molecules, are absent for atom–molecule collisions and perhaps it is naïve to search for analogies. However the great power of spectroscopy is in the wide range of initial channels that can be prepared and, more important, the large numbers of final channels which may be monitored following preparation of a given initial level. Thus information complementary to molecular beam work is available and some very interesting propensities have been observed, some of which were not readily apparent from molecular beam experiments.

Perhaps the best way to order this section is to start with the least 'averaged' experiments with the most variables specified, and then to move on to the more highly averaged 'cell' type experiments. Here also there is further sub-division depending on the degree of selection in preparing initial and monitoring final molecular eigenstates. We also separate double resonance spectroscopic experiments from the more conventional single resonance types.

It is well known that the process of state selection, done in molecular beams with electrostatic multipolar fields, may be achieved using radiation, and for atomic systems this method has been much employed using discharge lamps. Only the advent of the laser has made it feasible to populate significantly single rovibronic states of molecules and with the development of tunable lasers the possibility of exciting specified rotational levels exists. The combination of laser state selection with molecular beam methods represents the least averaged type of experiment and two main approaches are possible. First, one may prepare the molecule in a precisely defined excited state prior to collision, and then monitor the disposal of energy into final rotational levels from the resolved fluorescence spectrum. This involves crossing molecular and laser beams mutually orthogonally and is difficult experimentally. The fact that excited lifetimes are generally short and the molecules will travel only short distances before decaying radiatively is a major impediment in this process. A second method, exemplified recently by Wilcomb and Dagdigian[19] utilizes electrostatic selection of ground-state molecules, in this case lithium hydride, which are then scattered off argon. The populations of final j levels in LiH are then determined by laser-induced fluorescence[20] using a tunable dye laser to scan through the rotational levels of the ground state. The integral cross-sections for rotational transfer obtained from this particular experiment[19] were of interest since they were found not to agree with the predictions of the exponential gap law.[21] In principle the method of Wilcomb and Dagdigian[19] is capable of determining m-channel populations *via* the polarization of the resolved, laser-induced fluorescence and, although the theory for this more refined experiment has been very comprehensively presented by more than one group of workers,[22,23] experimental results are not yet forthcoming.

Further stages of averaging are necessary to treat results of experiments carried out in cells or bulbs, but the ease of operation of thermal cells compared to molecular beams has led to a far wider range of studies of relevance to this review. Probably the most direct way to study reorientation by elastic and inelastic collisions using laser methods is to excite initial $|jm\rangle$ states using polarized radiation and to monitor populations of the final $|j'm'\rangle$ levels, after collision, from the intensity and polarization of fully resolved levels. Atomic physicists of the 1950's made a very comprehensive study of collisional phenomena in atoms by techniques of this kind.[24] The problems in all but the lightest molecules are more severe however because of the close spacing of rotational levels and tunable single-mode lasers are almost essential for systematic work of this kind.

Atomic physicists have used density matrix methods to describe experiments involving atomic polarization[25,26] and these methods have proved to be very powerful in dealing with problems where, by the nature of the experiment, 'pure' quantum states are not prepared. The use of density matrix methods in optical experiments on molecules is not widespread[27] though more recently these techniques have been used to discuss angular momentum polarization in molecular

[19] B. E. Wilcomb and P. J. Dagdigian, *J. Chem. Phys.*, 1977, **67**, 3829.

[20] R. N. Zare and P. J. Dagdigian, *Science*, 1974, **185**, 739.

[21] J. Polanyi and K. Woodall, *J. Chem. Phys.*, 1972, **56**, 1563.

[22] M. H. Alexander, P. J. Dagdigian, and A. E. De Pristo, *J. Chem. Phys.*, 1977, **66**, 59.

[23] D. A. Case, G. M. McClelland, and D. R. Herschbach, *Mol. Phys.*, 1978, **35**, 541.

[24] J. A. Kastler, *J. Opt. Soc.*, 1957, **47**, 460.

[25] W. Happer, *Rev. Mod. Phys.*, 1972, **44**, 169.

[26] U. Fano, *Rev. Mod. Phys.*, 1957, **29**, 74.

[27] An exception to this is the work of Lehmann and colleagues, see *e.g.*, M. Broyer, G. Gouedard, J. C. Lehmann, and J. Vigue, *Adv. At. Mol. Phys.*, 1976, **12**, 165.

collisions.[23,28,29] The extension to molecules is particularly appropriate since pure states are rarely created, even with narrow line polarized laser radiation and this is due to the m degeneracy of most rotational levels not to mention the presence of close-lying hyperfine states. In most cases an incoherent superposition of a number of pure states is prepared, (experiments in which *coherent* superpositions of states are deliberately created will be discussed later in this section) and here the tensor representation of the density matrix is particularly valuable. A most thorough justification of the density matrix method is contained in the review article by Fano.[26] Since it is likely that optical experiments will use this method of analysis in future, a brief summary is given here to introduce the notation and to illustrate the advantage of using tensor methods in conjunction with the density matrix. The terminology is that used by Rowe and McCaffery in two recent papers[28,29] and has its origins in the early work of atomic physicists such as Barrat[30] and subsequent workers.[27]

In the normal formulation of quantum mechanics where a system is believed to be one of a range of pure states spanned by the functions $\psi_{m_1}, \psi_{m_2} \ldots$, the true state of the system is a linear superposition

$$\Psi = \sum_m a_m \psi_m \tag{2}$$

The mean value of an operator Q for this incoherent superposition is

$$\langle Q \rangle = \sum_{mm'} Q_{m'm} \sum_i a_m(i) a_{m'}^*(i) \tag{3}$$

where $Q_{m'm}$ is the matrix of Q. The density matrix $\rho_{mm'}$ is then defined as

$$\rho_{mm'} = \frac{1}{N} \sum_{i=1}^N a_m(i) a_{m'}(i)^* \tag{4}$$

Then $\langle Q \rangle$ becomes $Tr(Q\rho)$.

In dealing with molecular rotational states j having $(2j + 1)$ m components, additional labels are required, and the density matrix is written

$$^{jj'}\rho_{mm'} = \frac{1}{N} \sum_{i=1}^N a_m^j(i) a_{m'}^{j'*}(i) \tag{5}$$

It is very useful in optical experiments to carry out a multipole expansion of the density matrix using linear combinations that transform as irreducible representations of the full rotation group. This is the *tensor* density matrix and is defined by the linear combination

$$^{jj'}\rho_Q^K = \sum_{mm'} (-1)^{j-m} (2K + 1)^{\frac{1}{2}} \begin{pmatrix} j & j' & K \\ m & -m & -Q \end{pmatrix} {}^{jj'}\rho_{mm'} \tag{6}$$

[28] M. D. Rowe and A. J. McCaffery, *Chem. Phys.*, 1978, **34**, 81.
[29] M. D. Rowe and A. J. McCaffery, *Chem. Phys.*, 1979, **43**, 35.
[30] J. P. Barrat, *Proc. R. Soc. London, Ser. A*, 1961, **263**, 371.

where K is the rank and defines the state *multipole*. It has $(2K + 1) = Q$ components. For an individual rotational state, $j = j'$. The value of this multipolar form can be seen when a particular experiment is considered. It will be clear from arguments presented below that in optical absorption and emission, only the multipoles $K = 0$, 1, and 2 are present and thus the series may be truncated at $K = 2$ (note that this may not apply when the ground state is optically pumped). Distributions of m states corresponding to these three multipoles are shown in Figure 3. Here the names (normalized) population, orientation, and alignment are used for $K = 0$, 1, and 2 respectively, though others are in use such as trace, dipole, and quadrupole for 0, 1, and 2. This latter form is used in the analogous treatment of Raman spectroscopy of fluid media and dates back to the original work of Placzek.[31] Two other properties of the $^{jj'}\rho_Q^K$ are worth mention. Being represen-

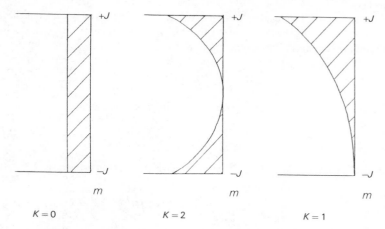

Figure 3 *Distribution of m states within a j manifold corresponding to the multipole population ($K = 0$), alignment ($K = 2$), and orientation ($K = 1$)*

tations of the full rotation group means that the multipoles are invariant to rotations of the co-ordinate frame. Further, under isotropic conditions, such as in a thermal cell, *each multipole polarization is decoupled from all other multipole polarizations and decays with its own characteristic decay time. This may be different for different multipoles*. This last point has relevence beyond the experiments discussed here, since laser radiation is invariably polarized and the balance between the multipoles created for different transitions may vary considerably. The assumption that only the population ($K = 0$) is excited and transferred in laser fluorescence investigations of energy transfer[32] may lead to incorrect interpretations. To determine fully all three multipoles, a combination of linear and circular polarization experiments is needed. This has rarely been attempted.

[31] G. Placzek, in 'Handbuch der Physik,' ed. E. Marx, Vol. 6, 1934, p. 205.
[32] See, *e.g.*, G. Ennen and C. Ottinger, *Chem. Phys.*, 1974, **31**, 404; or R. B. Kurzel, J. I. Steinfeld, D. A. Hatzenbuhler, and G. E. Leroi, *J. Chem. Phys.*, 1971, **55**, 4822.

On exciting an array of molecules with polarized light, multipoles are set up among the m states of the excited rotational level according to the formula

$$^{jj}\rho_Q^K(q_A) = \sum_{m'',m_1,m_2} (-1)^{j-m} (2K + 1)^{1/2} \begin{pmatrix} j & j & K \\ m & -m & Q \end{pmatrix} \langle j'' \, m'' | dq_A | jm_1 \rangle$$
$$\times \langle j'' \, m'' | dq_A | jm_2 \rangle^* \tag{7}$$

where d is the electric dipole moment operator having polarization q_A and in this formulation an unpolarized ($K = 0$) ground state is assumed. The procedure from this point depends on the nature of the experiment, but clearly some simplification may be achieved by determining polarization *ratios* for emission lines, since it will not then be necessary to evaluate the matrix elements fully. In the Reviewer's laboratory, circular polarization ratios are determined following excitation with circularly polarized laser radiation with the direction of propagation determining the quantization axis. Then only diagonal elements $^{jj}\rho_{mm}$ are created. Complications exist when hyperfine coupling increases the number of discrete states lying within the radiative width of the transition. This case is treated in some detail in ref. 28. The main effect of the hyperfine interaction is to cause an apparent depolarization at very low rotational levels of variable magnitude depending on radiative lifetime and the coupling parameters. Here the effect will generally be ignored unless it materially affects a particular experiment. For the above-mentioned experiment with circularly polarized light the multipoles may be simplified to[29]

$$^{jj}\rho_Q^K(q_A) = (-1)^{j''+j+K+q_A}(2K + 1)^{1/2} \begin{Bmatrix} j'' & j & 1 \\ K & 1 & j \end{Bmatrix} \begin{pmatrix} 1 & 1 & K \\ -q_A & q_A & Q \end{pmatrix} \tag{8}$$

which provides proof of the statement made earlier that only the multipoles $K = 0$, 1, and 2 are set up in the excited state. The relative magnitudes of the multipoles varies strongly with nature of transition as mentioned previously. Figure 4 illustrates this dramatically. Also of note is the marked variation as a function of j, though this is much less sharp above $j = 10$.

In the absence of collisions or other perturbations affecting the time evolution of the multipoles prior to emission, the emitted intensity of polarization q_E may be written

$$I_{q_E}^{q_A} = \sum_{KQ} [^{jj}\rho_Q^K(q_A)](-1)^{j'''+j+q_E}(2K + 1)^{1/2} \begin{Bmatrix} j''' & j & 1 \\ K & 1 & j \end{Bmatrix} \begin{pmatrix} 1 & 1 & K \\ -q_E & q_E & Q \end{pmatrix} \tag{9}$$

or, using the previous equation

$$I_{q_E}^{q_A} = \sum_{KQ} [^{jj}\rho_Q^K(q_A)][^{jj}\rho^K(q_E)](-1)^K \tag{10}$$

The closed forms for circular polarization ratios[33] may be obtained from these expressions.

Collisions will interfere with the evolution of the multipoles in the excited state and the tendency will be to revert to an unpolarized array at a rate determined by

[33] S. R. Jeyes, A. J. McCaffery, and M. D. Rowe, *Mol. Phys.*, 1978, **36**, 1865.

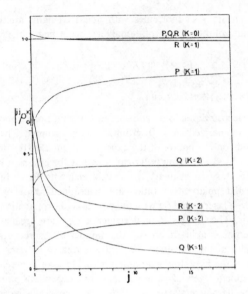

Figure 4 *A plot of the modulus of $^{jj}\rho_0^K$ vs. j for all K values of P, Q, and R transitions*
(Reproduced by permission from *Chem. Phys.*, 1979, **43**, 35)

the nature and frequency of the collisional interaction. Several experiments to test
this have been reported with varying degrees of resolution of the initial and final
states and with widely different theoretical analyses of the result. An early
experiment was performed by the Polish scientist Mrzowski,[34] whose study of the
polarization of unresolved total fluorescence from molecular iodine was the only
work available for many years, though Wood and colleagues[35] had established
the existence of the transfer spectrum as early as 1911. These results
appeared to indicate quite rapid depolarization as a function of foreign
gas pressure, the effect increasing as a function of foreign gas polarizability.
Some years later, Gordon[36] analysed the data of Mrzowski and emphasized
the value of the related fully resolved experiment. His analysis[26] presented
the fluorescence depolarization ratio as a Laplace transform of a correlation
function for molecular reorientation in a method similar to treatment
of Raman line-broadening and spin relaxation in gases. The experimental
depolarization ratio could then be expressed in terms of an average reorientation
angle, α, resulting from the collision. Kurzel and Steinfeld[37] studied the linear
polarization of resolved fluorescence from laser-excited I_2 using the 5145 Å line of
an argon ion laser. This excites the $B^3\Pi_{0_u}^+$ state some way below the dissociation
limit, and Kurzel and Steinfeld found that the resonance lines were very little
depolarized by up to a Torr (133.322 N m^{-2}) of helium. When j and v were changed

[34] S. Mrzowski, *Bull. Acad. Pol. Cracow*, 1933, 346.
[35] J. Franck and R. W. Wood, *Philos. Mag.*, 1911, **21**, 914.
[36] R. G. Gordon, *J. Chem. Phys.*, 1966, **45**, 1643.
[37] R. B. Kurzel and J. I. Steinfeld, *J. Chem. Phys.*, 1972, **56**, 3188.

by collision, however, the lines were less strongly polarized, and by use of Gordon's theory,[36] average reorientation angles of 50°—60° were calculated. Note though that this value is quite misleading. Jeyes *et al.*[33] have shown that an *m*-conserving transfer model will predict similar 'average reorientation angles' using the methods adopted by Kurzel and Steinfeld[37] where a sum of Δj transitions is taken. Clark and McCaffery[38,39] also made a study of this system and showed that a very large number of transitions are simultaneously excited by a multimode 5145 Å argon ion laser, and that some of these interfere strongly in the emission spectrum. They also found that elastic collisions did not depolarize the fully resolved resonance lines to very high foreign gas pressures although the emission intensity fell by over two orders of magnitude. Collision partners included the I_2 ground state molecule, rare gases, and oxygen and although the concept of collision frequency is not very precise when dealing with exponentially decaying excited states, particularly when collision-induced predissociation shortens the lifetime, the pressure range covered was very approximately equivalent to frequencies of up to one hundred per lifetime based on gas-kinetic cross-sections. Transfer lines were too strongly overlapped to yield useful results. The 5017 Å line of Ar^+ excites only a single rotational doublet and was found to be undepolarized[39] by foreign gases. This line was too close to the I_2^* dissociation limit for further pressure dependence studies.

An important point worth mentioning here is that circular polarizations generally have positive and negative signs whereas linear polarizations are mostly single-signed.[33] Overlapped transfer lines in circular polarization will therefore be very inaccurate though the inaccuracy may not be so great in linear polarization. Clearly though it is essential to excite single rotational levels for a proper study of reorientation by this technique and for heavy diatomics, a single-mode laser is needed. The first such study[40] on I_2 showed resonance and transfer features to be strongly polarized. Extensive rotational and vibrational transfer was seen, all highly polarized. This data and other similar results were interpreted by Katô[41] using a Born approximation, which has the effect of restricting access to some *m* channels on transfer. Given the current availability of tunable c.w. narrow line lasers capable of exciting single rotational levels of a diatomic, it is of interest to consider briefly what experiments may be performed and the information they yield. By resolving the resonance and transfer features one may study elastic and inelastic collisions separately as long as one works in the low-pressure region or carefully extrapolates back to zero pressure. A wide range of outgoing channels may be monitored by scanning the analysing monochromator and this is a major advantage in comparison to molecular beam methods. Studying the resonance line of a single rotational level at a single pressure is useful but not fully informative. Pressure dependence is needed to obtain cross-sections for reorientation by elastic collision, and *j'* dependence of polarization ratio is also important since at low *j'* values the fluorescence polarization is more sensitive to the reorientation process. Furthermore, individual rotational levels rarely yield theoretical polarizations even at very low pressures and the systematic variation with *j'* is needed. The difficulties in

[38] R. Clark and A. J. McCaffery, *Mol. Phys.*, 1978, **35**, 617.
[39] H. Katô, R. Clark, and A. J. McCaffery, *Mol. Phys.*, 1976, **31**, 943.
[40] H. Katô, S. R. Jeyes, A. J. McCaffery, and M. D. Rowe, *Chem. Phys. Lett.*, 1976, **39**, 573.
[41] H. Katô, *J. Chem. Phys.*, 1978, **68**, 86.

studying low j' values have been discussed earlier. If the molecule has nuclear spin there is depolarization, due to this effect, of variable magnitude.[28] It is sometimes difficult to 'find' the low j' values owing to crowding together near the band head and transitions to them often have inherently low intensity. Transfer lines, *i.e.*, fluorescence from levels populated by inelastic collisions, off low j' 'parents' are even weaker and thus the study of inelastic collisions on low rotational levels is hampered. Pressure dependence of circular polarization of transfer lines may be used to determine cross-sections for transfer of orientation, as shown by McCormack and McCaffery,[42] but mainly it is the Δj dependence of transfer features which is most useful for relating to theoretical concepts.

Despite the availability of tunable laser sources for studies of j' and v' dependence, relatively little has been published. Jeyes *et al.*[33] reported the polarization of resonance doublets from single rotational levels $j' = 2$ to $j' = 21$ of $v' = 16$, ($^3\Pi_{0_u}^+$) molecular iodine under conditions of several gas-kinetic collisions per lifetime. A marked reduction of the polarization ratio was observed to low j' values. The effect of nuclear hyperfine interactions was included using the method recently developed by Madden.[43] When this was properly included, there was no j'-dependent depolarization that could be attributed to the effect of collisions, I_2^*–I_2 in this case. This observation, in conjunction with the previously mentioned experiments on pressure dependence of polarization of resonance lines, led to the proposal of a selection rule on m in elastic collisions,[44] namely that $\Delta m = 0$. Li_2–rare-gas elastic collisions also show similar behaviour. A plot of polarization *vs.* pressure for $j' = 30$ of the $B^1\Pi_u$, $v' = 2$ level of Li_2 showed no depolarization with argon or helium out to 100 Torr (1.3 kN m^{-2}) of foreign gas pressure.[45] A more quantitative study on the $A^1\Sigma_u^+$, to be described shortly, extracted reorientation cross-sections for low j' states that were found to be very small indeed.[28]

That rotationally *inelastic* transfer is generally accompanied by transfer of orientation was established in I_2^* by Katô *et al.*[40] A careful experimental study of transfer polarization off $j' = 19$ of I_2^* was made by Jeyes *et al.*[45] who chose this relatively high j' value because of complications with hyperfine interactions at low j' and because of the intensity problems referred to earlier. Rotational transfer polarization to $\Delta j' = 30$ were recorded and it was noted that polarization ratios were still high after these substantial changes in j'. Two models were suggested to account for the results, one a θ-conserving model, where θ is the classical orientation angle, and the second an m-conserving model. Of the two, the latter gave a much closer fit to the experimental data. This led the authors to suggest a propensity rule based on the conservation of m in rotationally inelastic collisions,[45] which was subsequently modified following experiments on low rotational states of Li_2.[29] This aspect will be discussed later.

The interpretation of the experiments described above suffered from the assumption that the m populations initially formed in the excited level and

[42] J. McCormack and A. J. McCaffery, *Chem. Phys.*, 1980, in the press.
[43] P. A. Madden, *Chem. Phys. Lett.*, 1975, **35**, 521.
[44] S. R. Jeyes, A. J. McCaffery, M. D. Rowe, P. A. Madden, and H. Katô, *Chem. Phys. Lett.*, 1977, **47**, 550.
[45] S. R. Jeyes, A. J. McCaffery, and M. D. Rowe, *Mol. Phys.*, 1978, **36**, 1865.

determined by the Clebsch–Gordon coefficients (or $3j$ symbols) connecting the m components ground and excited states were invariants under the symmetry operations of the full rotational group. A more rigorous treatment requires the use of rotational invariants and, as mentioned earlier, the tensor density matrix method ensures that this symmetry is maintained. Two recent experimental studies have used this technique to extract numerical values of cross-sections for reorientation by elastic collisions[28] and to compare experimental results with the predictions of theory.[29]

In the first of these[28] it was found that the j' dependence of the resonance lines was unusual in that the depolarization at low j' values found in I_2^*, and explained in terms of nuclear hyperfine interaction,[43,45] was much less marked in the case of Li_2^*. This is because of the much shorter lifetime of $^1\Sigma_u^+$ Li_2, where the excited state does not live sufficiently long for the hyperfine coupling to have much effect. Quantum mechanically, this may be expressed by the inclusion of coherences between hyperfine components,[28] the excited state being formed in *coherent* time-dependent superposition of hyperfine levels, and which then evolve during the excited lifetime prior to emission. The Hanle effect[46] is perhaps the most familiar experiment in which coherent superpositions of pure states are deliberately created, and these are usually between individual magnetic sublevels of a particular rotational or hyperfine level. The coherences are destroyed by an external magnetic field of magnitude characteristic of the molecular g-factor and the excited lifetime. In the case of the $B^1\Sigma_u^+$ state of Li_2 the hyperfine coupling parameters were unknown but were adjusted to give a fit to the polarization *vs.* j' curve.[28] Pressure-dependence studies were carried out on $j' = 2$ and $j' = 10$ in Li_2, and as will be shown, an experimental plot of $1/C$ *vs.* pressure, where C is the circular polarization ratio, may be expressed as an equation for a straight line whose slope is related to the cross-section for reorientation by elastic collision, σ^1. For the case of $j' = 10$, this procedure gave[28] $\sigma^1 = 0.35 \pm 0.15$ Å^2 for Li_2^*–He, and 0.65 ± 0.3 Å^2 for Li_2^*–Ar. More limited experiments on $j' = 1$ also indicated very little pressure dependence of C for low rotational levels. These cross-sections are remarkably low. They are some two orders of magnitude smaller than the total rotationally inelastic (j-changing) cross-sections for Li_2[45,32] with rare gases. Clearly the previously quoted plots of C *vs.* pressure for I_2^*, which showed almost no depolarization to high foreign gas pressures,[37] would also give very low reorientation cross-sections, though these have not been numerically evaluated.

A theory suitable for dealing with the collision-induced perturbation of the state multipoles that are set up in optical experiments has been presented.[29] This was developed for calculating circular polarization ratios for resonance and transfer features in diatomics following initial excitation with circularly polarized light. It is suitable for dealing with most forms of optical experiment using polarized light since it is based on the tensor density matrix formalism described earlier and the basic ideas are given in outline here, taken from ref. 29. We follow on from equations (8) and (9) which express the intensity of light of polarization, q_E, emitted by a molecule following excitation with polarization q_A in terms of the state multipoles

[46] W. Hanle, *Z. Phys.*, 1924, **30**, 93.

$^{jj'}\rho_Q^K$ created on excitation; in this system the circular polarization ratio would be given by:

$$C = \frac{I_+^+ - I_-^+}{I_+^+ + I_-^+} \cdot I$$

To calculate C following collisional perturbation during the excited lifetime we must determine the values of the multipoles after collision. The method is based on the close coupled space frame treatment of Arthurs and Dalgarno[47] for the scattering of a diatomic molecule by an atom modified to allow for the laser propagation direction as initial and final quantization axes. Further changes from this original work include the collisions occurring isotropically, as is characteristic of cell experiments and then individual $jm \to j'm'$ transition cross-sections must be calculated before conversion to multipole cross-sections. The treatment is similar to the derivation by Kinsey *et al.*[48] for the collisional relaxation of nuclear spins in a gaseous medium.

Using the usual[47] angular momentum coupling of jm and lm_l to form a resultant total angular momentum J, M which is conserved during the collision, the reaction amplitude may be written:

$$f_{jm \to j'm'}(\theta\phi, \Theta\Phi) = 4\pi \sum_{Jm,\, lm_l,\, l'm_{l'}} i^{l-l'}(2J+1) \begin{pmatrix} j & l & J \\ m & m_l & -M \end{pmatrix} \begin{pmatrix} j' & l' & J \\ m' & m'_l & -M \end{pmatrix}$$

$$\times (-1)^{l+l'+j+j'} Y_{lm_l}^*(\Theta\Phi)\, Y_{lm_l}(\theta\phi)\, T_{jl,j'l'}^J \tag{11}$$

where the reaction amplitude f for a collision-induced transition $jm \to j'm'$ is a function of the angles θ, ϕ, specifying the rotor orientation, and $\Theta\Phi$ the angle of approach of the atom; the Y_{lm} are spherical harmonics representing ingoing and outgoing waves and $T_{jl,j'l'}^J$ is the transition matrix element.

The reaction amplitude may be then expressed as a cross-section by integrating over space co-ordinates $\theta\phi$ and averaging over incoming particle direction $\Theta\Phi$

$$\sigma_{jm \to j'm'} = \frac{1}{4\pi k^2} \int\int |f_{jm \to j'm'}(\theta\phi\Theta\Phi)|^2 \, d\Omega_{\theta\phi} \, d\Omega_{\Theta\Phi} \tag{12}$$

Using equation (11) and carrying out the integrations

$$\sigma_{jm,j'm'} = \frac{4\pi}{k^2} \sum_{J's\, M's\, l's\, ml's} (2J_1+1)(2J_2+1) \begin{pmatrix} l & j & J_1 \\ m_l & m & -M \end{pmatrix} \begin{pmatrix} J_1 & j' & l' \\ M_1 & -m' & -m'_l \end{pmatrix}$$

$$\times \begin{pmatrix} l' & j' & J_2 \\ m'_l & m' & -M_2 \end{pmatrix} \begin{pmatrix} J_2 & j & l \\ M_2 & -m & -m_l \end{pmatrix} T_{jl;\, j'l'}^{J_1} T_{jl;\, j'l'}^{J_2*} \tag{13}$$

[47] A. M. Arthurs and A. Dalgarno, *Proc. R. Soc. London, Ser. A*, 1960, **256**, 540.
[48] J. L. Kinsey, J. L. Reihl, and J. S. Waugh, *J. Chem. Phys.*, 1968, **49**, 5296.

This can be converted into a state multipole cross-section by taking linear combinations of m and m', as in equation (6) which, after some manipulation gives

$$
\sigma(^{jj}\rho_Q^K \to {}^{j'j'}\rho_{Q'}^{K'}) = \frac{4\pi}{k^2} \sum_{J_1 J_2 M_1 M_2} (-1)^{j+j'+l+l'} (2J_1 + 1)(2J_2 + 1) T_{j\bar{j}'\,l'}^{J_1} T_{j\bar{j}'\,l'}^{J_2*} \begin{Bmatrix} l & j & J_1 \\ K & J_2 & j \end{Bmatrix}
$$

$$
\times \begin{Bmatrix} l' & j' & J_1 \\ K' & J_2 & j' \end{Bmatrix} \begin{pmatrix} J_1 & J_2 & K \\ -M_1 & M_2 & Q \end{pmatrix} \begin{pmatrix} J_1 & J_2 & K' \\ -M_1 & M_2 & Q' \end{pmatrix}
$$

$$
\times (2K + 1)^{1/2} (2K' + 1)^{1/2} \tag{14}
$$

When M_1 and M_2 are summed out using the orthogonality property of the $3j$ symbols, it is clear that $K = K'$ and $Q = Q'$. Thus under isotropic conditions, *elastic or inelastic collisions transfer each multipole independently*. The multipole cross-section may be written as σ^K and (14) simplifies to

$$
\sigma^K = \sum_{J_1, J_2, l, l'} (-1)^{j+j'+l+l'} (2J_1 + 1)(2J_2 + 1)
$$

$$
\times \begin{Bmatrix} l & j & J_1 \\ K & J_2 & j \end{Bmatrix} \begin{Bmatrix} l' & j' & J_1 \\ K & J_2 & j' \end{Bmatrix} T_{jl;\,j'\,l'}^{J} T_{jl;\,j'\,l'}^{J*} \tag{15}
$$

From the symmetry restrictions on the $6j$ symbols, when $K = 0$, $J_1 = J_2$, when $K = 1$, $J_2 = J_1$, or $J_1 \pm 1$ and when $K = 2$, $J_2 = J_1 \pm 1$ or $J_1 \pm 2$.

In the experiment for which this theory was derived[29] only diagonal elements of $^{jj}\rho_{mm}$ are prepared by the light beam. In other experiments such as the Hanle effect, for example, where the quantization axis is not defined by the light beam, non-diagonal elements of $^{jj}\rho_{m_1 m_2}$ can be produced. The expression for the cross-section, equation (12), would then include terms in $f_{jm_1 \to j' m_1'} f_{jm_2 \to j' m_2'}^*$ giving terms with $M_1 \neq M_2$ and $Q \neq 0$ in equation (14). In either case, equation (15) is obtained and therefore it may be widely applied. In equation (15), initial- and final-state quantization axes are the same, and it is useful for the circular polarization experiments already mentioned. Emission polarization can be referred to any arbitrary axis by an appropriate rotation of the form

$$
{}^{j'j'}\rho_Q^K = \sum_Q {}^{j'j'}\rho_Q^K D_{QQ'}^K (\alpha\beta\gamma) \tag{16}
$$

We are now in a position to go back to equation (10) and modify this for the effect of collisional perturbation of the excited state. The intensity of polarized emission is given by equation (17) and the circular polarization ratio by equation

$$
I_{qE}^{q_A} = \sum_{KQ} [^{jj}\rho_Q^K(q_A)] [^{jj}\rho_Q^K(q_E)] \sigma^K (-1)^K \tag{17}
$$

(18), where σ^0, σ^1, and σ^2 are the cross-sections for the transfer of population, orientation, and alignment respectively in a rotationally inelastic collision-induced transition. The effect of elastic collisions before and after inelastic transitions may

$$C = \frac{\begin{Bmatrix} j'' & j & 1 \\ 1 & 1 & j \end{Bmatrix} \begin{Bmatrix} j''' & j' & 1 \\ 1 & 1 & j' \end{Bmatrix} \sigma^1}{\frac{2}{3} \begin{Bmatrix} j'' & j & 1 \\ 0 & 1 & j \end{Bmatrix} \begin{Bmatrix} j''' & j' & 1 \\ 0 & 1 & j' \end{Bmatrix} \sigma^0 + \frac{1}{3} \begin{Bmatrix} j'' & j & 1 \\ 2 & 1 & j \end{Bmatrix} \begin{Bmatrix} j''' & j' & 1 \\ 2 & 1 & j' \end{Bmatrix} \sigma^2}$$

(18)

be introduced *via* their effect on the density matrices. For example $^{jj}\rho_Q^K(q_A)$ becomes $^{jj}\rho_Q^K(q_A)/(\Gamma + N\sigma^K \bar{v})$, where N is the number density of the collision partner, $\bar{v}$ is the average velocity, and Γ is the natural linewidth of the excited state.

The equation corresponding to (18) for elastic collisions becomes

$$C = \frac{\begin{Bmatrix} j'' & j & 1 \\ 1 & 1 & j' \end{Bmatrix} \begin{Bmatrix} j''' & j' & 1 \\ 1 & 1 & j' \end{Bmatrix} (\Gamma + N\sigma^1 \bar{v})^{-1}}{\frac{2}{3} \begin{Bmatrix} j'' & j & 1 \\ 0 & 1 & j \end{Bmatrix} \begin{Bmatrix} j''' & j' & 1 \\ 0 & 1 & j' \end{Bmatrix} (\Gamma + N\sigma^0 \bar{v})^{-1} + \frac{1}{3} \begin{Bmatrix} j'' & j & 1 \\ 2 & 1 & j' \end{Bmatrix} \begin{Bmatrix} j''' & j' & 1 \\ 2 & 1 & j' \end{Bmatrix} (\Gamma + N\sigma^{2l} \bar{v})^{-1}}$$

(19)

Quenching collisions may be introduced by including a cross-section σ^Q additively.

An example of the use of equation (19) to obtain a reorientation cross-section comes from ref. 28, where the explicit form of equation (19) is given for an $R\!\uparrow P\!\downarrow$ transition. The expression for C may be approximated to

$$C = -\frac{5}{7} \frac{(\Gamma + KP\sigma^0)}{[\Gamma + KP(\sigma^1 + \sigma^0)]}$$

(20)

which may be written

$$C^{-1} = C_0^{-1} \frac{(1 + KP\sigma^1)}{(\Gamma + KP\sigma^0)}$$

(21)

where $C_0 = C$ at zero foreign gas pressure. From equation (21) it can be seen that a plot of C^{-1} *vs.* P is a straight line if it can be assumed that $KP\sigma^0$ is negligible compared to Γ, and in this case σ^1 can be obtained from the slope. This was done[28] for $j = 10$ of $Li_2{}^*$ *vs.* He and Ar with the results $\sigma^1 = 0.35 \pm 0.15$ Å^2 for Li_2^*–He and $\sigma^1 = 0.65 \pm 0.3$ Å^2 for Li_2^*–Ar.

This development of the nature of the multipoles set up by polarized radiation and their evolution in time under the influence of collisions indicates that laser fluorescence studies of energy transfer are quite complex. Laser radiation is invariably polarized and the nature of the multipoles set up depends on the type of transition and on the polarization of the light. An example will clarify this. Circularly polarized excitation of a diatomic *via* a P or R transition produces a distribution of excited m states in which the dominant multipole is the orientation ($K = 1$). When excitation is *via* a Q transition with circularly polarized light the dominant multipole is the alignment. Rates of relaxation of these multipoles depend on the relative magnitude of the σ^K in equations (18) and (19) and although there is some relationship between these cross-sections in the angular momentum coefficients of equation (15), the T matrices are not necessarily related.

A very interesting feature of rotationally inelastic transfer is demonstrated in the treatment of ref. 29, and is seen if transitions among m-channels are assumed to have equal probability in the collision-induced transition $jm \rightarrow j' m'$. In both the close-coupled space frame of Arthurs and Dalgarno or in an uncoupled space frame, the aforementioned assumption gives an equation for σ^K in which only the population ($K = 0$) carries through the collision: both orientation and alignment are destroyed.[29] This would produce total depolarization, as can be seen in equation (19), since the orientation cross-section is zero. There are numerous experiments[29,40,44,45] showing that collisional transfer lines are in fact quite strongly polarized in a range of atom–diatomic molecule collisions. In other words the $K = 1, 2$ multipoles *are* transferred on collision and *thus there are restrictions on the number of channels accessible to a particular m state in the scattering process.*

In the last section of this Report we look at some of the theoretical approaches to atom–diatomic molecule scattering, which limit the number of accessible m channels in elastic and rotationally inelastic transfer. Before this, several other relevant experiments both of the 'single resonance' type already discussed and some double resonance spectroscopic work on both diatomic and polyatomic molecules will be reviewed.

The bare essentials of the Hanle effect experiment were mentioned earlier and that coherent superpositions of states are created on excitation. Mostly this experiment has been confined to resonance lines and used to determine excited lifetimes. The observation of a Hanle effect in collisionally populated transfer lines is a strong indication of transfer of coherence in inelastic collisions. Such an experiment was performed by Caughey and Crossley[49] on S_2. Individual transfer lines were not fully resolved and an average over a small number of Δj quanta was evaluated. Caughey and Crossley[49] calculated from the experimental Hanle width that $85 \pm 15\%$ of the initial alignment was carried over in rotationally inelastic collisions of S_2 with He. Two polarization studies on Li_2 using laser fluorescence have been reported. Bormann and Poppe[50] and Vidal[51] have both used fixed lines of the argon ion laser to study a limited range of rotational levels. The former authors[50] use Gordon's[35] method to evaluate classical reorientation angles in inelastic collisions.

Most of the work described in the foregoing has been on homonuclear diatomics and it is of interest to consider the effect of a molecular dipole moment on reorientation. The next section on double resonance methods deals with dipolar molecules where the general view is that reorientation is a relatively easy process. However a polarized fluorescence study on NaK,[52] with rare gases as collision partners, gave reorientation cross-sections (σ^1) of a few tenths of a square Å for $j' = 30$, though very low j values were not investigated. Rotationally transferred lines were quite strongly polarized, but a very unusual alternation was observed in these.[53] The intensities were found to fall-off with Δj in an unexceptional fashion,

[49] T. A. Caughey and D. Crosley, *Chem. Phys.*, 1977, **20**, 467.
[50] J. Bormann and D. Poppe, *Z. Naturforsch., Teil A*, 1976, **31**, 739.
[51] C. R. Vidal, *Chem. Phys.*, 1978, **35**, 218.
[52] J. McCormack, A. J. McCaffery, and M. D. Rowe, *Chem. Phys.*, 1980, **48**, 121.
[53] J. McCormack and A. J. McCaffery, *Chem. Phys.*, 1980, in the press.

but the polarization ratios alternated strongly with odd Δj transitions less polarized than even Δj.[53]

The last group of spectroscopic measurements to be considered is the double resonance techniques using microwave, i.r., and optical radiation. There is an excellent review of double resonance methods and analysis of results by Steinfeld and Houston[54] which contains a detailed account of all aspects of both double resonance and coherence spectroscopy. Only results of relevance to this Report are discussed here, without extended cover of experimental methods or analysis. The experimental method is straightforward, in principle, though generally non-trivial in practice. Perhaps the most relevant of the studies to this Report, which is mostly about diatomics, is the optical–optical double resonance (OODR) experiment of Silvers *et al.*[55] on BaO. In this a linearly polarized tunable dye laser prepares the $m = 0$ sublevel of $j' = 1$, $A^1\Sigma^+(v' = 1)$, BaO, and a second polarized tunable laser monitors other m sublevels of $j' = 1$ and $j' = 2$ by inducing transitions to $C^1\Pi$ from which fluorescence is observed. On monitoring $j' = 1$ substantial signals were observed when CO_2 was used as collision partner, indicating that m changes had been induced by elastic collisions, but argon produced no signal. Inelastic transfer with both Ar and CO_2 populates sublevels other than $m = 0$ indicating again that m transitions were taking place. This method is clearly very useful for studying m transitions and has the advantage that velocity selection is possible through pumping different regions of the Doppler width. It does however rely heavily on there being truly 'single collision' conditions in the thermal cell, particularly for the elastic studies.

By use of microwave and i.r. radiation, a wider range of molecules has been studied by double resonance. The earliest of these was a microwave–microwave double resonance experiment on OCS by Cox *et al.*,[56] later repeated by Unland and Flygare,[57] which showed that the $j = 1$, $m = 0$ level was not elastically relaxed to $m = \pm 1$ by collisions, and thus m was conserved. The interpretation of this is not fully clear-cut however since OCS retains approximate parity, and a parity selection rule prohibits m transfer in dipolar molecules when long-range dipolar forces govern the collision dynamics.[58] There have been several double resonance studies of polyatomic molecules using i.r. lasers and the results obtained by Shoemaker *et al.*[59] on CH_3F are of interest. Using two frequency-locked CO_2 lasers they studied the $(v_3 JK) = (0,4,3) \rightarrow (1,5,3)$ methyl fluoride transition and saw satellite resonances due to collisional coupling between M levels. Shoemaker *et al.*[59] found that reorientation took place with preserved velocity distribution and that cross-sections for this were very high, of the order of 100 Å^2. Reorientation shows a very steep J dependence however and by the (12,2) level this value had dropped to less than 1 Å. Johns *et al.*[60] also studied i.r. Lamb dips in this way, but widened the range of molecular systems to include NH_3 and H_2CO as well as methyl fluoride. A range of

[54] J. I. Steinfeld and P. R. Houston, in 'Laser and Coherence Spectroscopy,' ed. J. I. Steinfeld, Plenum, New York, 1978.

[55] S. J. Silvers, R. A. Gottscho, and R. W. Field, *J. Chem. Phys.*, to be published.

[56] A. P. Cox, G. W. Flynn, and E. B. Wilson, *J. Chem. Phys.*, 1965, **42**, 3094.

[57] M. L. Unland and W. H. Flygare, *J. Chem. Phys.*, 1966, **45**, 2421.

[58] T. Oka, *Adv. At. Mol. Phys.*, 1973, **9**, 127.

[59] R. L. Shoemaker, S. Stenholm, and R. G. Brewer, *Phys. Rev. A*, 1974, **10**, 2037.

[60] J. W. C. Johns, A. R. W. McKellar, T. Oka, and M. Rohmeld, *J. Chem. Phys.*, 1975, **62**, 1488.

rotational levels was studied and the observations of Shoemaker *et al.*[59] were confirmed. Johns *et al.*[60] felt that parity conservation in collisional transfer was an important factor and accounted for the large cross-sections seen in CH_3F since in this molecule levels with $K = 0$ are doubly degenerate with $+$ and $-$ parity. Those of NH_3 and H_2CO have single parity until the Stark field used in these experiments is sufficient to overcome the restrictions. However, substantial reorientation was observed in single collisions of H_2CO with other molecules.

The use of Stark fields is avoided in the experiment by Leite *et al.*[61] who studied laser-induced line-narrowing in NH_3. Berman[62] had shown that the lineshape in i.r. double resonance is a sensitive function of relaxation processes and the (8,7) transition[61] in v_2 was saturated and probed using N_2O lasers to study m changing collisions. The experiment suggested that alignment destroying collisions were very probable and that m changing collisions accounted for a substantial fraction of the alignment relaxation rate. Frankel and Steinfeld[63] investigated reorientation following *vibrationally* inelastic transitions in BCl_3 and SF_6, concluded that the molecular alignment created by CO_2 laser excitation must remain partially unchanged in the collision process.

It is worth further emphasizing the difference in approach between the molecular beam workers and the spectroscopists, this time in their treatment of data. The spectroscopic results have mostly been used to test theoretical models for elastic and inelastic collisional processes without direct recourse to the intermolecular potential, whereas molecular beam results have almost always been used to relate directly to the potential *via* an assumed model. Clearly it is important for both aspects of scattering to be examined. The spectroscopic results are now probably sufficiently refined that it is possible to decide on a model, or limited range of models, as discussed later, and to extract information on the intermolecular potential. The essential similarities of the scattering and spectroscopic approaches should then become more strikingly apparent.

4 Scattering Theory and Approximation Methods

The final section of this Report concerns atom–diatomic molecule scattering theory, and the approximations which have been made to the exact theory to render it tractable for general use. The reason that theoretical developments and experiment are so closely interwoven stems from the observations, described earlier, that in the scattering process all m channels are *not* open, and that some restrictions exist at the very least in a substantial number atom–diatom systems. It is not the intention to review scattering theories exhaustively since this has been done in several recent excellent accounts.[18,64] Here discussion is limited to areas of direct relevance to the experiments mentioned earlier. The basic scattering theory formalism for structureless atom–rigid-rotor collisions was first expounded by Arthurs and

[61] J. R. R. Leite, M. Ducloy, A. Sanchez, D. Seligson, and M. S. Feld, *Phys. Rev. Lett.*, 1977, **39**, 1465.

[62] P. R. Berman, *Phys. Rev.*, 1976, **13**, 2191.

[63] D. S. Frankel and J. I. Steinfeld, *J. Chem. Phys.*, 1975, **62**, 3358.

[64] (*a*) See, *e.g.*, contributions to 'Atom–Molecule Collision Theory: A Guide to the Experimentalist,' ed. R. B. Bernstein, Plenum, New York, 1977; (*b*) H. Rabitz, in 'Modern Theoretical Chemistry,' ed. W. H. Miller, Plenum, New York, 1976; (*c*) A. E. De Pristo and H. Rabitz, to be published.

Dalgarno.[47] In this the space-fixed formulation of the Schrodinger equation leads to a set of coupled differential equations

$$\frac{\hbar^2}{2\mu}\left[-\frac{d^2}{dr^2}+\frac{l'(l'+1)}{r^2}-k_{j'l'}^2\right]$$

$$\times u_{j'l'}^{JjI}(r)+\sum_{j''}\sum_{l''}\langle j''\,l'';J|v|\,j'\,l';J\rangle\mu_{j''l''}^{JjI}(r)=0 \qquad (22)$$

where $k_{j'j}$ is the channel wavenumber

$$k_{j'j}^2=\frac{2\mu}{\hbar^2}\left[E_j=\frac{\hbar}{2I}j'(j'+1)\right] \qquad (23)$$

and rotational angular momentum j,m is coupled with orbital angular momentum l,m_l to form the resultant JM which is conserved in the collision. The $U_{j'l}^{JjI}(r)$ are radial functions. The scattering amplitude for a transition between initial rotor level $|jm\rangle$ to final level $|j'm'\rangle$ can then be written[47]

$$f_{j_1m\to j'm'}(\theta)=\sum_{JMlml'm'}i^{l_2-l_1}\pi^{1/2}(-1)^{j_1+j_2}[J][l_1]\begin{pmatrix}j_1&l_1&J\\m_1&0&-M\end{pmatrix}$$

$$\times\begin{pmatrix}j_1&l_2&J\\m_2&m&-M\end{pmatrix}T_{jl'l'}^J Y_{lm}(\hat R) \qquad (24)$$

where the quantization axis is the initial relative velocity vector.[47] $\hat R$ represents the vector joining the atom with the molecular centre of mass. An extension of this equation to arbitrarily oriented initial and final axes has been given by Alexander *et al.*[65] whilst Rowe and McCaffery[29] have generalized to the case of all orientations of the incoming wave as in a thermal cell. The fully state-selected cross-section is related to the scattering amplitude by equation (25). From a practical point of view the difficulty in a quantum calculation comes in solving the infinite set of coupled equations in (22) and a substantial amount of effort has been directed at finding acceptable ways of reducing the dimension of this problem. Several full quantal calculations, referred to as close-coupling (CC) studies, in which orientation effects were specifically considered, have been carried out for H_2–Kr (ref. 66), He–HCl (ref. 67), He–CO (ref. 68), and Ar–N_2 (ref. 69). Of these the He–HCl cross-sections[67] show very dominant diagonal values which agree well with some of the optical experiments described earlier.

$$\sigma_{jm\to j'm'}=\frac{1}{4\pi k^2}\iint|f_{jm\to j'm'}(\theta)|^2\,d\Omega \qquad (25)$$

[65] M. H. Alexander, P. J. Dagdigian, and A. E. De Pristo, *J. Chem. Phys.*, 1977, **66**, 59.
[66] M. Jacobs and S. Stolte, *Chem. Phys.*, 1977, **25**, 425.
[67] L. Monchick, *J. Chem. Phys.*, 1977, **67**, 4626.
[68] L. Monchick, *J. Chem. Phys.*, 1979, **71**, 578.
[69] M. H. Alexander, *J. Chem. Phys.*, 1977, **66**, 59.

For many problems the numerical solution of equation (22) is too time-consuming and several attempts have been made to resolve this problem, notably by the use of 'effective Hamiltonian' procedures.[64b] These involve a reduction in, or elimination of, some of the dynamical degrees of freedom before calculation. Some of these methods involve pre-averaging over degenerate m states and thus the information relevant to this Report is lost though the dimensionality is reduced. In other approximations, accessible m channels are restricted to achieve simplification, either through the dynamics or the kinematics or some combination of both. It is the relationship between these and the previously described experimental results which will now be discussed.

The most widely used procedure is the coupled states (CS) approximation put forward by McGuire and Kouri[70] and by Pack[71] which effectively achieves a decoupling of l and j in the quantum formulation. The full details of this approximation and its further refinements are given in chap. 9 of ref. 64b, but basically it involves replacement of the centrifugal barrier in equation (22) by the effective value $\bar{l}(\bar{l} + 1)$. There has been some controversy over the choice of the value of $\bar{l}$, as described in the review by De Pristo and Rabitz,[64c] which has been identified with total angular momentum J, initial orbital angular momentum l, and final orbital angular momentum l'. Differential cross-sections are quite sensitive to this decision and, for example, choosing $\bar{l} = l'$ leads to the conclusion that changes in m are forbidden[72] whilst $\bar{l} = l$ does not. Although there were claims for conservation of m in inelastic transfer based on polarized fluorescence work on I_2,[45] this now seems unlikely in the light of the more recent study on Li_2 described above.[29] Monchick and Kouri[73] found that $\bar{l} = l$ gave a better agreement with exact calculations on He–HCl.

The sudden approximations such as the infinite order sudden (IOS)[74] and the l-conserving energy sudden[75] approximations achieve further simplification to the coupled equations by a closure approximation on the rotor states. These have implications regarding m transitions and multipolar cross-sections which will be discussed briefly. Some interesting physical implications arise from the decoupled l-dominant (DLD) approximation introduced by De Pristo and Alexander[76] in which the centrifugal potential plays a dominant role. This is expected to be the case when the long-range part of the intermolecular potential contributes strongly to the inelasticity. In this long-range region the potential varies slowly, and therefore different values of orbital angular momentum l lead to very different values of the classical turning point. McGuire and Kouri had noted[77] in a coupled states and close coupled study of Li^+, H_2 collisions that the T matrix was dominated by the lowest values of l, i.e. $l = J - j$ and $l' = J - j'$. This yields the lowest centrifugal barrier and the smallest classical turning point enabling more of the potential to be

[70] P. McGuire and D. Kouri, *J. Chem. Phys.*, 1974, **60**, 2488.
[71] R. T. Pack, *J. Chem. Phys.*, 1974, **60**, 633.
[72] Y. Shimoni and D. J. Kouri, *J. Chem. Phys.*, 1977, **66**, 2841.
[73] D. J. Kouri and L. Monchick, *J. Chem. Phys.*, 1978, **69**, 3262.
[74] R. T. Pack, *J. Chem. Phys.*, 1974, **60**, 633.
[75] S. I. Chu and A. Dalgarno, *Proc. R. Soc. London, Ser. A*, 1975, **342**, 191.
[76] A. E. De Pristo and M. H. Alexander, *J. Chem. Phys.*, 1976, **64**, 3009.
[77] P. McGuire and D. J. Kouri, *Chem. Phys. Lett.*, 1974, **29**, 414.

sampled. De Pristo and Alexander[76] have formally incorporated this into scattering theory in which the centrifugal potential is included exactly while the potential is approximated. This contrasts with the CS approximations where the reverse is the case. In both instances there is a consequent reduction in complexity to one coupled equation per rotor state.

The physical implication of the DLD approximation is that the preferred geometry for the collision is a coplanar one, with the atom approaching in the plane of rotation of the diatomic rotor. Intuitively this has some appeal since it fits in well with the idea of slow reorientating, but relatively rapid j-changing, collisions.

Perturbation theory approximations such as the Born approximation, the distorted-wave Born approximation (DWA),[11,78] and the exponential distorted-wave approximation[5] have been widely used. Much of the work by Reuss described earlier uses the DWA extensively, though more recently Stolte and Reuss[79] have recommended a space-fixed sudden approximation for treating elastic reorientation in an intensive examination of the available models.

The question which of these approximate models gives a best fit to the experimental observation of channel restrictions is partially answered in the study by Rowe and McCaffery[29] who took limiting forms of the main approximations and, after casting them into appropriate form to calculate multipolar cross-sections, compared the results with j-dependent fluorescence polarization curves for Li_2^*–He. The Born approximation gave the least satisfactory results. High-order terms need to be included to approach the experimental curves, the P_2, P_4, and P_6 terms being necessary to fit the $\Delta j = 2$ curve for example. By contrast the j_z-conserving CS method gave a good fit to the data up to high rotational inelasticities. Ref. 29 shows that for $|\Delta m| < \Delta j$ the multipole cross-sections are independent of Δm and for this reason it is impossible to distinguish selection or propensity rules on Δm more precisely than $\Delta m \leqslant \Delta j$ since this and greater restrictions (even $\Delta m = 0$) yield identical results in the CS approximation. The DLD approximation also yields a good fit to $\Delta j = 2$ and $\Delta j = 4$ curves by incorporating quite low values of the parameter λ where $l = J - j + \lambda$ in the DLD theory; Figure 5 indicates this clearly. T matrices were not evaluated in the foregoing comparisons and thus further work needs to be done before one model may be chosen as 'best' for this purpose. However, some valuable indicators have emerged and the results indicate profitable avenues for further research.

In conclusion, one might ask if there are some convenient rules of thumb to tell us if reorientation will be slow in atom–molecule encounters, as found for rare gas–alkali-metal dimers and several other cases, or if it will be a fast process as observed in *e.g.*, NH_3–NH_3 collisions. It would be valuable to be in a position to make qualitative judgments without having to go through the process of a scattering calculation of even the most truncated kind. Some key to this comes from the review by Oka[58] who derived selection and propensity rules on rotational and magnetic quantum numbers based on the leading multipolar moments of the molecules involved in the collision. Of course it is not always the permanent moments which form the major part of the long-range attractive intermolecular

[78] R. E. Roberts and J. Ross, *J. Chem. Phys.*, 1970, **52**, 5011.
[79] J. Reuss, in ref. 64a.

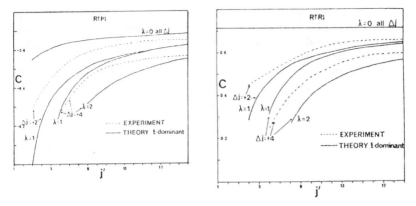

Figure 5 *Theoretical plots of C vs. j' based on the l-dominant approximation for R↑P↓ Δj = +2 and +4 with values of λ as indicated. Scaled experimental curves are also shown for comparison*

(Reproduced by permission from *Chem. Phys.*, 1979, **43**, 35)

interaction, and the surface is composed of contributions from dipole–dipole, dipole–induced-dipole and induced-dipole–induced-dipole interactions, any one of which might dominate in a particular situation. What little is currently known about molecular reorientation does seem to suggest that when dipolar forces, including the induction force, dominate, then reorientation may be expected to be relatively rapid, and simple angular momentum arguments support Oka's allowed Δm transitions. The dipole–dipole force can very simply be shown to be the major influence in NH_3–NH_3 and CH_3F–CH_3F collisions and when those molecules collide with rare gases the dipole–induced-dipole effect dominates. In BaO–CO_2 this also holds, though it is difficult to estimate excited molecular state polarizabilities accurately. In the homonuclear diatomics, and also in heteronuclear alkali-metal dimers such as NaK, this is most certainly not the case, and it is the dispersion force which dominates (by over an order of magnitude in the case of NaK). In these instances reorientation appears to be slow and the dipolar or multipolar rules on m and j probably should not be expected. This of course leads to the next question, namely: why should an interaction dominated by dispersion forces lead to gas kinetic cross-sections for rotational transfer but considerably smaller values for reorientation? It cannot be that the intermolecular interaction is isotropic since a glance at the full form for the molecular polarizability separated into its constituent parts, as given by Buckingham for instance,[80] indicates that the dispersion force for a diatomic molecule would be very markedly anisotropic. For the rare-gas atom-alkali-metal dimer interaction the greatest (most negative) value of the dispersion energy would be for approach along the bond direction in ground state molecules, and the anisotropy is expected to be very much greater when the molecule is excited.[75] It is difficult to resist drawing a parallel between this preferential direction

[80] A. D. Buckingham, *Adv. Chem. Phys.*, 1967, **12**, 107.

of approach, resulting from anisotropic dispersion forces, and the ideas that led to the formulation of the dynamic *l*-dominant approximation already described which we have seen predicts slow m changes but reasonably rapid j-changes in atom–diatomic molecule energy transfer collisions.

The next few years should see many developments in experimental measurements of reorientation by collision. The theoretical models, and the author's rule of thumb suggestion, described here, will be subjected to stringent examination as a wider range of molecular systems is exposed to experimental investigation.

Acknowledgments. The author thanks Drs. De Pristo, Rabitz, and Field for preprints of relevant papers. The theoretical development of multipolar cross-sections given here is due to Dr. M. D. Rowe.

3

Infrared Multiple Photon Excitation and Dissociation: Reaction Kinetics and Radical Formation

BY M. N. R. ASHFOLD AND G. HANCOCK

1 Introduction

The interaction of gas-phase molecules with intense infrared laser fields is one of the most rapidly developing areas of experimental and theoretical study in present day chemical physics. Since the first suggestions, in the early 1970's, that collisionless dissociation of a molecule could take place following multiple photon absorption of infrared light from a pulsed laser,[1-3] and the subsequent (and, at the time, startling) observation that this process could take place in an isotopically selective fashion,[4,5] the number of publications in the field has increased dramatically, and shows little signs of levelling off from its current rate of well over 100 major articles per year. The subject has attracted the attention of investigators in a wide range of disciplines, from theoreticians studying the physics of the multiple photon absorption process itself, to experimentalists measuring the products of organic reactions brought about by i.r. lasers and comparing the results with those from more conventional chemical synthesis.

The effects of the presence of internal energy in a molecule have been of interest to chemical kineticists for many years. The unimolecular decomposition of molecules excited either thermally or chemically to levels above their dissociation limit has been a particularly active area of experimental and theoretical research; chemical reactions of small molecules with specific excitation in low vibrational levels is a subject of appreciable current interest. Multiple photon absorption of i.r. light provides a unique way of preparing molecules with considerable internal energy, and the study of the behaviour of these species, towards dissociation, collisional relaxation or chemical reaction is a relatively new and potentially exciting field. The purpose of this Report is to present a chemical kineticist's view

[1] N. R. Isenor and M. C. Richardson, *Appl. Phys. Lett.*, 1971, **18**, 224; *Optics Commun.*, 1971, **3**, 360.
[2] V. S. Letokhov, E. A. Ryabov, and O. A. Tumanov, *Optics Commun.*, 1972, **5**, 168; *Sov. Phys. JETP Engl. transl.*, 1973, **36**, 1069.
[3] N. R. Isenor, V. Merchant, R. S. Hallsworth, and M. C. Richardson, *Can. J. Phys.*, 1973, **51**, 1281.
[4] R. V. Ambartzumian, Yu. A. Gorokhov, V. S. Letokhov, and G. N. Makarov, *JETP Lett. Engl. transl.*, 1974, **21**, 171; 1975, **22**, 43.
[5] J. L. Lyman, R. J. Jensen, J. Rink, C. P. Robinson, and S. D. Rockwood, *Appl. Phys. Lett.*, 1975, **27**, 87.

of the current understanding of i.r. multiple photon absorption (IRMPA) and dissociation (IRMPD), and to outline applications (and point out limitations) of these processes to studies involving chemical change, particularly those involving kinetic measurements.

The main features of the currently accepted model of IRMPA and IRMPD will be briefly discussed, with emphasis placed upon details of the theoretical treatments that can be experimentally verified. Then follows a description of the relatively few direct experimental investigations into IRMPA, and the far more numerous accounts of IRMPD, indicating how studies of the dissociation products have helped to unravel details of the absorption process itself, and of the behaviour of molecules excited above their dissociation limit. The effects of collisions upon the multiple photon processes are then discussed, together with an account of how IRMPD compares with more conventional thermal dissociation processes. Finally, applications of the technique are described, with special reference to aspects not covered in previous reviews, namely the kinetics and spectroscopy of reactive intermediates formed by IRMPD.

Literature is cited up to October 1979, with emphasis upon the most recently published investigations, which inevitably involves omission of some notable early work. Space limitations preclude discussion of some recent creditable publications. This is the price paid by authors working in a popular and exciting field, to whom we tender our apologies.

2 The Model for Infrared Multiple Photon Absorption (IRMPA) and Dissociation (IRMPD)

The vast majority of the experimental IRMPA and IRMPD studies to date have been carried out using the pulsed radiation from CO_2 lasers, tunable over the discrete vibration–rotation transitions of the $(001 \rightarrow 100)$ and $(001 \rightarrow 020)$ bands around 10.5 and 9.5 μm, respectively. SF_6 has been the most extensively studied molecule, and the following mechanism was originally developed to account for its dissociation at wavelengths around 10.6 μm.[6] The concepts, however, are general and can be applied to the IRMPD of other species.

The model envisages absorption taking place through three successive regions as the total energy deposited in the molecule increases. At low levels of excitation (corresponding to absorption of a few i.r. photons), vibrational states are essentially discrete. Absorption of a single i.r. photon, raising molecules from their ground vibrational state to $v = 1$ in a given vibrational mode, is a straightforward process, but successive transitions to higher vibrational levels are immediately out of resonance with the monochromatic i.r. laser pulse, owing to the molecules' vibrational anharmonicities. In order to achieve further excitation within a single mode, mechanisms compensating for these anharmonicities must be in operation. This 'region I' of IRMPA is illustrated in Figure 1. The density of all vibrational states within a polyatomic molecule increases with energy, the rapidity of this increase depending upon the molecular complexity and the magnitudes of the frequencies of the individual modes. At some value of the molecules' internal energy

6 J. G. Black, E. Yablonovitch, N. Bloembergen, and S. Mukamel, *Phys. Rev. Lett.*, 1977, **38**, 131.

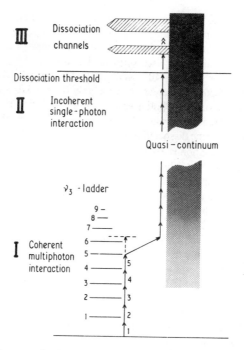

Figure 1 *Schematic energy-level diagram as originally applied to the IRMPD of* SF$_6$*. In region I, 'coherent multiphoton interaction' describes one of the mechanisms (intensity-dependent power-broadening effects) whereby non-resonant absorption can take place in a sparse region of vibrational levels. Other compensating mechanisms are discussed in the text. In region II, the quasicontinuum, resonant absorption steps are always possible, and can be treated as stepwise (incoherent) excitation processes. Region III lies above the dissociation threshold*
(Reproduced by permission from *Phys. Rev. Lett,* 1977, **38**, 1131)

these levels are considered to merge forming what is termed the 'quasicontinuum': the main feature of this region (II) is that resonant absorption processes can always take place at the monochromatic laser frequency (and indeed at essentially all other frequencies, although there may be structure within the quasicontinuum). If a molecule can be excited through the discrete levels (region I), then an absorption path is always available through the quasicontinuum of vibrational states. Eventually molecules acquire enough energy by absorption to be able to dissociate into fragments. Randomization of internal energy in this region (III), above the dissociation limit, is considered to take place on a timescale that is short in comparison with the lifetime of the dissociating species, so that statistical theories of unimolecular decomposition, *e.g.*, RRKM theory, may be used to relate this lifetime to the energy absorbed in excess of that needed for dissociation, and to estimate (in the absence of dynamical constraints as the fragments separate) the distribution of available energy within the products.

The absorption of the first few i.r. photons within the same vibrational mode in region I has been explained, for the IRMPA of SF$_6$, by allowing particular rotational transitions to compensate for the vibrational anharmonicity[7,8] and by recognizing that rotation–vibration interactions can produce level splittings which enhance the probability of finding resonant absorption steps.[9] Furthermore, at the reasonably high laser *intensities* used (measured in laser power per unit area, and generally of the order of 10—100 MW cm^{-2} for typical CO_2 laser pulses) power-broadening effects can increase considerably the overlap between laser field and molecular absorptions.[6,10] Photon absorption within region I will depend upon laser intensity and frequency, and clearly is responsible for the high isotopic selectivity observed in some dissociation yields.

Within region II, excitation is expected to take place in consecutive single photon absorption steps, with the absorption coefficients being independent of laser intensity, and the overall number of molecules driven through the quasicontinuum being dependent upon the integral of the laser intensity, termed the fluence (J cm^{-2}). Above the dissociation limit, the up-pumping rate must compete with unimolecular decomposition, and the laser intensity will determine the excess energy present in the molecule and available for partitioning in the dissociation fragments. Furthermore, if more than one dissociation channel is energetically accessible, then the product branching ratios may depend upon laser intensities (see later).

The overall effects on the IRMPD yield of parameters such as fluence, intensity, and frequency of the laser source thus depend upon the properties of each of the regions of absorption illustrated in Figure 1. For example, laser frequency effects within region II have been (as expected) demonstrated to be of less importance than for the crucial absorption steps in region I.[11,12] Laser intensity effects are often obscured by the relatively high fluences needed to pump a measurable yield of molecules through region II, these fluences arising from laser pulses comprising intensities large enough to overcome the intensity dependent absorption steps of region I. The effect of collisions upon the IRMPA and IRMPD processes will be discussed later.

This simple division of the absorption and dissociation processes into three regions leaves unanswered many important questions, notable amongst these being how collisionless coupling takes place between the pumped mode and the rest of the molecule (the transition between regions I and II).[13] Indeed, this division has been removed in theoretical approaches which treat molecular eigenstates (*via* which photon absorption occurs) as mixtures of normal modes, removing the necessity of considering intramolecular energy transfer.[8,14] This idea has also been adopted in the phenomenological approach of treating the absorption processes as sets of

[7] (a) R. V. Ambartzumian, in 'Tunable Lasers and Applications,' ed. A. Mooradian, T. Jaeger, and P. Stokseth, Springer, Berlin, 1976, p. 150; (b) D. M. Larsen and N. Bloembergen, *Optics Commun.*, 1976, **17**, 254.

[8] R. V. Ambartzumian, Yu. A. Gorokhov, V. S. Letokhov, G. N. Makarov, and A. A. Puretzky, *Sov. Phys. JETP Engl. transl.*, 1976, **44**, 231; *JETP Lett.* 1976, **23**, 22.

[9] I. N. Knyazev, V. S. Letokhov, and V. V. Lobka, *Optics Commun.*, 1978, **25**, 337.

[10] N. Bloembergen, C. D. Cantrell, and D. M. Larsen, in ref 7a, p. 102.

[11] V. S. Letokhov, *Ann. Rev. Phys. Chem.*, 1977, **28**, 133.

[12] V. S. Letokhov and C. B. Moore, in 'Chemical and Biochemical Applications of Lasers,' ed. C. B. Moore, Academic Press, New York, 1977, **3**, 1.

[13] (a) S. Mukamel and J. Jortner, *J. Chem. Phys.*, 1976, **65**, 5204; (b) E. Yablonovitch, *Optics Lett.*, 1977, **1**, 87; (c) S. Mukamel, *Chem. Phys.*, 1978, **31**, 327; *J. Chem. Phys.*, 1979, **70**, 5834.

[14] J. R. Ackerhalt and H. W. Galbraith, *J. Chem. Phys.*, 1978, **69**, 1200; *Optics Lett.*, 1978, **3**, 109.

linked rate equations, with absorption parameters deduced from experimental observations.[15,16] For transitions through the quasicontinuum, the validity of this rate equation method has been confirmed theoretically,[17,18] and the introduction of the absorption steps within the initially pumped discrete levels into this treatment has been demonstrated.[18] The observation of IRMPD in molecules with few vibrational frequencies or low densities of vibrational states [such as OCS, O_3 (ref. 19) and NH_3 (ref. 20)] causes some problems for an explanation of the process in terms of discrete and continuous vibrational energy levels. In these species, the distinction between regions I and II takes place much closer to the dissociation limit than for molecules such as SF_6, and coherent multiphoton absorption steps must play a more important role than for larger molecules. Recent review articles[21-23] have considered these and other theoretical aspects of IRMPA and IRMPD in far more detail than is possible here.

3 Investigations of The Nature of The Process

In this section we will attempt to provide an overview of some of the recent experimental studies undertaken with a view to establishing a more detailed understanding of the mechanism of IRMPD. Absorption studies on the excited molecules themselves, then experiments in which dissociation products have been investigated, will be considered in turn. Whereas the former type of study only provides information relating to regions I and II of the generalized model for the excitation process, the latter and far more numerous type of investigation can be applied to studies of the excitation mechanism both below (regions I and II) and above (region III) the dissociation threshold.

Studies in Absorption.—The vast majority of investigations of this kind have been directed at SF_6, and have recently been reviewed in considerable detail by Schulz *et al.*[21] In the 'discrete' region, the spectroscopy of the ground and first excited vibrational levels of the v_3 pumped mode around 948 cm^{-1} has been studied at very high resolution,[24] and Doppler-broadened rotational–vibrational structure has been observed in the $2 \leftarrow 1$[25,26] and $3 \leftarrow 0$[27] vibrational absorption bands. From the

[15] J. L. Lyman, *J. Chem. Phys.*, 1977, **67**, 1868.

[16] E. R. Grant, P. A. Schulz, Aa. S. Sudbo, Y. R. Shen, and Y. T. Lee, *Phys. Rev. Lett.*, 1978, **40**, 115.

[17] I. Schek and J. Jortner, *J. Chem. Phys.*, 1979, **70**, 3016.

[18] M. Quack, *J. Chem. Phys.*, 1978, **69**, 1282.

[19] D. Proch and H. Schroder, *Chem. Phys. Lett.*, 1979, **61**, 426.

[20] Hancock, R. J. Hennessy, and T. Villis, in 'Laser Induced Processes in Molecules,' ed. K. L. Kompa and S. D. Smith, Vol. 6, Springer-Verlag, Berlin, 1978, p. 190.

[21] P. A. Schulz, Aa. S. Sudbo, D. J. Krajnovich, H. S. Kwok, Y. R. Shen, and Y. T. Lee, *Ann. Rev. Phys. Chem.*, 1979, **30**, 379.

[22] C. D. Cantrell, S. M. Freund, and J. L. Lyman, in 'Laser Handbook IIIb,' ed. M. Stitch, North-Holland, Amsterdam, 1979, p. 485.

[23] V. N. Panfilov and Yu. N. Molin, *Russ. Chem. Rev.*, 1978, **47**, 503.

[24] J. P. Aldridge, H. Filip, H. Flicker, R. F. Holland, R. S. McDowell, N. G. Nereson, and K. Fox, *J. Mol. Spectrosc.*, 1975, **58**, 165.

[25] P. F. Moulton, D. M. Larsen, J. N. Malpole, and A. Mooradian, *Optics Lett.*, 1977, **1**, 51.

[26] C. C. Jensen, T. G. Anderson, C. Reiser, and J. I. Steinfeld, *J. Chem. Phys.*, 1979, **71**, 2648.

[27] J. R. Ackerhalt, H. Flicker, H. W. Galbraith, J. King, and W. B. Person, *J. Chem. Phys.*, 1978, **69**, 1461, and refs. therein.

spectroscopic point of view, detailed analyses of the region around the v_3 fundamental and overtone absorption bands are inevitably complicated by the fact that, under normal experimental conditions (300 K), the vibrational 'hot band' populations in SF_6 have been estimated to be such that only *ca.* 30% of the molecules are actually in the $v_3 = 0$ level.[28,29]

From the general mechanistic viewpoint the potentially more informative experiments are those designed to investigate the properties of molecules excited into their quasicontinuum (region II). Studies of this kind fall into two broad categories, namely, those which have attempted to measure the induced absorption spectrum of these vibrationally excited molecules, and those which have concentrated on providing an estimate of V–V intramolecular energy transfer rates within the quasicontinuum. Each type of experiment is fraught with experimental difficulties: their performance remains few and far between, and those time-dependent studies which have been reported yield widely disparate conclusions.

For example, Deutsch and Brueck[30] and Fuss and Hartmann[31] have carried out time-resolved i.r.–i.r. double resonance experiments on the v_3 absorption region in SF_6 for a variety of pumping laser fluences and have observed an immediate and marked red-shift in the induced v_3 absorption spectrum following the pump CO_2 laser pulse. Such an observation, both in induced absorption[30–32] and in the dissociation yield (the so called 'excitation' spectrum [7a, 10, 11]) has frequently been reported in studies of IRMPD. It arises as an inevitable consequence of the vibrational anharmonicity of the 'discrete' levels involved in the first few resonant absorption steps; less energetic photons are required for the excitation of vibrationally excited molecules. The induced absorption was subsequently observed to revert to shorter wavelengths over a period of several microseconds. This was interpreted[30] in terms of 'collisionless' cooling of the v_3 vibrational mode, and a time constant of 3 ± 2 μs was deduced for the rate of complete vibrational energy randomization amongst the various vibrational modes following an initial average excitation level of 3 CO_2 photons per molecule. A similar instantaneous red-shift in the absorption spectrum of the v_3 mode was observed for higher average excitation energies (*ca.* 10 CO_2 photons per molecule), but the eventual 'relaxed' spectrum (which developed with a very similar time constant) was found to be considerably broader than the original room-temperature absorption profile.[30] In a very early study, Ambartzumian[7a] suggested a comparable microsecond timescale for the rate of intramolecular energy transfer from the pumped v_3 mode of OsO_4 to the observed, i.r.-inactive v_1 vibrational mode. Furthermore, it has been argued that the results of a recent i.r.–visible double-resonance study of vibrational relaxation following multiple photon excitation of propynal ($HC\dot{.}C \cdot CHO$) under the rigorously collision-free conditions of a molecular beam demonstrate that, at an average excitation level of 3—4 CO_2 photons per molecule, the initially pumped v_6 mode of this molecule does not equilibrate with the lower vibrational levels of

[28] S. S. Alimpiev, N. V. Karlov, B. G. Sartakov, and E. M. Khokhlov, *Optics Commun.*, 1978, **26**, 45.

[29] W. Tsay, C. Riley, and D. O. Ham, *J. Chem. Phys.*, 1979, **70**, 3558.

[30] T. F. Deutsch and S. R. J. Brueck, *Chem. Phys. Lett.*, 1978, **54**, 258; *J. Chem. Phys.*, 1979, **70**, 2063.

[31] W. Fuss and J. Hartmann, *J. Chem. Phys.*, 1979, **70**, 5468.

[32] W. Fuss, J. Hartmann, and W. E. Schmid, *Appl. Phys.*, 1978, **15**, 297.

non-pumped modes,[33] at least within the 10 μs timescale of these experiments.

Other experimental investigations have been reported, however, which suggest that the rate of intramolecular V–V transfer is very much faster. Frankel and Manuccia[34] observed i.r. emission at 16 μm from the ν_4 vibrational mode of SF_6 following excitation of the ν_3 absorption feature. Under essentially collisionless conditions the 16 μm fluorescence was found to have a risetime faster than the microsecond time-resolution of the experiment. This more rapid timescale for vibrational energy flow has been supported with a vengeance by the recent experimental results of Kwok and Yablonovitch,[35] which have been interpreted in terms of a collisionless intramolecular relaxation time for SF_6 molecules (at an average level of excitation corresponding to the absorption of one CO_2 laser photon per molecule) in the range 1—30 ps. The experiment employed 30 ps CO_2 laser pulses initially to excite the SF_6 molecules, and used a second, weak probe pulse of the same frequency (split from, and delayed with respect to, the excitation pulse) to monitor the partial recovery of the ν_3 absorption as vibrational energy was redistributed amongst the various modes. These latter workers[35] also measured a *collisional* relaxation time of 13 ns Torr (2.4×10^{-9} cm^3 mol^{-1} s^{-1}) for SF_6 which has led Schulz *et al.*[21] to question the 'collisionless' nature of the earlier i.r.–i.r. double-resonance absorption measurements.[30] To date, however, there appears to be insufficient data to ascertain precisely what processes are being measured in these two experiments and whether, in fact, either of the suggested rates for vibrational energy randomization, differing by at least *five* orders of magnitude, provides a measure of the true rate of collisionless V–V energy transfer in SF_6.

One experimental parameter that may have received insufficient attention in these experiments,[30,35] the interpretation of both of which depend to some extent on an estimate of the average excitation level per molecule, is the laser *intensity* which is important in determining the initial population distribution over vibrational states, especially at relatively low laser fluences. Optoacoustic measurements of the (collisional) i.r. multiple photon excitation of vinyl chloride at a pressure of 133 N m^{-2} have revealed that the average number of CO_2 photons absorbed at a given laser fluence has a marked dependence upon the pulse duration.[36] More recently, Sudbo *et al.*[37] have carried out photoionization studies on multiphoton excited SF_6 molecules under molecular beam conditions and shown that at low laser intensities, comparable to those employed by Deutsch and Brueck,[30] only a small fraction (<20%) of the molecules are excited out of the discrete levels into the quasicontinuum. Kwok[38] has also observed this 'bottleneck' effect of the discrete levels in SF_6 at low laser fluences. He found that the average number of CO_2 photons absorbed by SF_6 at a constant (low) fluence increased significantly when the laser pulsewidth was decreased from 150 to 30 ps, but that the effect became much less significant at higher fluences (>1 J cm^{-2}) in accord with a number of

[33] D. M. Brenner and K. Brezinsky, *Chem. Phys. Lett.*, 1979, **67**, 36.
[34] D. S. Frankel, jun., and T. J. Manuccia, *Chem. Phys. Lett.*, 1978, **54**, 451.
[35] H. S. Kwok and E. Yablonovitch, *Phys. Rev. Lett.*, 1979, **41**, 745.
[36] F. M. Lussier, J. I. Steinfeld, and T. F. Deutsch, *Chem. Phys. Lett.*, 1978, **58**, 277.
[37] Aa. S. Sudbo, P. A. Schulz, D. J. Krajnovich, Y. T. Lee, and Y. R. Shen, *Optics Lett.*, 1979, **4**, 219.
[38] H. S. Kwok, Ph.D. Thesis, Harvard University, Cambridge, Mass., 1978, reported in ref. 21.

earlier 'high' fluence absorption[39] and dissociation[40-44] studies on SF_6. This hitherto frequently underestimated role of possible laser intensity effects in any overall description of the IRMPD process represents a common theme, which will recur quite frequently in this Report.

Studies of Dissociation Relevant to Excitation Regions I and II.—As mentioned previously, investigations of the wavelength dependence of the IRMPD yield have been reported for a number of polyatomic molecules in which the peak of the measured dissociation yield or excitation spectrum is observed to be red-shifted with respect to the single-photon i.r. spectrum of the parent molecules.[45-47] That this effect arises as a consequence of the anharmonicity of the first few 'discrete' vibrational levels through which the quasicontinuum is accessed, was first conclusively demonstrated in a number of elegant experiments on the dissociation of OsO_4 and SF_6, using two lasers of different frequencies.[7a, 12] In these experiments, the first weak laser was tuned to the maximum of the v_3 absorption feature, its function being simply to excite molecules through the 'discrete' vibrational levels in region I. The wavelength of the second, more powerful, laser was then varied. A much improved dissociation efficiency (as compared to using a single laser of fixed frequency) was found when the second laser operated at frequencies to the red of the first pumping wavelength. In effect, the second, and dissociating laser was used to map out an excitation spectrum of molecules prepared in the quasicontinuum by the first *weak* laser pulse; the observed red-shift is analogous to the red-shifted induced absorption spectra of SF_6 described previously.[30, 31]

More recently, Tiee and Wittig have used two pulsed i.r. lasers of different frequencies to effect the IRMPD of SeF_6[48] and UF_6,[49] neither of which dissociate to any significant extent under the influence of CO_2 laser irradiation alone, although SeF_6 does possess a very weak resonant absorption feature centred around 923 cm^{-1} associated with the $v_2 + v_6$ combination band. In the first instance[48] a pulsed NH_3 laser operating at 780.5 cm^{-1} was used to pump the resonant v_3 mode of SeF_6. A 50-fold increase in the dissociation yield, and no loss of isotopic selectivity, was observed when this first, *weak* (50 mJ per pulse) resonant excitation pulse was followed after 0.5 μs by a more powerful (1.5 J per pulse) CO_2 laser pulse; furthermore, the dissociation rate was found to be insensitive to the precise frequency of the second laser pulse. In the latter study[49] the pulsed output of a CF_4 laser was used to pump the v_3 absorption of UF_6 at 615 cm^{-1}; comparable enhancements in the overall dissociation probability were observed when this *weak* resonant excitation source was used in conjunction with the more powerful output

[39] J. L. Lyman, B. J. Feldmann, and R. A. Fischer, *Optics Commun.*, 1978, **25**, 391.
[40] J. L. Lyman, J. W. Hudson, and S. M. Freund, *Optics Commun.*, 1977, **21**, 112.
[41] J. L. Lyman, S. D. Rockwood, and S. M. Freund, *J. Chem. Phys.*, 1979, **67**, 4545.
[42] M. C. Gower and K. W. Billman, *Appl. Phys. Lett.*, 1977, **30**, 514.
[43] P. Kolodner, C. Winterfield, and E. Yablonovitch, *Optics Commun.*, 1977, **20**, 119.
[44] J. G. Black, P. Kolodner, M. J. Schultz, E. Yablonovitch, and N. Bloembergen, *Phys. Rev. A.*, 1979, **19**, 704.
[45] R. V. Ambartzumian and V. S. Letokhov, in ref. 12, p. 166.
[46] G. Hancock, R. J. Hennessy, and T. Villis, *J. Photochem.*, 1978, **9**, 197; 1979, **10**, 305.
[47] M. Rossi, J. R. Barker, and D. M. Golden, *Chem. Phys. Lett.*, 1979, **65**, 523.
[48] J. J. Tiee and C. Wittig, *J. Chem. Phys.*, 1978, **69**, 4756.
[49] J. J. Tiee and C. Wittig, *Optics Commun.*, 1978, **27**, 377.

of a CO_2 laser. Both experimental results emphasize yet again the potential selectivity introduced by the first few resonant absorption steps. The increase in the vibrational-state density once the molecules are excited into the quasicontinuum is such that suitable levels are available for sequential excitation by i.r. photons of any frequency that lies within the quite extensive induced absorption profile of these vibrationally-excited molecules; hence the driving influence of the CO_2 laser in these two experiments, which merely serves to provide the necessary *fluence* to pump the molecules to their dissociation limit. These results clearly demonstrate the breadth of the induced absorption spectrum of molecules in their quasicontinuum but, of course, do not necessarily imply that the induced absorption is uniformly 'black' to all i.r. frequencies; indeed, for SF_6, structure has been observed both in the induced absorption[32] and in the dissociation yield excitation spectrum.[11] Although the *peak* of the induced spectrum is red-shifted as a consequence of vibrational anharmonicity, the overall spectrum also broadens considerably. Deutsch and Brueck[30] have actually observed weak absorption associated with a 'blue' quasicontinuum extending for over 100 cm^{-1} to the high-frequency side of the v_3 absorption band in SF_6.

A number of other interesting results have appeared recently in which the frequency dependence of the CO_2 laser-induced dissociation yield from a variety of molecular ions has been investigated. Woodin *et al.*[50] used a low-power (4 W cm^{-2}) CW CO_2 laser to irradiate proton-bound dimers of diethyl ether, $[(Et_2O)_2H]^+$, produced in an ion cyclotron resonance (ICR) spectrometer, and observed no wavelength dependence for the primary production of $[Et_2OH]^+$ ions over the range 925—1090 cm^{-1}. A similar wavelength independence has been reported for the collisionless dissociation probability of $MeOHF^-$ ions following pulsed CO_2 laser irradiation.[51] It has been suggested[21,51] that this may be a consequence of the use of an ICR production technique in which the parent ions are initially formed with a considerable, but rather ill-defined, amount of internal excitation. Under these conditions the observed dissociation may arise from molecules which are already in the quasicontinuum as a result of chemical (rather than laser) activation. This suggestion must remain somewhat speculative however, since the single-photon i.r. spectrum of these two species are not known. Furthermore, and apparently in complete contrast, Woodin *et al.*[52] have subsequently demonstrated that the collision-free excitation spectrum for production of $C_2F_4^+$ ions following CW excitation of $C_3F_6^+$ parent molecular ions shows a marked wavelength dependence.

One final study of this kind must be mentioned, in which the IRMPD of EtCl using the pulsed output of a Nd:YAG (YAG = yttrium aluminium garnet) laser-pumped $LiNbO_3$ parametric oscillator around 3.3 μm has been investigated as a function of the exciting frequency.[53] The shape and structure of the C_2H_4 product-yield spectrum was found to reproduce faithfully many of the features not only of the fundamental absorption band around 3000 cm^{-1} (associated with the

[50] R. L. Woodin, D. S. Bomse, and J. L. Beauchamp, *J. Am. Chem. Soc.* 1978, **100**, 3248.
[51] R. N. Rosenfeld, J. M. Jasinski, and J. I. Brauman, *J. Am. Chem. Soc.* 1979, **101**, 3999.
[52] R. L. Woodin, D. S. Bomse, and J. L. Beauchamp, *Chem. Phys. Lett.*, 1979, **63**, 630.
[53] H.-L. Dai, A. H. Kung, and C. B. Moore, *Phys. Rev. Lett.*, 1979, **43**, 761; A. H. Kung, H.-L. Dai, M. R. Berman, and C. B. Moore, in 'Laser Spectroscopy IV' ed. H. Walther and K. W. Rothe, Springer-Verlag, Berlin, 1979, p. 309.

various C—H stretching frequencies), but also the first and second overtone bands. It has been suggested[53] that the unusually close frequency correlation between the absorption and product-yield spectra ($\lesssim 0.4$ cm^{-1}) arises because of the presence of five near-degenerate C—H stretching frequencies in the molecule in close resonance with the 3.3 μm pumping wavelength. These effectively compensate for vibrational anharmonicity and thus allow transitions through the discrete levels of region I to occur at or near resonance.

The crucial influence of the first few discrete absorption steps in the overall excitation scheme has been further demonstrated in a variety of other experimental studies. At relatively *low* laser intensities, power-broadening effects are not very significant and only a small fraction of all the molecules are likely to be in the few vibrational-rotational levels which can undergo resonant excitation through the discrete levels to the quasicontinuum and eventually dissociate. A number of workers have demonstrated that under suitable pressure conditions the multiple photon absorption[36,54] and dissociation[54-60] probabilities may be significantly enhanced through collisions, notably with the rare gases. Clearly the effect of these collisions is to reduce the 'bottleneck' effect introduced by the first few resonant transitions. The mechanisms most frequently advanced to account for this increased excitation cross-section[36,54-59] virtually all reflect the idea that collisions encourage rotational equilibration within the pumped vibrational mode of the absorber molecule and thus allow the pumping of molecules which were originally outside the frequency width of the pumping laser line: so called rotational 'hole filling'.[61] Studies of the IRMPD of SF$_6$ at higher laser fluences have revealed that the dissociation yield shows an inverse dependence upon the pressure of added diluent.[59,62] Under the conditions of these experiments, 'bottleneck' effects are overcome by the comparatively high laser intensities employed (see later). Thus the measured decrease in dissociation yield is due to the fairly efficient vibrational de-excitation of SF$_6^\ddagger$ molecules already in the quasicontinuum. A similar V–T relaxation mechanism has been suggested to account for the observed decline in the IRMPD yield from CH$_2$:CHCl,[36] C$_2$H$_4$,[55] CDF$_3$,[58] and CF$_2$HCl[60] at high buffer gas pressures.

Evans *et al.*[63] have demonstrated another method of reducing the severity of this early bottleneck in the multiple photon excitation of small molecules. They employed a pulsed DF laser (*ca.* 2700 cm^{-1}) to effect the IRMPD of H$_2$CO and observed that the dissociation yield obtained when a sample was simultaneously irradiated with multiple DF lines showed a 100-fold increase over that obtained through the use of a single line of the same overall energy. It has been suggested[63] that the range of excitation frequencies provided by the 'broadband' source enables many more molecules to find a suitable sequence of resonant absorption steps for

54 C. P. Quigley, *Optics Lett.*, 1978, **3**, 106; *J. Photochem.*, 1978, **9**, 177.
55 N. C. Peterson, R. G. Manning, and W. Braun, *J. Res. Natl. Bur. Stand.*, 1978, **83**, 117.
56 C. R. Quick, jun., and C. Wittig, *Chem. Phys.*, 1978, **32**, 75; *J. Chem. Phys.* 1978, **69**, 4201.
57 J. C. Stephenson, D. S. King, M. F. Goodman, and J. Stone, *J. Chem. Phys.*, 1979, **70**, 4496.
58 I. P. Herman and J. B. Marling, *Chem. Phys. Lett.*, 1979, **64**, 75, and refs. therein.
59 R. Duperrex and H. van den Bergh, *Chem. Phys.*, 1979, **40**, 275.
60 R. Duperrex and H. van den Bergh, *J. Chem. Phys.*, 1979, **71**, 3613.
61 J. D. Campbell, G. Hancock, J. B. Halpern, and K. H. Welge, *Chem. Phys. Lett.*, 1976, **44**, 404.
62 P. Bado and H. van den Bergh, *J. Chem. Phys.*, 1978, **68**, 4188.
63 D. K. Evans, R. D. McAlpine, and F. K. McClusky, *Chem. Phys. Lett.*, 1979, **65**, 226.

access into the quasicontinuum. As an alternative explanation, however, the increased dissociation yield may simply reflect (in part at least) the fact that the induced absorption spectrum of the vibrationally excited levels is different from that of molecules in the ground state.

Numerous studies of the fluence-dependence of the 'collisionless' dissociation yield from a variety of molecules, ranging in size from the triatomics ^{19}OCS and O_3 to the large, volatile uranium complex, $UO_2(hfacac)_2 \cdot thf^*$,[64] which contains 44 atoms and some 126 normal modes, have been reported. As usual, however, the most extensively studied molecule is SF_6,[40-44,65-69] the results for which have been used as a test-bed for some of the more recent theoretical models of the multiple photon excitation mechanism.[15,18,70,71] The general form of the yield *vs.* laser energy plot for SF_6 (conventionally plotted on a double logarithmic scale as in Figure 2) is typical of that observed for the IRMPD of many other molecules. The results from some of these IRMPD studies on SF_6 in which the mode quality,[40] pulse shape,[41] and duration[42-44] of the CO_2 laser pulse were varied have long been cited as demonstrations of the fact that the laser *fluence* and not the laser *intensity* is the important parameter in the IRMPD process. In passing, it is worth mentioning that one of the observable characteristics of a truly multiphoton excitation (*i.e.* a process involving the simultaneous absorption of *n* photons) is that the observed product yield exhibits an *n*th power dependence upon the intensity of the exciting laser; by and large though, the temptation to interpret the slope of these IRMPD yield plots (which usually have a gradient, $m \sim 3$—5) in terms of such a high-order *intensity* dependent process has (rightly) been avoided. As already described briefly in Section 2, it has now been shown theoretically[15,18,22,70] that the form of the observed fluence dependences can be modelled accurately by a simple treatment in which the rates of the transitions induced by the laser radiation between energy levels (both in absorption and stimulated emission) are assumed to depend linearly upon the laser intensity. In this approach, the total dissociation yield (the integral of the dissociation rate) becomes intensity independent, but can retain a greater than linear dependence upon fluence. By assuming that steady-state conditions prevail once *ca.* 30% of the precursor molecules have dissociated, Quack[70] has shown that it is possible to estimate unimolecular rate constants for the IRMPD of any molecule for which adequate experimental information (namely, the fluence dependence of the collision-free dissociation yield, once a sufficiently high level of dissociation has occurred) is available.

---

* Uranyl bis(hexafluoroacetylacetonate)–tetrahydrofuran.

[64] A. Kaldor, R. B. Hall, D. M. Cox, J. A. Horsley, P. Rabinowitz, and G. M. Kramer, *J. Am. Chem. Soc.*, 1979, **101**, 4465.

[65] R. V. Ambartzumian, Yu. A. Gorokhov, V. S. Letokhov, G. N. Makarov, and A. A. Puretzky, *JETP Lett. Engl. transl.*, 1976, **23**, 26.

[66] J. D. Campbell, G. Hancock, and K. H. Welge, *Chem. Phys. Lett.*, 1976, **43**, 581.

[67] W. Fuss and T. P. Cotter, *Appl. Phys.*, 1977, **12**, 265.

[68] F. Brunner, T. P. Cotter, K. L. Kompa, and D. Proch, *J. Chem. Phys.* 1977, **67**, 1547.

[69] F. Brunner and D. Proch, *J. Chem. Phys.*, 1978, **68**, 4936.

[70] M. Quack, *Ber. Bunsenges. Phys. Chem.*, 1978, **82**, 1252; 1979, **83**, 757; *J. Chem. Phys.*, 1979, **70**, 1069.

[71] J. A. Horsley, J. Stone, M. F. Goodman, and D. A. Dows, *Chem. Phys. Lett.*, 1979, **66**, 461, and refs. therein.

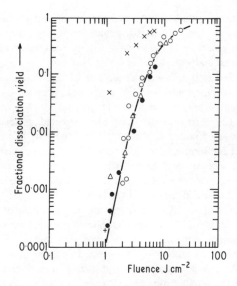

Figure 2 *Experimentally measured and modelled fluence-dependence of the product yield from the IRMPD* $SF_6 \rightarrow SF_5 + F$ *using the P(20) (001—100) CO_2 laser line. Experimental data points taken from △ ref. 41, ● ref. 43, + ref. 66, ○ ref. 67, and × ref. 68. The calculated dependence (full line) connects the experimental bulk data for SF_6 initially at 300 K, whereas the beam data (×) corresponds to an initial temperature of ca. 150 K.*
(Reproduced by permission from *Ber. Bunsenges. Phys. Chem.*, 1979, **83**, 757)

In contrast to the earlier SF_6 results,[40-44] recent studies of the collisionless IRMPD of $MeNH_2$[20,72] and NH_3[20] in our own laboratory have provided a further powerful illustration of the fact that the laser intensity as well as the laser fluence can be significant in determining the rate of multiple photon absorption and dissociation. In the more recent studies[72] the fluence-dependence of the NH_2 product yield from $MeNH_2$ was measured for a variety of times during the CO_2 laser pulse (see Figure 3). Three different exciting frequencies (resonant with three different regions of the single photon absorption spectrum of the parent molecule) were investigated; in each case, the measured dissociation yield for a given fluence was greatest when that fluence was delivered in the shortest time, *i.e.*, at the highest average intensity. The magnitude of the measured intensity-dependence was observed to decrease with increasing fluence, and to be more pronounced in the 'peak' of the CO_2 laser pulse than in the less energetic 'tail'; overall, however, more dissociation was observed during the 'tail' of the pulse than during the 'peak' as expected purely from fluence considerations. These results provide further support for the concept of an intensity-dependent 'bottleneck' at the beginning of the excitation process. It has been suggested[72] that, particularly at relatively low energies, the intense initial spike of the CO_2 laser pulse plays an important role in exciting molecules through this bottleneck; these 'prepared' molecules can then be

[72] M. N. R. Ashfold, G. Hancock, and G. W. Ketley, *Faraday Discuss. Chem. Soc.*, 1979, **67**, 204.

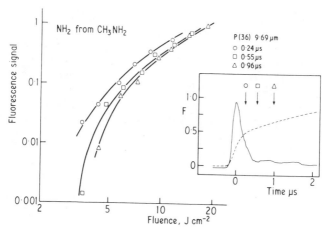

Figure 3 *Relative dissociation yield of* $NH_2(\tilde{X})$ *from the IRMPD of* $MeNH_2$ *using the P(36) (001—020)* CO_2 *laser line plotted as a function of fluence for the three delay times shown. Inset is the temporal profile of the* CO_2 *laser pulse (solid line) and the fraction, F, of the total pulse energy deposited (vertical scale and dashed line)* (Reproduced by permission from *Faraday Discuss. Chem. Soc.*, 1979, **67**, 204).

driven through the quasicontinuum by the sequential (intensity-independent) absorption of more photons from the 'peak' or the 'tail' of the pulse.

In an analogous experimental study, King and Stephenson[73] observed a major intensity-dependence in the yield of CF_2 radicals produced through the IRMPD of CF_2HCl. At the lowest fluence employed (1 J cm^{-2}) they report that a 6-fold increase in the average pulse intensity gave rise to a 400-fold increase in product yield, indicating that in this particular dissociation (which represents by far the most dramatic intensity effect reported to date) the intensity dependence of the IRMPD yield is of comparable importance to the fluence dependence. As in the previous study[72] the magnitude of the intensity effect was observed to decrease at higher fluences, and to disappear altogether when an inert gas was added.[73] Intensity-dependent product yields have also been reported in the collisionless IRMPD of CF_3I[47] and CF_3COCF_3[74] and in the collisional dissociation of HDCO,[75] in the multiple photon excitation of chromyl chloride, CrO_2Cl_2,[74] and in the multiple photon isomerization of *trans*-1,2-dichloroethylene.[74] Quantitative estimates of intensity effects are complicated by the very uneven intensity of the 'spiky' multimode CO_2 laser pulses used in the vast majority of IRMPD experiments reported to date. Thus, the ingenious yet very simple experimental arrangement used by Naaman and Zare,[74] through which they obtain pulses of equal fluence but different intensity, deserves special mention. The temporal profile of a normal multimode CO_2 laser pulse resembles a 'comb' of spikes superimposed on a broad background; the spikes (each of a few ns duration and separated from each other

[73] D. S. King and J. C. Stephenson, *Chem. Phys. Lett.*, 1979, **66**, 33.
[74] R. Naaman and R. N. Zare, *Faraday Discuss. Chem. Soc.*, 1979, **67**, 242.
[75] G. Koren and U. P. Oppenheim, *Optics Commun.*, 1978, **26**, 449.

by the cavity round-trip time) are caused by beating together of the various longitudinal modes within the laser cavity. However, as Naaman and Zare[74] have pointed out, if a pulse of this kind is split into two (Figure 4), and one half of it is delayed (by a time corresponding to approximately *half* the cavity round-trip time) before recombination with the other half of the original pulse, the net result is a pulse of the same total fluence but with only about half the intensity of the original. The magnitude of the visible emission resulting from the multiple photon excitation of CrO_2Cl_2 (see later discussion) using the recombined beam has been found to be a

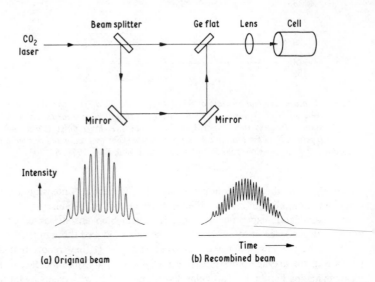

Figure 4 *Experimental configuration for carrying out i.r. multiple photon excitation studies with either a, the original beam, or b, the recombined beam of a CO_2 laser*
(Reproduced by permission from *Faraday Discuss. Chem. Soc.*, 1979, **67**, 242)

factor of eight smaller than that obtained using a 'normal' pulse of exactly the same fluence.[74]

Studies in Dissociation Relevant to Region III.—Over the last few years a wealth of relevant experimental results has appeared, which provides information regarding the nature of the i.r. multiple photon absorption process once the lowest dissociation threshold has been reached (region III). Virtually none of them would seem to be inconsistent with the hypothesis that complete energy randomization within the various vibrational modes occur before dissociation. Thus, apart from the few early and now often quoted erroneous reports of 'mode selective' i.r. laser dissociation,[76] the predominant products observed in all these IRMPD studies are those derived from the statistically predicted dissociation channel.

[76] D. F. Dever and E. Grunwald, *J. Am. Chem. Soc.*, 1976, **98**, 5055.

In a classic series of experiments, Lee and co-workers[77-80] used a molecular beam–mass spectrometric detection technique to investigate the IRMPD of a variety of molecules including SF_6, N_2F_4, and a number of halogenated hydrocarbons. For all these molecules dissociation was observed to proceed *via* the thermodynamically favoured channel, which in all cases involved a single bond-rupture (Table 1).[21, 79] By measuring the angular and velocity distributions of the dissociation fragments these workers are able to derive the fragment translational energy distributions and thus gain considerable information regarding the dynamics of the dissociation process. Zero-peaking translational energy distributions were obtained for all the simple bond-cleavage reactions studied.[79] It was found that the measured distributions could all be modelled using RRKM theory in conjunction with intuitively reasonable assumed lifetimes for the

Table 1 *Infrared multiple photon dissociation (IRMPD) products from simple bond-rupture reactions**

Molecule	Obs. fragments	BDE†/kJ mol⁻¹	Calc. lifetime of dissociating level/ns
N_2F_4	$2NF_2$	93.2	1
CF_3I	$CF_3 + I$	213	1
CF_3Br	$CF_3 + Br$	278	2
CF_3Cl	$CF_3 + Cl$	341	5
CF_2Br_2	$CF_2Br + Br$	255	5
CF_2Cl_2	$CF_2Cl + Cl$	326	5
$CFCl_3$	$CFCl_2 + Cl$	303	12
SF_6	$SF_5 + F$	385	10—20
C_2F_5Cl	$C_2F_5 + Cl$	347	60

* From refs. 21 and 79. † Bond dissociation energy.

vibrationally excited activated complex. For example, in the case of the IRMPD of SF_6 [equation (1)] the calculated dissociation rate was found to approach the laser

$$SF_6 \rightarrow SF_5 + F \qquad (1)$$

excitation rate at levels corresponding to the absorption of 7—11 CO_2 photons (*ca.* 113 kJ mol⁻¹) above the dissociation threshold; the experimental data can be reproduced by assuming an average lifetime of 10—20 ns for these vibrationally excited levels [21, 28] (see Figure 5). This translational energy distribution reinforces the results of earlier scavenging experiments[81,82] from which it was concluded that

[77] M. J. Coggiola, P. A. Schulz, Y. T. Lee, and Y. R. Shen, *Phys. Rev. Lett.*, 1977, **38**, 17.
[78] E. R. Grant, M. J. Coggiola, Y. T. Lee, P. A. Schulz, Aa. S. Sudbo, and Y. R. Shen, *Chem. Phys. Lett.*, 1977, **52**, 595.
[79] Aa. S. Sudbo, P. A. Schulz, E. R. Grant, Y. R. Shen, and Y. T. Lee, *J. Chem. Phys.*, 1978, **68**, 1306; 1979, **70**, 912.
[80] Aa. S. Sudbo, P. A. Schulz, Y. R. Shen, and Y. T. Lee, *J. Chem. Phys.*, 1978, **69**, 2312.
[81] C. R. Quick, jun., and C. Wittig, *Chem. Phys. Lett.*, 1977, **48**, 420.
[82] J. M. Preses, R. E. Weston, jun., and G. W. Flynn, *Chem. Phys. Lett.*, 1977, **48**, 425.

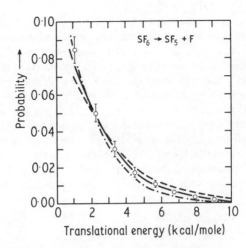

Figure 5 *Experimental centre of mass recoil energy distribution of the* SF_5 *fragments produced through IRMPD of* SF_6 *using the P(20) (001—100)* CO_2 *laser line at a fluence of* 10 J cm^{-2} *(open circles). Also shown are the statistical rate theory predictions for* SF_6 *molecules having absorbed seven (–·–·–), nine (————), and eleven (————) photons above the dissociation threshold (1 cal $\equiv$ 4.18 J)*
(Reproduced by permission from *Chem. Phys. Lett.*, 1977, **52**, 595)

the IRMPD of SF_6 gave rise to translationally 'cool' F atoms. Thus the balance of the excess of excitation energy must remain as vibrational energy in the SF_5 fragments; this level of internal excitation ensures that the SF_5 fragments are initially formed in their own quasicontinuum of states and can readily undergo further stepwise excitation leading to secondary dissociation products.[78] Clearly, secondary fragmentations of this kind constitute a potential complication in investigations of the primary collisionless dissociation process, most especially when real time resolution is not employed.

Product analysis may be further complicated when a molecule possesses two (or more) low-lying dissociation channels of sufficiently similar energy that the thresholds for both (or all) of them lie within the significantly populated excited levels. Under such circumstances the branching ratio between the various competing channels will be dependent upon the laser *intensity*; indeed Quack,[18,70] and Duperrex and van den Bergh[59] have pointed out that even within the framework of a purely statistical dissociation theory it is possible to envisage situations in which products formed *via* a higher-energy dissociation channel could dominate those derived from the low-energy channel. Such a situation for a molecule with two low-lying decomposition pathways is represented schematically in Figure 6. If the unimolecular rate constant for dissociation *via* the higher-energy channel, $k_2(E)$, increases with energy much more rapidly than that for the less energetic route then clearly, in the limit of fast optical pumping (high intensities), neither $k_1(E)$ nor $k_2(E)$ will compete effectively with the optical up-pumping rate until most of the molecules are well above the threshold for the high-energy decomposition channel, whereupon $k_2(E) > k_1(E)$ and the higher-energy products will predominantly, although never

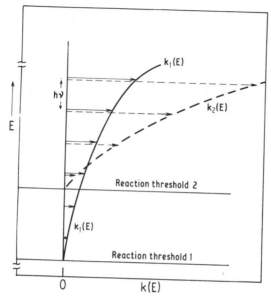

Figure 6 *Schematic illustration of a situation whereby products from the more energetic of two competing dissociation channels can predominate in an IRMPD process. Unimolecular rate constants for the lower energy channel $k_1(E)$ increase slowly with increasing energy above reaction threshold 1, while those for the more energetic pathway $k_2(E)$ increase rapidly with energy above reaction threshold 2* (Reproduced by permission from *Chem. Phys.*, 1979, **40**, 275)

exclusively,[18,70] result. Such intensity-dependent behaviour in region III has been observed in the IRMPD of ethyl vinyl ether[83,84] and (probably) cyclopropane.[85]

The two low-lying dissociation channels for ethyl vinyl ether are, respectively, the retro-ene molecular elimination (2) and the bond rupture (3) in which the primary radical products subsequently disproportionate to yield stable products. At low pressures (0.27 N m^{-2}) and low CO_2 laser fluence (0.9 J cm^{-2}) Brenner[84] has observed that butyraldehyde is the almost exclusive product of reaction (3). The relative importance of the two dissociation channels can therefore be estimated simply by measuring the butyraldehyde–acetaldehyde product ratio. Brenner[84] found that laser pulses of equal fluence but different duration gave rise to different product branching ratios; the higher-energy channel involving radical formation is favoured by the use of a short (0.2 μs), more intense laser pulse.

$$CH_3CH_2OCH{:}CH_2 \xrightarrow{\ k_2\ } CH_3CHO + CH_2{:}CH_2 \qquad (2)$$

$$\log k_2 = 11.6 - \frac{9.6 \times 10^3}{T}$$

[83] R. N. Rosenfeld, J. I. Brauman, J. R. Barker, and D. M. Golden, *J. Am. Chem. Soc.*, 1977, **99**, 8063.
[84] D. M. Brenner, *Chem. Phys. Lett.*, 1978, **57**, 357.
[85] R. B. Hall and A. Kaldor, *J. Chem. Phys.*, 1979, **70**, 4027.

$$CH_3CH_2OCH{:}CH_2 \xrightarrow{\ k_3\ } CH_3CH_2^{\bullet} + CH_2{:}CHO^{\bullet} \qquad (3)$$

$$\log k_3 = 15.0 - \frac{14.2 \times 10^3}{T}$$

Similar intensity effects in dissociation may have been unwittingly observed in a recent study of the IRMPD of cyclopropane,[85] the results of which have been claimed to provide evidence for a mode-selective excitation process. The two reaction channels of interest in this example are isomerization (4) and fragmentation (5). Hall and Kaldor[85] observed that excitation of the C—H asymmetric

$$\triangle \xrightarrow{\ k_4\ } CH_2{:}CHCH_3 \qquad (4)$$

$$\log k_4 = 15.2 - \frac{14.2 \times 10^3}{T}$$

$$\triangle \xrightarrow{\ k_5\ } CH_2{:}CH_2 + {:}CH_2 \qquad (5)$$

$$\log k_5 \sim 15 - \frac{21.8 \times 10^3}{T}$$

stretching mode of cyclopropane at 3.22 μm (using the output of a Nd:YAG laser pumped LiNbO$_3$ optical parametric oscillator) gave rise to exclusive formation of the isomerization product, propylene, while CO$_2$ laser excitation of the CH$_2$ 'wagging' mode at 9.5 μm yielded a 50:50 mixture of isomerization and fragmentation products. Although interpretation of these results must involve a certain amount of speculation, owing to the rather unpredictable effect of collisions at the high pressures employed in the experiment, it is not unlikely that it is the higher intensity of the CO$_2$ laser pulse that allows access to the higher energy (fragmentation) reaction channel rather than any 'mode selective' excitation process. Additional support for this proposal is provided by the fact that an increased relative yield of the low-energy channel (isomerization) product was observed when the CO$_2$ laser pulse duration was increased (*i.e.*, its average intensity was decreased).[85] Yet a further complication, is that all the foregoing discussion concerning the IRMPD of cyclopropane has necessarily assumed equal *absorption* rates at the two excitation wavelengths. Alternatively, it is possible to envisage the enhanced product yield from the higher-energy channel following excitation at 9.5 μm as arising as a consequence of the parent molecular absorption cross-section at this wavelength being much greater than at 3.22 μm. Since the overall excitation rate is determined by the product of the laser intensity multiplied by the absorption cross-section, such an observation would be expected under these circumstances even if the two laser pulses were of very comparable fluence and intensity. Clearly the case for 'mode selective' absorption in the excitation of this molecule remains, at best, debatable.

There have been a number of studies of the IRMPD of CF$_2$Cl$_2$, not only because of its inherent selectivity with respect to both ^{13}C[86-88] and the two Cl isotopes,[89] but also for mechanistic reasons. Morrison and Grant[90] have measured the fractional yield of CF_2 fragments to provide a measure of the relative importance of the two lowest-lying fragmentation channels:

$$CF_2Cl_2 \rightarrow CF_2 + Cl_2 \qquad \Delta H = 309 \text{ kJ mol}^{-1} \qquad (6)$$

$$\rightarrow CF_2Cl + Cl \qquad \Delta H = 326 \text{ kJ mol}^{-1} \qquad (7)$$

They found that the relative yield of CF_2 increases as the laser fluence (and hence intensity) is decreased, which tends to rule out the possible secondary dissociation (8) as a significant source of the CF_2 radical. Schulz *et al.*[21] reached a similar

$$CF_2Cl \rightarrow CF_2 + Cl \qquad \Delta H = 226 \text{ kJ mol}^{-1} \qquad (8)$$

conclusion regarding the increasing importance of the three-centre molecular elimination channel at low fluences by considering the results of a number of previous studies[79,91,92] in which the product branching ratio was estimated. Schulz *et al.*[21] have argued that the critical configuration in the atomic elimination pathway (7) should be characterized by a considerably stretched C−Cl bond and by a reduction in vibrational frequencies as compared with those in the parent molecule (a 'loose' complex), whereas the transition state for the three-centre molecular elimination (6) should be represented by a 'rigid' activated complex involving relatively high vibrational frequencies. Consequently, the rate of atomic elimination might be expected to increase more rapidly with excitation than the rate of molecular elimination; precisely the behaviour observed in the IRMPD of CF_2Cl_2.

There have also been a number of studies of the IRMPD of molecules in which the elimination of a hydrogen halide molecule constitutes the least energetic dissociation channel. In the collisionless regime, Sudbo *et al.*[80] have applied their molecular beam technique to a study of the translational energy distributions of the HCl fragments produced in the IRMPD of CH_3CCl_3, CH_3CF_2Cl, $CHClCF_2$, CHF_2Cl, and $CHFCl_2$ *via* either three- or four-centre molecular elimination reactions, all of which are characterized by the presence of an activation barrier to the reverse recombination reaction. Apart from the reaction involving the 3-centre elimination of HCl from $CHClCF_2$, all these dissociations yielded non-zero peaking (*i.e.*, non-statistical) translational energy distributions, in marked contrast to the previously described behaviour exhibited by the simple bond-rupture reactions[79] in which the energy required for dissociation is roughly equivalent to the difference between the enthalpies of formation of the product fragments and the parent molecule itself, *i.e.*, there is essentially no energy barrier for the back reaction. (As will be discussed in the succeeding paragraph, however, the observation of non-statistical translational energy partitioning in the molecular elimination reactions[80] in no way necessarily implies that there is a non-statistical

[86] J. L. Lyman and S. D. Rockwood, *J. Appl. Phys.*, 1976, **47**, 595.
[87] J. J. Ritter, *J. Am. Chem. Soc.*, 1978, **100**, 2441.
[88] D. S. King and J. C. Stephenson, *J. Am. Chem. Soc.*, 1978, **100**, 7151.
[89] R. E. Huie, J. T. Herron, W. Braun, and W. Tsang, *Chem. Phys. Lett.*, 1978, **56**, 193.
[90] R. J. S. Morrison and E. R. Grant, *J. Chem. Phys.*, 1979, **71**, 3537.
[91] D. S. King and J. C. Stephenson, *Chem. Phys. Lett.*, 1977, **51**, 48.
[92] J. W. Hudgens, *J. Chem. Phys.*, 1978, **68**, 777.

energy distribution within the vibrationally excited parent molecules immediately before dissociation, since the observed non-statistical energy partitioning within the fragments can be qualitatively rationalized in terms of a dynamical partitioning of the energy associated with the barrier to the back reaction). As an obvious consequence, however, it is not possible to obtain a detailed description of the dissociation dynamics of these molecular eliminations solely from experimentally determined fragment translational energy distributions and the judicious application of RRKM theory; additional information concerning the partitioning of the potential energy of the exit channel barrier amongst the various internal degrees of freedom is also required. Such information can on occasion be provided by a variety of spectroscopic techniques such as laser-induced fluorescence or, most simply, by observation of spontaneous emission from vibrationally (or electronically) excited products.

Quick and Wittig[56] have demonstrated the collisionless formation of vibrationally excited $HF^{\ddagger}$ molecules in the IRMPD of vinyl fluoride and a variety of fluorinated ethylenes and ethanes through time-resolved analysis of the spontaneous $HF^{\ddagger}$ i.r. chemiluminescence. Subsequent analysis of the vibrational energy partitioning amongst the vibrationally excited $HF^{\ddagger}$ fragments from the collisional IRMPD of CH_2CHF, CH_3CH_2F, CH_3CHF_2, and CH_2F_2 revealed a non-statistical distribution[93] somewhat comparable to that reported for the HF molecules produced from these same molecules by other excitation methods, *e.g.*, triplet mercury sensitization[94] or chemical activation techniques.[95] Similarly, it has been shown that the IRMPD of CF_3CH_3 yields some vibrationally excited $HF^{\ddagger}$ and $CH_2CF_2^{\ddagger}$ amongst the products, and that both $HF^{\ddagger}$ and $HCl^{\ddagger}$ are formed in the IRMPD of CF_2ClCH_3.[96] Both sets of workers[93,96] have been at lengths to point out that such an observation is not necessarily incompatible with the accepted statistical nature of the IRMPD process; indeed it is beginning to appear that the lesson learnt from the earlier premature proposals regarding mode-selective product yields has led to a quite widespread reluctance even to suggest the possibility of a genuinely non-statistical energy partitioning in the excited parent molecules before dissociation. Quick and Wittig[93] have suggested that the overall $HF^{\ddagger}$ product distribution observed reflects a superposition of the statistically partitioned 'excess' excitation energy and the dynamically partitioned 'fixed' energy (associated with the activation barrier to the back reaction) that is released as the transition state evolves into products. Thus, it is reasonably argued, the observed non-Boltzmann distribution within the $HF^{\ddagger}$ fragments originates primarily from the 'fixed' energy partitioning, since the small HF fragment would receive only a minor share of the 'excess' energy apportioned by a statistical partitioning. While such an explanation is wholly consistent with the currently accepted mechanism of the IRMPD process (which assumes complete energy randomization at excitation levels approaching the dissociation threshold) it should be pointed out that there is little *direct* experimental evidence that unequivocably demonstrates statistical

[93] C. R. Quick, jun., and C. Wittig, *J. Chem. Phys.*, 1980, **72**, 1694.
[94] P. N. Clough, J. C. Polanyi, and R. T. Taguchi, *Can. J. Chem.*, 1970, **48**, 2912.
[95] E. R. Sickin and M. J. Berry, *IEEE J. Quant. Electron.*, 1974, **QE10**, 701.
[96] G. A. West, R. E. Weston, jun., and G. W. Flynn, *Chem. Phys.*, 1978, **35**, 275.

energy partitioning within the activated complex associated with a dissociation channel for which there exists an appreciable barrier to the back reaction.

Several groups have attempted to measure the dissociation rate of molecules excited to levels above their dissociation threshold by i.r. multiple photon absorption. Information of this kind, when considered in conjunction with measurements of the product energy distributions over a range of laser excitation energies, should provide a much clearer insight into the nature of the actual dissociation process. So far, the most extensive studies have involved the IRMPD of CF_2HCl.[57,60,97] Duperrex and van den Bergh[60,97] have used kinetic absorption spectroscopy to monitor the build-up in the product CF_2 radical concentration as a function of time during and after the CO_2 laser pulse and, for incident fluences in the range 2—8 J cm^{-2}, have deduced an average decomposition rate constant of *ca.* 10^6 s^{-1} for the vibrationally excited parent molecules; this rate constant is found to increase slowly with increasing laser fluence (and hence intensity). A similar value for this dissociation rate may be derived from the complementary study of Stephenson and co-workers,[57] which was carried out at a higher fluence (24 J cm^{-2}) and employed time-resolved laser-induced fluorescence radical detection methods. It should be noted in passing that these observed average excited state lifetimes of *ca.* 1 μs are roughly an order of magnitude longer than has been previously suggested[80] on the basis of experimentally measured velocity distributions and the application of RRKM theory, again demonstrating the potential pitfalls associated with the use of a statistical description for the product energy distributions obtained from decompositions for which there is an activation barrier to the back reaction.

One potential problem in any time-resolved study of this kind is the uncertainty regarding the contribution made by the relatively weak tail of a conventional CO_2 laser pulse to the observed product build-up. Thus, the experimental technique employed in a recent study of the IRMPD of 1,1-difluoroethane by Wittig and co-workers[98] represents a significant advance. By using a 'chopped' CO_2 laser pulse, *i.e.*, one with a very fast falling (< 10 ns) trailing edge, and monitoring the time evolution of the HF* chemiluminescence produced both during and immediately after the pulse, these workers were able to provide a direct estimate of the rate of unimolecular decomposition from the distribution of vibrationally excited states produced in the IRMPD of CHF_2CH_3. The results of this pioneering study reveal that while the laser is *on* CHF_2CH_3 molecules are pumped very rapidly, and that dissociation occurs primarily from highly excited vibrational states characterized by short lifetimes (*ca.* 3 ns); however, once the laser is *off* the continuing HF* yield observed is due to the much slower dissociation of less highly excited parent molecules. It is argued[98] that the measured unimolecular lifetime of *ca.* 150 ns represents an average lifetime for the 'centre of gravity' in the vibrational state distribution of those molecules excited above the dissociation threshold.

Laser-induced fluorescence (LIF) provides a very sensitive method for the determination of both the internal and translational energy distributions in small photofragments. Stephenson and King[91,99] have applied the technique to a study of

[97] R. Duperrex and H. van den Bergh, *J. Mol. Struct.*, 1980, **61**, 291.
[98] C. R. Quick, jun., J. J. Tiee, T. A. Fischer, and C. Wittig, *Chem. Phys. Lett.*, 1979, **62**, 435.
[99] J. C. Stephenson and D. S. King, *J. Chem. Phys.*, 1978, **69**, 1485.

the energy partitioning in the $CF_2(\tilde{X})$ fragments produced through the collisionless IRMPD of CF_2HCl, CF_2Br_2, and CF_2Cl_2. For each one of these three molecules they found that the vibrational energy distribution within the v_1 symmetric stretching mode and the v_2 bending mode of the resulting $CF_2(\tilde{X})$ photofragments could both be characterized by a particular temperature (Table 2). Further evidence for the statistical nature of the dissociation process is provided by the fact that in the IRMPD of CF_2Cl_2 the fragment vibrational temperature, T_V, was found to be independent of the initially pumped mode in the parent molecule.[91] These workers have also estimated the energy partitioned into $CF_2(\tilde{X})$ fragment rotation in the IRMPD of these three molecules by computer modelling the observed spectral band contours assuming a thermal rotational state distribution.[99] The validity of such an assumption has been reinforced by several recent studies in which LIF measurements have revealed well resolved, essentially Boltzmann, rotational state distributions in a number of diatomic[102, 103] or lighter triatomic[100] dissociation products (see Table 2).

Campbell *et al.*[104] have employed an optical time-of-flight technique to measure the translational energy of $C_2(a^3\Pi_u)$ fragments formed (*via* a sequence of dissociation steps) in the collisionless IRMPD of C_2H_3CN. In this study, the parallel pump and probe laser beams are arranged to be 1 cm apart; spatial selection is achieved by use of a specially designed skimmer assembly, which ensures that only those fragments that move out of the photolysis zone at right

Table 2 *Summary of results obtained using LIF to monitor energy disposal in fragments produced through collisionless i.r. multiple photon dissociation (IRMPD)*

Radical	Precursor	T_V/K (E_V/kJ mol^{-1})	T_R/K (E_R/kJ mol^{-1})	T_T/K (E_T/kJ mol^{-1})
$CF_2{}^{a, b}$	CF_2Cl_2	1050 (12.1)	550 (6.7)	510[c] (6.3)
	CF_2Br_2	790 (7.5)	450 (5.9)	570[c] (7.1)
	CF_2HCl	1160 (15.1)	2000 (25.1)	(14.6)[d]
NH_2	$MeNH_2$	900 ± 100[e]	400 ± 20[e]	760 ± 150[f] ~300[e]
CN	C_2H_3CN	~1000[h]	450 ± 20[g] 1000 ± 200[h]	
C_2	C_2H_3CN	~700[h]	700 ± 200[h]	~500 (4.6 ± 1.3)[i]

[a] Ref. 91; [b] ref. 99; [c] assuming a Boltzmann distribution of translational velocities; [d] assuming a δ-function velocity distribution; [e] ref. 100; [f] ref. 101; [g] ref. 102; [h] ref. 103; [i] ref. 104.

[100] M. N. R. Ashfold, G. Hancock, and A. J. Roberts, *Faraday Discuss. Chem. Soc.*, 1979, **67**, 247.
[101] R. Schmiedl, R. Bottner, H. Zacharias, U. Meier, and K. H. Welge, *Optics Commun.*, 1979, **31**, 329; H. Zacharias, R. Schmiedl, R. Bottner, M. Geilhaupt, U. Meier, and K. H. Welge in ref. 53, p. 329.
[102] C. M. Miller and R. N. Zare, *Faraday Discuss. Chem. Soc.*, 1979, **67**, 245.
[103] M. R. Levy, M. Mangir, H. Reisler, M. H. Yu, and C. Wittig, *Faraday Discuss. Chem. Soc.*, 1979, **67**, 243; M. H. Yu, M. R. Levy, and C. Wittig, *J. Chem. Phys.*, 1980, **72**, 3789.
[104] J. D. Campbell, M. H. Yu, M. Mangir, and C. Wittig, *J. Chem. Phys.*, 1978, **69**, 3854.

angles to the CO_2 laser axis are subsequently probed by the dye laser. The time-of-flight spectrum is obtained directly, simply by measuring the laser-induced fluorescence intensity (proportional to the C_2 radical concentration) over a range of delay times. Making the assumption that the C_2H_3CN precursor molecules are randomly oriented with a room-temperature Boltzmann velocity spread before dissociation, and that the dissociation itself occurs isotropically with respect to the probing laser axis, it is a relatively straightforward computational exercise to extract the required C_2 fragment velocity (and, hence, translational energy) distribution in the centre-of-mass frame of reference from the measured laboratory time-of-flight spectrum.

LIF has also found use in two other less-direct methods by which fragment translational energies have been estimated. King and Stephenson[99] using an antiparallel, collinear pump and probe laser geometry have estimated the average translational velocity of nascent $CF_2(\tilde{X})$ fragments produced in the collisionless IRMPD of CF_2HCl, CF_2Br_2, and CF_2Cl_2 simply by monitoring the rate of decrease of the LIF signal due to their free flight out of the probing laser beam. As an alternative, Welge and co-workers[101] have used a probing dye laser of sufficiently narrow bandwidth to permit the measurement of the Doppler linewidths of $NH_2(\tilde{X})$ fragments produced in the IRMPD of $MeNH_2$ under molecular beam conditions. By assuming a Gaussian line profile the experimentally observed rotational linewidths were transformed into a most probable speed. As with the method of King and Stephenson[99] the translational energy partitioning can only be estimated if a Boltzmann fragment velocity distribution is assumed. In the case of the IRMPD of CF_2HCl the previously described molecular beam studies[79] have shown this assumption to be invalid; the fragment translational energy distribution is sharply peaked owing to the presence of an activation barrier in the exit channel.

The internal and translational energy distributions for fragments formed by IRMPD listed in Table 2 reveal that although the distribution of energy within a given degree of freedom may, in general, be described as Boltzmann and thus represented by a characteristic temperature T, this value may be different for each degree of freedom. On the basis of this observation it has been suggested[100] that the IRMPD process may not give rise to complete energy randomization amongst the various fragment degrees of freedom since it may reflect the influence of any dynamical constraints during the fragmentation process (as in the IRMPD of CF_2HCl). Reference to Table 2 reveals a marked discrepancy in the rotational temperature measured for the $CN(X^2\Sigma^+)$ fragments produced through collisionless IRMPD of C_2H_3CN in two different laboratories, and in the translational distributions of NH_2 $(\tilde{X}^2B_1)$ fragments measured by two different techniques. The most probable explanation for this has been provided by two recent experimental studies,[104a, 104b] which have shown the important effect that the i.r. laser *intensity* has on the energy available for partitioning in the resulting dissociation fragments. In the absence of collisions, molecules excited above their dissociation threshold can undergo either of two competing processes throughout the remaining duration of the i.r. laser pulse; namely, unimolecular decomposition (through one, or more,

[104a] C. M. Miller and R. N. Zare, *Chem. Phys. Lett.*, 1980, **71**, 376.
[104b] M. N. R. Ashfold, G. Hancock and M. L. Hardaker, *J. Photochem.*, 1980, **14**, 85.

dissociation channels) or continued up-pumping as a result of further i.r. photon absorption. The rate of the former process increases markedly with internal energy content of the molecule above the dissociation barrier, whilst that of the latter depends on the laser intensity. Hence the relative rates of these two processes will determine the average level of excitation attained by the pumped molecules before dissociation and thus the energy available for partitioning into the fragments.[104b] Consistent with this view is the observation[104a, 104b] that CN radicals formed in the collisionless IRMPD of C_2H_3CN during the intense initial spike of the CO_2 laser pulse can be ascribed a much hotter rotational temperature (~ 1000 K) than those formed at longer times as a result of dissociation by the low intensity tail. Similarly, Stephenson *et al.*[104c] have reported that the vibrational temperature of the CF_2 fragments formed in the collisionless IRMPD of C_2F_3Cl at a fixed delay time decreases from ~ 1860 to ~ 1100 K as the fluence (and thus the intensity) is reduced six-fold from 30 to 5 J cm^{-2}. Different laser intensities are thus almost certainly responsible for the different CN and NH_2 temperatures noted in Table 2. For comparison of results obtained in different laboratories, and for modelling of energy partitioning in dissociation fragments, the intensity will clearly need to be specified, a task virtually impossible unless well shaped CO_2 laser pulses (produced, for example, by pulse switching with electro-optic crystals) with no mode-beating effects are used.

By its very nature the collisionless IRMPD process is unlikely to yield a high proportion of electronically excited *fragments*. In support of this conclusion, Stephenson and King[99] have estimated that the ratio of $CF_2(\tilde{A})$ to $CF_2(\tilde{X})$ produced in the IRMPD of CF_2Cl_2 and CF_2Br_2 is $< 10^{-8}$ (corresponding to an 'electronic temperature' of < 2900 K); a comparably low branching ratio for the formation of electronically excited NH_2 in the low-pressure IRMPD of NH_3 had previously been reported by Campbell *et al.*[105] Furthermore, Merchant[106] has recently reinvestigated the IRMPD of SiF_4 (one of the very first molecules shown to undergo multiple photon excitation[1,3]), and has concluded that even the so called 'instantaneous' component in the observed luminescence has collisional origins. Reisler and Wittig[107] have considered the kinetics of a variety of possible collisionless and collisional mechanisms whereby electronic emission could reasonably be expected to accompany the IRMPD of gas-phase molecules and, as a result, have proposed that as a *minimum* requirement for demonstrating that an excited product arises through a truly collisionless dissociation process, the observed product luminescence should be shown to exhibit a linear dependence upon precursor pressure at low pressures ($\leqslant 1$ N m^{-2}). However, the i.r. multiple photon excitation of a number of different molecules has recently been shown to produce visible luminescence under unambiguously collisionless conditions; indeed, such an observation might be expected for parent molecules which possess one (or more) bound excited electronic states lying at energies below the ground-state dissociation limit. Such *parent* molecular emission arising as a result of this 'inverse

[104c] J. C. Stephenson, S. E. Bialkowski and D. S. King, *J. Chem. Phys.*, 1980, **72**, 1161.

[105] J. D. Campbell, G. Hancock, J. B. Halpern, and K. H. Welge, *Optics Commun.*, 1976, **17**, 38.

[106] V. S. Merchant, *Optics Commun.*, 1978, **25**, 259.

[107] H. Reisler and C. Wittig, *Adv. Chem. Phys.*, in the press.

electronic relaxation'[108] has recently been observed from CrO_2Cl_2,[109] F_2CO,[110] and OsO_4.[111] Uncharacterized visible emission has been observed in the collisionless multiple photon excitation of UF_6,[49] tetramethyldioxetane,[112] C_2H_3CN,[113] C_2H_4,[107] C_3H_6,[107] and C_2H_3CHO;[107] it has been suggested[107] that, in the last-mentioned cases at least, this emission arises from primary collisionless dissociation products that have been further excited by the i.r. laser. One final and interesting example of a vibrationally driven electronic process has recently been provided by Rosenfeld *et al.*[114] who have reported the i.r. multiple photon induced detachment of an electron from the benzyl radical anion; at least seven CO_2 photons are required to overcome the electron affinity of the radical.

Thus it would seem that the currently accepted model of the i.r. multiple photon absorption process can so far successfully accommodate virtually all the experimental findings. Experimental evidence for the excitation mechanism being both *intensity* and *fluence* dependent in regions I and III is accumulating rapidly. It can be expected that considerable further experimental effort will be devoted to:

(*a*) unambiguously collision-free time resolved i.r.–i.r. double-resonance studies of molecules (and especially small molecules) involving two different laser frequencies to provide further estimates of the time-scale for V–V intramolecular energy transfer at various levels of excitation within the quasicontinuum region, and

(*b*) time-resolved dissociation studies using shaped laser pulses of the kind pioneered by Wittig and co-workers.[98] When combined with the additional sensitivity provided by LIF detection methods this technique should allow a much more detailed investigation of the mechanism and dynamics of the actual fragmentation process at a variety of 'average excitation levels' above the dissociation threshold.

4 Laser-induced Chemistry: Effect of Collisions

This Section reviews some of the many experimental studies (most of which have been carried out under anything but collision-free conditions), in which attempts have been made to distinguish between thermal and i.r. laser-induced chemistry. In their recent review Cantrell *et al.*[22] have argued that many of the early reports of 'non-thermal laser chemistry' can in fact be attributed to the effects of high-temperature thermochemistry in the region of the laser focus, to reactions with secondary photolysis products, and to wall reactions. Karny and Zare[115] have shown that heterogeneous effects can significantly influence the i.r. laser-induced *trans–cis* isomerization of 1,2-dichloroethylene and increase the product yield from

[108] A. Nitzan and J. Jortner, *Chem. Phys. Lett.*, 1978, **60**, 1.

[109] Z. Karny, A. Gupta, R. N. Zare, S. T. Lin, J. Niemann, and A. M. Ronn, *Chem. Phys.*, 1979, **37**, 15.

[110] J. W. Hudgens, J. L. Durant, jun., D. J. Bogan, and R. A. Coveleskie, *J. Chem. Phys.*, 1979, **70**, 5906.

[111] R. V. Ambartzumian, G. N. Makarov, and A. A. Puretzky, *JETP Lett. Engl. transl.*, 1979, **28**, 647.

[112] Y. Haas and G. Yahav, *Chem. Phys. Lett.*, 1977, **48**, 63; G. Yahav and Y. Haas, *J. Am. Chem. Soc.*, 1978, **100**, 4885; *Chem. Phys.*, 1978, **35**, 41.

[113] M. H. Yu, H. Reisler, M. Mangir, and C. Wittig, *Chem. Phys. Lett.*, 1979, **62**, 439.

[114] R. N. Rosenfeld, J. M. Jasinski, and J. I. Brauman, *J. Chem. Phys.*, 1979, **71**, 1030.

[115] Z. Karny, and R. N. Zare, *Chem. Phys.*, 1977, **23**, 321.

the IRMPD of cyclopropane and propylene. Conversely, Shaub and Bauer[116] have proposed that observed differences between laser-induced and thermal reactions may arise as a consequence of the fact that conventional high-temperature chemistry often has a dominant heterogeneous component that will not be present when an i.r. laser is used to induce 'homogeneous' high-temperature chemistry well away from the vessel walls. Whilst the common concensus at present is that i.r. laser excitation does not provide a means of achieving 'mode-selective' *dissociation*, the prospects for the more general concept of 'mode-selective' chemistry (*i.e.*, chemical transformations brought about as a result of vibrational energy being localized in a particular part of a molecule) do not appear to have been ruled out. In addition, many experiments have been reported that unequivocally demonstrate that it is possible to obtain non-thermal products distributions by using the non-equilibrium excitation conditions provided by a pulsed i.r. laser.

One of the most successful methods of demonstrating that a particular set of products arises from a truly laser-induced transformation, and not through a 'thermal' reaction resulting from redistribution of the absorbed energy *via* intermolecular V–V and V–T,R energy-transfer processes, involves the addition of a 'chemical thermometer' to the system. In a pioneering study of this kind Danen *et al.*[117] studied the CO_2 laser-induced decomposition [equations (9) and (10)] of ethyl acetate in the presence of isopropyl bromide. The latter performs the role of an internal standard for measuring thermal effects. It does not possess an absorption feature resonant with the exciting CO_2 laser frequency and consequently it is possible to estimate thermal effects by monitoring the eliminated HBr. If possible the internal standard is chosen so that both decompositions have similar activation energies, and the resultant ratio of the two thermal rate constants is then essentially independent of temperature. At low fluences (<1 J cm^{-2}) the measured propylene–ethylene ratio was found to be in accord with the thermally calculated value, but at higher fluences (8 J cm^{-2}) the relative yield of ethylene was observed to increase dramatically, indicating a significant contribution from the laser-induced non-thermal chemical reaction. Not surprisingly, the relative importance of the laser-induced component in the overall decomposition was found to be greatest at low pressure where collisional thermalization effects are minimized.

$$CH_3COOEt \xrightarrow{k_9} CH_3COOH + CH_2:CH_2 \qquad (9)$$

$$\log k_9 = 12.59 - \frac{10.5 \times 10^3}{T}$$

$$CH_3CHBrCH_3 \xrightarrow{k_{10}} CH_3CH:CH_2 + HBr \qquad (10)$$

$$\log k_{10} = 13.60 - \frac{10.4 \times 10^3}{T}$$

[116] W. M. Shaub and S. H. Bauer, *Int. J. Chem. Kinet.*, 1975, **7**, 509.
[117] W. C. Danen, W. D. Munslow, and D. W. Setser, *J. Am. Chem. Soc.*, 1977, **99**, 6961.

Braun and co-workers[118-120] have reported very similar results for this system and have found that the specificity in dissociation of the initially absorbing species is retained at pressures as high as 6.6 kN m^{-2} and that the addition of an excess of helium gives rise to non-thermal product ratios even at pressures approaching 1 atmosphere. To account for the first observation it has been suggested[119,120] that there is a low-lying bottleneck in the absorption process so that at these high pressures the bulk of the molecules constitute a 'cold' bath gas and that the relatively small number that are excited through the bottleneck are disproportionately 'hot'; non-absorbing isopropyl bromide molecules can only become involved in the overall excitation and dissociation scheme as a result of energy transfer arising through collisions with the 'hot' absorbing molecules. In passing, it may be worth pointing out that such a decoupling of the photolysed molecule and bath gas temperature may account for the non-thermal behaviour reported for the collisional IRMPD of cyclopropane,[121] $CHF_2 Cl$,[122] and C_2F_3Cl.[123] In order to account for the observed non-thermal product yield, even under the highly collisional conditions imposed by the addition of nearly one atmosphere of helium, it has been suggested that in this situation, the diluent provides a sink for the excitation energy of the absorbing molecule and greatly reduces the probability of intermolecular energy transfer to the non-absorbing molecule. Thus, although collisions undoubtedly reduce the net *intra*-molecular excitation rate, and hence the dissociation probability of the initially pumped ethyl acetate molecules, *inter*-molecular energy transfer to the non-absorbing internal standard is virtually ruled out altogether and what little isopropyl bromide dissociation is observed, probably arises as a consequence of thermal heating of the bulk gas by the laser.

Richardson and Setser[124] have applied the same experimental technique to a study of the CO_2 laser-induced decomposition of CH_3CF_3 and CH_3CH_2F. As before, the relative yield of the laser-induced reaction product was observed to increase significantly at lower sample pressures. Subsequent studies of the laser-induced elimination of HF from CH_3CF_3 in the presence of an excess of helium or nitrogen have demonstrated that the extent of laser-induced reaction has a marked dependence upon the laser intensity.[125] For a short intense CO_2 laser pulse the HF product yield is observed to decrease (as compared with that obtained from the same pressure of CH_3CF_3 precursor in the absence of diluent) relatively slowly with increasing pressure of diluent. For longer pulses however the dissociation quenching effect is far more pronounced. From a comparison of the relative quenching efficiencies of the various diluents it has been proposed[125] that the essential steps in the quenching of laser-induced reactions involve collisional deactivation of highly vibrationally excited molecules within their quasicontinuum.

[118] D. Gutman, W. Braun, and W. Tsang, *J. Chem. Phys.*, 1977, **67**, 4291.
[119] W. Tsang, J. A. Walker, W. Braun, and J. T. Herron, *Chem. Phys. Lett.*, 1978, **59**, 487.
[120] W. Braun, J. T. Herron, W. Tsang, and K. Churney, *Chem. Phys. Lett.*, 1978, **59**, 492.
[121] M. L. Leseicki and W. A. Guillory, *J. Chem. Phys.*, 1977, **66**, 4317.
[122] R. C. Slater and J. H. Parks, *Chem. Phys. Lett.*, 1979, **60**, 275.
[123] K. Nagai, M. Katayama, H. Mikuni, and M. Takahashi, *Chem. Phys. Lett.*, 1979, **62**, 499.
[124] T. H. Richardson and D. W. Setser, *J. Phys. Chem.*, 1977, **81**, 2301.
[125] D. W. Setser, *Faraday Discuss. Chem. Soc.*, 1979, **67**, 241; J. C. Jang and D. W. Setser, *J. Phys. Chem.*, 1979, **83**, 2809.

There have been numerous other reports of i.r. laser-induced dissociations involving the selective excitation of one component in a relatively low-pressure mixture of reactants, including the vast majority of multiple photon isotope separations. These are considered in somewhat greater detail amongst the various applications of i.r. multiple photon excitation but, in order to illustrate the essential principle of the method, one example involving an isotope of commercial importance will be described briefly. The example concerns the pulsed IRMPD of natural fluoroform, for which it has been shown[58,126] that excitation at 10.2 μm (a wavelength at which CHF_3 is essentially transparent, but which is resonant with the ν_5 rocking vibration in CDF_3) can lead to, at least, a 1000-fold deuterium enrichment in the resulting HF product.

In the following paragraphs we have attempted to provide a fairly exhaustive summary of the various other reported i.r. laser-induced organic reactions. An interesting study of a type not far removed from those described previously has investigated the effect of pulsed CO_2 laser radiation on the hydrogen bromide catalysed bimolecular dehydration of an equimolar mixture of ethanol and propan-2-ol.[127] In the conventional thermal bimolecular process HBr serves both as a reacting partner and as a catalyst for the dehydration; the activation energies for catalysed dehydration of the two alcohols are such that heating such a mixture gives rise to >98% propylene. In complete contrast, Danen[127] has observed that by using a pulsed CO_2 laser at a frequency resonant only with a vibrational mode of ethanol it is possible to induce selectively the unimolecular elimination of H_2O from ethanol in preference to either the lower-energy bimolecular HBr-catalysed dehydration, or the thermodynamically more facile dehydration of propan-2-ol. The selectivity of the product yield decreases as collisions begin to dominate and at total pressures $\gtrsim$330 N m^{-2} the 'thermal' product distribution is observed.

At relatively lower overall levels of excitation, i.r. multiple photon absorption can be used to promote selective isomerization reactions. An early example of this type of laser-induced chemistry was provided by Yogev and Loewenstein-Benmair[128] who found that the pulsed CO_2 laser irradiation of an equimolar mixture of the *cis*- and *trans*-isomers of but-2-ene led to the mixture becoming enriched with respect to the *cis*-isomer as a result of selective decomposition of the *trans*-isomer. Subsequently it has been shown[115,129-131] that the pulsed CO_2 laser excitation of *trans*-1,2-dichloroethylene actually leads to the formation of some of the *cis*-isomer and that both the *trans–cis* and the *cis–trans* interconversions can be effected in a thermal fashion by adding SF_6 to a mixture of the isomers to act as a photosensitizer. In the direct laser-induced *trans–cis* interconversion the pumped molecule must absorb *ca.* 20 CO_2 photons to overcome the activation energy for isomerization; the activated molecule can subsequently decay to either the *cis* or *trans* ground state with the relative decay probabilities determined solely by

[126] S. A. Tuccio and A. Hartford, jun., *Chem. Phys. Lett.*, 1979, **65**, 234.
[127] W. C. Danen, *J. Am. Chem. Soc.*, 1979, **101**, 1187.
[128] A. Yogev and R. M. J. Loewenstein-Benmair, *J. Am. Chem. Soc.*, 1973, **95**, 8487.
[129] R. V. Ambartzumian, N. V. Chekalin, V. S. Doljikov, V. S. Letokhov, and V. N. Lokhman, *Optics Commun.*, 1976, **18**, 220; *J. Photochem.*, 1976/7, **6**, 55.
[130] K. Nagai and M. Katayama, *Chem. Phys. Lett.*, 1977, **51**, 329.
[131] C. Reiser, F. M. Lussier, C. C. Jensen, and J. I. Steinfeld, *J. Am. Chem. Soc.*, 1979, **101**, 350.

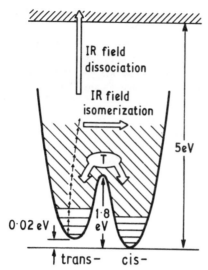

Figure 7 *Schematic representation of the competition between the laser-induced isomerization and dissociation, as well as the thermal isomerization of* trans-*1,2-dichloroethylene in an intense i.r. field (after ref. 129)*

Boltzmann statistics and not by the particular isomeric form present initially (see Figure 7). As confirmation of this mechanism, Knyazev *et al.*[132] have recently reported the selective isomerization of the *trans*-isomer using a weak CO_2 laser to provide the initial absorption selectivity and a KrF laser (249 nm) to excite these prepared molecules above the barrier to isomerization. Buechele *et al.*[133] have studied the CO_2 laser-induced multiple photon isomerization of the hexadienes [equation (11)] at sufficiently low laser fluences for the extensive formation of fragmentation products to be avoided. In accord with the previous studies these workers have demonstrated that it is possible to convert selectively and non-destructively an absorbing into a non-absorbing isomer; furthermore, on the basis of product analysis following excitation by a single CO_2 laser pulse, it is possible for two (or more) successive isomerizations to occur within the time duration of the pulse.[133]

$$2,4 \quad \rightleftharpoons \quad \rightleftharpoons \tag{11}$$

trans-,trans-   cis-,trans-   cis-,cis-

1,3

trans-   cis-

[132] I. N. Knyazev, Y. A. Kudryavtsev, N. P. Kuz'mina, and V. S. Letokhov, *Sov. Phys. JETP Engl. transl.*, 1978, **47**, 1049.
[133] J. L. Buechele, E. Weitz, and F. D. Lewis, *J. Am. Chem. Soc.*, 1979, **101**, 3700.

As a follow up to their initial work with but-2-ene, Yogev and co-workers[134, 135] have reported a number of other interesting laser-induced isomerization reactions. For example, the pulsed CO_2 laser excitation of a relatively low-pressure (665 N m^{-2}) sample 2,3,3,4,4,5-hexadeuteriohexa-1,5-diene (A) can cause its almost total conversion into the product (B) *via* a Cope rearrangement [equation (12)].[134] The pulsed CO_2 laser-induced isomerization[135] of hexafluorocyclobutene to the

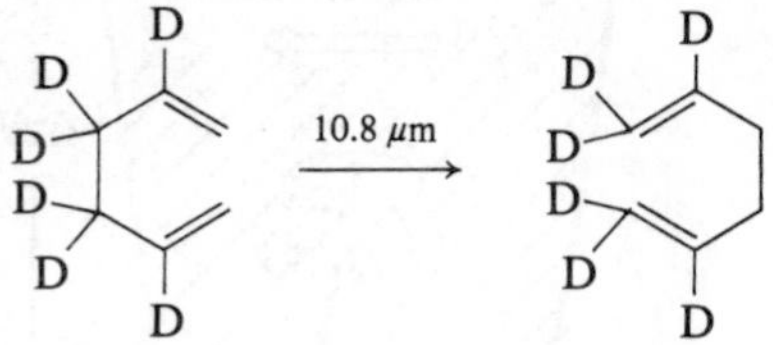

$$10.8 \ \mu m$$

(12)

(A)   (B)

thermodynamically less-stable hexafluorobutadiene [equation (13)] has given conversions of *ca.* 30% observed by use of the unfocussed laser output. Although

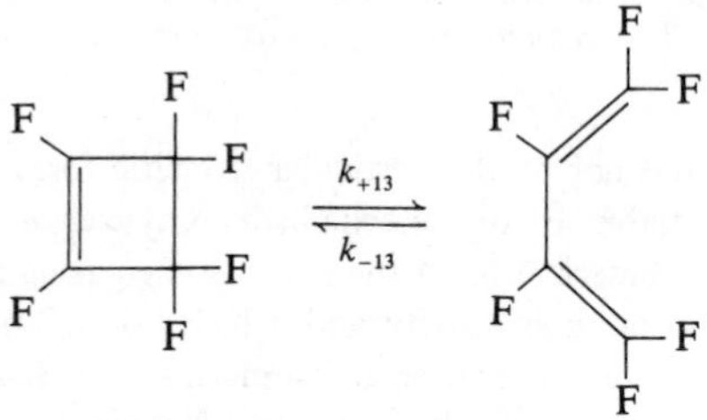

$$k_{+13}$$
$$k_{-13}$$

(13)

$$\log k_{13} = 14.00 - \frac{10.3 \times 10^3}{T}; \qquad \log k_{-13} = 12.03 - \frac{7.7 \times 10^3}{T}$$

the addition of a large excess of helium was found to reduce the initial reaction rate (k_{13}) it also had the effect of completely quenching the reverse reaction (k_{-13}); thus quantitative conversion to the less-stable non-cyclic isomer could be obtained. It was also observed that the laser-induced isomerization process in the presence of an excess of helium could be isotopically selective with respect to carbon-13.[135] Somewhat analogously, it has been reported that at sufficiently low pressures ($<$270 N m^{-2}) the multiple photon isomerization of CH_3NC and C_2H_5NC in their ground electronic states is isotopically selective for nitrogen.[136]

From the stereochemical viewpoint, Danen *et al.*[137] have investigated the CO_2

[134] I. Glatt and A. Yogev, *J. Am. Chem. Soc.*, 1976, **98**, 7087.
[135] A. Yogev and R. M. J. Benmair, *Chem. Phys. Lett.*, 1977, **46**, 290; 1979, **63**, 558.
[136] A. Hartford, jun., and S. A. Tuccio, *Chem. Phys. Lett.*, 1979, **60**, 431.
[137] W. C. Danen, D. F. Koster, and R. N. Zitter, *J. Am. Chem. Soc.*, 1979, **101**, 4281.

laser-induced ring-opening of *cis*-3,4-dichlorocyclobutene and found the exclusive product to be *cis,trans*-1,4-dichlorobuta-1,3-diene:

$$\log k_{14} \sim 13.4 - \frac{6.6 \times 10^3}{T}; \qquad \log k_{15} \sim 14.0 - \frac{9.8 \times 10^3}{T}$$

The observed product, which arises through a thermally allowed (ground electronic state) conrotatory electrocyclic ring-opening process [equation (14)], is precisely that predicted by the Woodward–Hoffmann rules. Despite the use of a range of CO_2-laser fluences and intensities these workers were unable to access any other, higher-energy reaction channel (involving, for example, a biradical intermediate or a low-lying excited electronic state of the parent molecule) that might give rise [equation (15)] to the other two isomeric forms of the product. Since it has been estimated[138] that the vertical energy difference between the transition states for the Woodward–Hoffmann allowed and non-allowed product channels is only $\gtrsim 63$ kJ mol^{-1} this experimental non-observation of any products from the higher-energy disrotatory product channel apparently contrasts with the previously described results of Brenner[84] for the IRMPD of ethyl vinyl ether. In accord with the earlier discussion, Danen *et al.*[137] have suggested that in the IRMPD of ethyl vinyl ether at moderate laser intensities the higher-energy bond rupture channel, involving a 'loose' transition state and a correspondingly high A factor, is able to compete successfully because its specific rate constant increases much more rapidly with energy than does that of the lower energy channel. In the laser-induced reaction of *cis*-3,4-dichlorocyclobutene however, the estimated A factor for the higher-energy product channels is only slightly greater than that for the Woodward–Hoffmann allowed process and is therefore quite unable to compensate for the difference in activation energies.

Steinfeld and co-workers[36,131,139] have carried out a series of experiments designed to probe the stereochemistry and dissociation dynamics of the i.r. laser-induced multiple photon reactions of a series of chlorine-substituted ethylenes.

[138] J. I. Brauman and W. C. Archie, jun., *J. Am. Chem. Soc.*, 1972, **94**, 4262.
[139] F. M. Lussier and J. I. Steinfeld, *Chem. Phys. Lett.*, 1977, **50**, 175.

In an early study [139] it was shown that the primary step in the IRMPD of vinyl chloride was elimination of HCl; the actual dissociation probability was found to be independent of the symmetry of the initially pumped vibrational mode in the parent molecule. HCl elimination has subsequently been shown to represent a dominant reaction channel in the IRMPD of all the mono-, di-, and trichloro-substituted ethylenes investigated.[131] However, the fact that isomerization of the reactant molecule was also observed to occur in the multiple photon excitation of *trans*-1,2-dichloroethylene and β-1-deuteriovinyl chloride has led to the suggestion that free rotation around the C=C double bond precedes the HCl elimination process.[131] Although products were often observed different from those reported in classical pyrolysis studies of these molecules, it has been demonstrated that the observed product branching ratios arising in the i.r. laser-induced chemistry of each of these molecules (and the pressure dependence of these ratios) can be accurately reproduced by a statistical, non-mode-selective model which presupposes unhindered vibrational energy flow throughout the molecule.[131]

The fact that molecules such as SiF_4, SF_6, and NH_3 readily undergo multiple photon absorption in the 10 μm region has led to their use as sensitizing agents; intermolecular energy transfer from the vibrationally excited sensitizer can be used to bring about reaction of a non-absorbing species. For example, Keehn and co-workers[140-142] have demonstrated that pulsed CO_2 laser photosensitization methods can be employed not only to promote the 'high-temperature' thermal reaction of non-absorbing parent molecules but also to provide a cleaner product yield than conventional pyrolysis techniques. By use of SiF_4 as sensitizer this 'cold pyrolysis' technique was used initially to effect the isomerization [equation (16)] of

$$CH_2=C=CH_2 \xrightarrow{\ SiF_4^{\ddagger}\ } CH_3C{\equiv}CH \tag{16}$$

allene to methyl acetylene,[140, 141] and more recently to promote a number of laser-induced retro-Diels–Alder reactions,[142] [equations (17)—(19)]. In both the

$$\tag{17}$$

Norbornadiene   Cyclopentadiene   Acetylene

$$\tag{18}$$

Cyclohexene   Butadiene   Ethylene

$$\tag{19}$$

D-Limonene   Isoprene

[140] K. J. Olszyna, E. Grunwald, P. M. Keehn, and S. P. Anderson, *Tetrahedron Lett.*, 1977, 1609.
[141] C. Cheng and P. Keehn, *J. Am. Chem. Soc.*, 1977, **99**, 5808.
[142] D. Garcia and P. Keehn, *J. Am. Chem. Soc.*, 1978, **100**, 6111.

pyrolysis and the u.v. photodecomposition of norbornadiene the isomeric form cycloheptatriene is found amongst the products; in contrast, none is observed in the sensitized i.r. decomposition. The observed product yield from cyclohexene decomposition is very much simpler than that obtained from the corresponding pyrolysis at 500 °C. Similarly, it is found[142] that the yield of isoprene from the sensitized decomposition of D-limonene is much cleaner than that obtained in the CW CO_2 laser irradiation of this molecule.[143] Steel *et al.*[144] have employed essentially the same technique in a study of the i.r. laser-induced decomposition of cyclobutanone using NH_3 as the sensitizing agent [equations (20) and (21)] and

$$\text{(20)}$$

Isomerization → Propylene

$$\log k_{20} = 14.37 - \frac{12.7 \times 10^3}{T}$$

$$\text{(21)}$$

$$CH_2CO + C_2H_4$$

$$\log k_{21} = 14.56 - \frac{11.4 \times 10^3}{T}$$

have suggested that analysis of the product ratios can provide information about the effective temperature reached by the reacting system and the timescale for which the system remains at this limiting high temperature.

Another potentially important application of CO_2 lasers to organic synthesis is in the preparation of isotopically labelled compounds.[87,145] For example, Ritter[87] has shown that the i.r. multiple photon excitation of a CF_2Br_2–isobutylene mixture yields the adduct 1,1-difluoro-2,2-dimethylcyclopropane as a major constituent in the product [equation (22)]. By irradiating at frequencies that overlap with

$$Me_2C{=}CH_2 + CF_2Br_2 \longrightarrow Me_2C{-}CH_2 \quad \text{(22)}$$

absorption features specific to the ^{13}C halogenocarbon it is possible selectively to prepare[87,88] the CF_2 reactive intermediate enriched in ^{13}C and thus provide a synthetic route for specifically labelled fluorinated compounds.

The CO_2 laser has also found use in a number of inorganic syntheses involving boron compounds. For example, Rockwood and Hudson[146] have demonstrated that

[143] A. Yogev, R. M. J. Loeweinstein, and D. Amar, *J. Am. Chem. Soc.*, 1972, **94**, 1091.
[144] C. Steel, V. Starov, F. Leo, P. John, and R. G. Harrison, *Chem. Phys. Lett.*, 1979, **62**, 121.
[145] J. J. Ritter and S. M. Freund, *J. Chem. Soc., Chem. Commun.*, 1976, 811.
[146] S. D. Rockwood and J. W. Hudson, *Chem. Phys. Lett.*, 1975, **34**, 542.

the laser-induced reaction (23) of BCl_3 with molecular hydrogen yields exclusively

$$BCl_3 + H_2 \xrightarrow{10.6\,\mu m} BHCl_2 + HCl \qquad (23)$$

pure $BHCl_2$. By contrast, the catalysed thermal reaction yields a complex mixture of products including B_2H_6, B_2H_5Cl, $BHCl_2$, *etc*. Interestingly, it has sub-sequently been reported that in the presence of a catalyst of powdered titanium metal the laser-induced reaction gives rise to the dimeric form of $BHCl_2$ and is isotopically selective for boron.[147,148]

Thus there is now a wealth of experimental evidence to show that non-thermal chemistry can be achieved by i.r. multiple photon excitation. In contrast to conventional high-temperature chemistry, the use of an i.r. laser allows selected components in a mixture of reactants to be singled out and rapidly excited in the absence of heterogeneous wall effects. On the other hand, no conclusive experimental evidence for the most attractive and exciting concept of 'mode-selective' laser-induced chemistry has appeared to date. Here, again, it is important to reiterate the distinction between laser-induced dissociations and possible laser-induced chemical reactions in which one reactant carries a relatively low, and very possibly localized, level of excitation. In the former category there have been no unequivocal demonstrations of bond-selective multiple photon dissociations; even now, it is still not clear whether it is valid to generalize and assume that this conclusion applies to the multiple photon dissociation of all molecules, especially very large, spatially extensive molecules containing a number of well defined and well separated groups. For example, in order to account for the estimated laser-driven dissociation rate of $UO_2(hfacac)_2 \cdot thf$, the largest molecule investigated to date, Kaldor *et al.*[64] have suggested that complete energy randomization in this molecule does not occur within the timescale of the dissociation process, but that the initial vibrational excitation actually remains localized in a bonding sphere centred around the uranyl moiety. Further experimental studies of the i.r. laser-induced dissociation of large molecules aimed at substantiating or disproving this suggestion can be expected in the near future.

Similarly, considerable experimental effort is likely to be focussed upon the far from exhausted possibility of 'mode-selective' chemical reactions involving vibrationally excited molecules. Current models of the multiple photon absorption process (see earlier) indicate that in region I the initial absorption occurs within a specific vibrational mode and that intramode coupling does not become important until several photons have been absorbed: the onset of the quasicontinuum. Thus within this discrete absorption region it should prove possible to prepare vibrationally excited molecules which contain considerable vibrational excitation, but have a markedly non-random internal energy distribution, and which, as a consequence, can undergo non-thermal 'bond specific' chemical reactions.

One final type of laser-induced chemical reaction worthy of mention is typified by the experimental study of Hsu and Manuccia[149] in which a low-power CW CO_2

[147] C. T. Lin, T. D. Z. Atvars, and F. B. T. Pessine, *J. Appl. Phys.*, 1977, **48**, 1720.
[148] C. T. Lin and T. D. Z. Atvars, *J. Chem. Phys.*, 1978, **68**, 4233.
[149] D. S. Y. Hsu and T. J. Manuccia, *Appl. Phys. Lett.*, 1978, **33**, 915.

laser has been used at a wavelength that selectively excites only the CH_2D_2 molecules (in their ν_7 vibrational mode) in a CH_2D_2–CH_4 mixture. Interisotopic V–V scrambling is minimized (at the expense of photon usage) by the addition of an excess of argon diluent; a steady-state situation develops in which only the CH_2D_2 molecules are vibrationally excited (almost certainly to the $\nu_7 = 1$ level only). These subsequently react with chlorine atoms and molecules present in the initial reaction mixture, at a much faster rate than the vibrationally unexcited species, leading to formation of deuterium-enriched stable methyl chloride product. The same general technique, based on efficient V–T relaxation and the fact that even one quantum of vibrational excitation in the selected isotopic form of the reactant polyatomic can lead to a significantly enhanced rate of product formation, has been similarly applied to Br [150] and C [151] isotope enrichment through the respective reactions (24) and (25) (isotope in square brackets). In essence, these observed isotope enrichments serve to provide another illustration of the fact that reagent internal energy can significantly promote the rate of bimolecular reactions.

$$
\left.
\begin{aligned}
CH_3[Br] + Cl &\longrightarrow CH_2[Br] + HCl \\
CH_2[Br] + Cl_2 &\longrightarrow CH_2[Br]Cl + Cl
\end{aligned}
\right\} \tag{24}
$$

$$
\left.
\begin{aligned}
[C]H_3F + Br &\longrightarrow [C]H_2F + HBr \\
[C]H_2F + Br_2 &\longrightarrow [C]H_2FBr + Br
\end{aligned}
\right\} \tag{25}
$$

5 Chemical Applications of Infrared Lasers

The advent of the i.r. laser has provided a significant boost to studies in which the dependence of reaction rates upon the initial levels of vibrational excitation in the reactants have been investigated. The bimolecular reaction of O_3 with NO provides the classic example,[152] while the isotope enrichment work of Manuccia and co-workers,[149,150] described in the concluding paragraph of the previous section, illustrates an interesting application of these studies. Several recent reviews have considered laser enhancement of bimolecular reactions in detail,[153–156] and since single photon excitations do not fall strictly within the scope of this review, studies of this kind will not be considered further here.

However, a number of recent studies have demonstrated that the introduction of a comparatively low level of vibrational excitation (typically 2—4 CO_2 laser photons) to the electronic ground state of a molecule can lead to a substantial modification of its visible and u.v. absorption and dissociation characteristics. Figure 8 illustrates some possible origins of the observed change in the electronic

[150] T. J. Manuccia and M. D. Clark, *J. Chem. Phys.*, 1978, **68**, 2271.

[151] Yu. N. Molin, V. N. Panfilov, and V. P. Strunin, *Chem. Phys. Lett.*, 1978, **56**, 557.

[152] See, *e.g.*, K. K. Hui and T. A. Cool, *J. Chem. Phys.*, 1978, **68**, 1023, and refs. therein.

[153] J. H. Birely and J. L. Lyman, *J. Photochem.*, 1975, **4**, 269.

[154] J. Wolfrum, *Ber. Bunsenges. Phys. Chem.*, 1977, **81**, 114; M. Kneba and J. Wolfrum, *Annu. Rev. Phys. Chem.*, 1980, 31.

[155] I. W. M. Smith, in 'Physical Chemistry of Fast Reactions', vol. II, 'Reaction Dynamics,' ed. I. W. M. Smith, Plenum Press, New York, 1979, p. 1.

[156] C. B. Moore and I. W. M. Smith, *Faraday Discuss. Chem. Soc.*, 1979, **67**, 146.

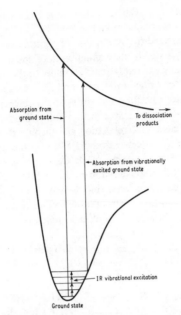

Figure 8 *Schematic illustration of the modifying influence of i.r. laser-induced vibrational excitation on electronic transition probabilities*

absorption profile. As a result of the ground state molecules being vibrationally excited the Franck–Condon factors for excitation to the, in this case dissociative, upper state are altered and the mean electronic excitation frequency is lowered by an amount dependent upon the magnitude of the vibrational energy introduced, leading to a red shift in the electronic absorption profile. In addition, the symmetry of a polyatomic molecule can be lowered by the excitation of bending modes so that, for example, transitions that are symmetry forbidden in a linear ground state molecule may be allowed in the vibrationally excited distorted configuration. Pummer *et al.*[157] have demonstrated that the long wavelength wing of the near-u.v. absorption band of CF_3I is significantly enhanced following initial vibrational excitation with a pulsed low-powered CO_2 laser. It has been suggested[157] that this much broadened spectral profile could be advantageous in the production of electronically excited I* atoms and iodine laser systems from the broad-band flash photodissociation of CF_3I. Similarly, low-powered CO_2 laser excitation has been used to create a vibrationally induced broadening of the u.v. absorption profile of the intense $^1\Sigma^+-^1\Sigma^+$ transition in OCS[158] and $OCSe$.[159] In both instances subsequent excitation with an ArF laser at 193 nm excites a region with a vibrationally enhanced absorption coefficient and results in a substantially increased yield of excited metastable $S(^1S)$ and $Se(^1S)$ atoms amongst the respective dissociation products.

[157] H. Pummer, J. Eggleston, W. K. Bischel, and C. K. Rhodes, *Appl. Phys. Lett.*, 1978, **32**, 427.
[158] D. J. Kligler, H. Pummer, W. K. Bischel, and C. K. Rhodes, *J. Chem. Phys.*, 1978, **69**, 4652.
[159] W. K. Bischel, J. Bokor, J. Dallarosa, and C. K. Rhodes, *J. Chem. Phys.*, 1979, **70**, 5593.

Somewhat comparable i.r.–u.v. double resonance experiments have been used to monitor energy transfer processes in biacetyl.[160, 161] In the more recent study[161] the 457.9 nm line of an Ar^+ laser was used to excite a vibrational hot-band transition and thus prepare biacetyl in both its first excited singlet state $\tilde{A}^1A_u$ and, through rapid intersystem crossing, the triplet state $\tilde{a}^3A_u$. Subsequent excitation with a weakly focussed CO_2 laser resulted in the triplet phosphorescence intensity being reduced both in intensity and duration; simultaneously, a burst of 'delayed fluorescence' was observed in the green. The findings have been interpreted by assuming that the CO_2 laser excites biacetyl molecules already prepared in the *triplet* manifold by the optical excitation process. Thus the more rapid phosphorescent decay is assumed to be characteristic of more highly excited vibrational levels of the triplet state and the 'delayed' fluorescence can be attributed to 'triplet' molecules that have been excited above the origin of the upper singlet state and which therefore possess some degree of excited singlet (^1A_u) character.

By far the most widely reported application of the i.r. multiple photon excitation process is in the field of isotope separation. Numerous reviews have described in detail[11, 12, 22, 162–164] the physical principles, experimental methods, and commercial feasibility of the various processes. Table 3 therefore merely attempts first to

Table 3 *Isotopic selectivity and enrichment achieved through i.r. multiple photon excitation*

Element	Parent molecule (refs. in parentheses)
B	BCl_3 (86, 165, 166)
Br	MeBr (150)
C	CF_3Cl (167), CF_3Br (167–169), CF_3I (167, 170–172), CCl_4 (173), CF_2Cl_2 (86–88), MeF (151), $(CF_3)_2CO$ (174)
Cl	CCl_4 (173), CF_2Cl_2 (89), $C_2F_2Cl_2$ (86)
H	CH_2Cl_2 (175), CH_2DCH_2Cl (176), H_2CO (63, 177), HCO_2H (178), MeOH (179), CHF_3 (58, 126), CF_3CHCl_2 (180), CH_4/CH_2D_2 (149)
Mo	MoF_6 (181)
N	MeNC (136), EtNC (136)
O	$(CF_3)_2CO$ (174)
Os	OsO_4 (4, 182, 183)
S	SF_6(4, 5, 41, 69, 184–187), S_2F_{10} (188), SF_5Cl (189), SF_5NF_2 (190)
Se	SeF_6 (48)
Si	SiF_4 (86)
U	$U(OMe)_6$ (191), UO_2 (hfacac)$_2 \cdot$ thf (192)

[160] B. J. Orr, *Chem. Phys. Lett.*, 1976, **43**, 446.

[161] I. Burak, T. J. Quelly, and J. I. Steinfeld, *J. Chem. Phys.*, 1979, **70**, 334.

[162] J. P. Aldridge, III, J. H. Birely, C. D. Cantrell, and D. C. Cartwright, in 'Laser Photochemistry, Tunable Lasers and Other Topics: Physics of Quantum Electronics,' vol. 4, ed. S. F. Jacobs, M. Sargent, III, M. O. Scully, and C. T. Walker, Addison-Wesley, Reading, Pa., 1976, p. 157.

[163] V. S. Letokhov and C. B. Moore, *Sov. J. Quant. Electron.*, 1976, **6**, 129 and 259.

[164] V. S. Letokhov, *Nature (London)*, 1979, **277**, 605.

[165] R. V. Ambartzumian, V. S. Letokhov, E. A. Ryabov, and N. V. Chekalin, *JETP Lett. Engl. transl.*, 1974, **20**, 273.

[166] S. M. Freund and J. J. Ritter, *Chem. Phys. Lett.*, 1975, **32**, 255.

[167] M. Drouin, M. Gauthier, R. Pilon, P. A. Hackett, and C. Willis, *Chem. Phys. Lett.*, 1978, **60**, 16.

[168] M. Gauthier, P. A. Hackett, M. Drouin, R. Pilon, and C. Willis, *Can. J. Chem.*, 1978, **56**, 2227.

[169] W. A. Jalenak and N. S. Nogar, *Chem. Phys.*, 1979, **41**, 407.

summarize the current state of the art as regards those elements whose isotopes have proved amenable to selective laser-induced chemistry and, secondly, to illustrate the extensive range of parent molecules for which isotopic enrichment has been observed as a result of i.r. multiple photon excitation.

The wavelength selectivity introduced by the first few absorption steps of the IRMPD process has also found application in the purification of gases. For example, Ambartzumian *et al.*[193] have demonstrated that it is possible selectively to dissociate the small amount of 1,2-dichloroethane and carbon tetrachloride impurity in commercial $AsCl_3$ by use of the pulsed output of a CO_2 laser tuned in turn to the absorption features of each impurity molecule. Subsequent removal of the photolysis products (notably C_2Cl_4 and C_2Cl_6) is then relatively easy. In a somewhat comparable study Merritt and Robertson[194] have demonstrated that the phosgene impurity in a commercial sample of BCl_3 can be removed using the 'chopped' output from a low-powered CW CO_2 laser. It would appear that in this latter study, however, the principle role of the laser is that of an expensive bunsen burner (although the exclusion of heterogeneous wall effects may be significant) since the initial excitation is actually in the v_3 mode of BCl_3. This then acts as an i.r. sensitizer for the phosgene dissociation. Again, subsequent removal of the resulting products is relatively simple.

From the gas kineticist's viewpoint one of the most interesting applications of the

[170] S. Bittenson and P. L. Houston, *J. Chem. Phys.*, 1977, **69**, 4819.
[171] V. N. Bagratashvili, V. S. Doljikov, V. S. Letokhov, and E. A. Ryabov, in ref. 20, p. 179.
[172] I. N. Knyazev, Yu. Kudriavtzev, N. P. Kuz'mina, V. S. Letokhov, and A. A. Sarkisian, *Appl. Phys.*, 1978, **17**, 427.
[173] R. V. Ambartzumian, Yu. A. Gorokhov, V. S. Letokhov, G. N. Makarov, and A. A. Puretzky, *Phys. Lett.*, 1976, **56A**, 183.
[174] P. A. Hackett, M. Gauthier, C. Willis, and R. Pilon, *J. Chem. Phys.*, 1979, **71**, 546; P. A. Hackett, C. Willis, and M. Gauthier, *ibid.*, p. 2682.
[175] A. Yogev and R. M. J. Benmair, *J. Am. Chem. Soc.*, 1975, **97**, 4430.
[176] A. J. Colussi, S. W. Benson, R. J. Hwang, and J. J. Tiee, *Chem. Phys. Lett.*, 1977, **52**, 349.
[177] G. Koren, U. P. Oppenheim, D. Tal, M. Okon, and R. Weil, *Appl. Phys. Lett.*, 1976, **29**, 40.
[178] D. K. Evans, R. D. McAlpine, and F. K. McClusky, *Chem. Phys.*, 1978, **32**, 81.
[179] R. D. McAlpine, D. K. Evans, and F. K. McClusky, *Chem. Phys.*, 1979, **39**, 263.
[180] J. B. Marling and I. P. Herman, *Appl. Phys. Lett.*, 1978, **34**, 439.
[181] S. M. Freund and J. L. Lyman, *Chem. Phys. Lett.*, 1978, **55**, 435.
[182] R. V. Ambartzumian, N. P. Furzikov, Yu. A. Gorokhov, V. S. Letokhov, G. N. Makarov, and A. A. Puretzky, *Optics Lett.*, 1977, **1**, 22.
[183] R. V. Ambartzumian, V. S. Letokhov, G. N. Makarov, and A. A. Puretzky, *Optics Commun.*, 1978, **25**, 69.
[184] R. V. Ambartzumian, N. P. Furzikov, Yu. A Gorokhov, V. S. Letokhov, G. N. Makarov, and A. A. Puretzky, *Optics Commun.*, 1976, **18**, 517.
[185] G. Hancock, J. D. Campbell, and K. H. Welge, *Optics Commun.*, 1976, **16**, 177.
[186] M. C. Gower and K. W. Billman. *Optics Commun.*, 1977, **20**, 123.
[187] S. T. Lin, S. M. Lee, and A. M. Ronn, *Chem. Phys. Lett.*, 1978, **53**, 260.
[188] J. L. Lyman and K. M. Leary, *J. Chem. Phys.*, 1978, **69**, 1858.
[189] K. M. Leary, J. L. Lyman, L. B. Asprey, and S. M. Freund, *J. Chem. Phys.*, 1978, **68**, 1671.
[190] J. L. Lyman, W. C. Danen, L. C. Nilsson, and A. V. Nowak, *J. Chem. Phys.*, 1979, **71**, 1206.
[191] S. S. Miller, D. D. DeFord, T. J. Marks, and E. Weitz. *J. Amer. Chem. Soc.* 1979, **101**, 1036.
[192] D. M. Cox, R. B. Hall, J. A. Horsley, G. M. Kramer, P. Rabinowitz, and A. Kaldor, *Science*, 1979, **205**. 390.
[193] R. V. Ambartzumian, Yu. A. Gorokhov, S. L. Grigorovich, V. S. Letokhov, G. N. Makarov, Yu. A. Malanin, A. A. Puretzky, E. P. Filippov, and M. P. Furzikov, *Sov. J. Quant. Electron. Engl. transl.*, 1977, **7**, 96.
[194] J. A. Merritt and L. C. Robertson, *J. Chem. Phys.*, 1977, **67**, 3545.

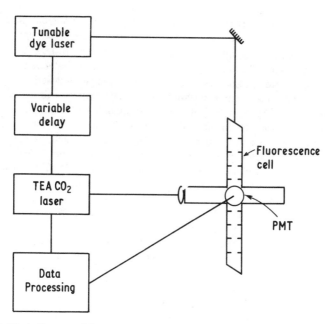

Figure 9 *Block diagram of the conventional experimental arrangement for measurements of the kinetic removal rate of fragments produced through IRMPD (after ref. 195)*

IRMPD process is as a source of reactive free radicals; the remainder of this section will therefore be devoted to a survey of the various relevant kinetic and spectroscopic studies reported. The basic experimental arrangement employed in virtually all the kinetic investigations is shown in Figure 9. The focussed output of a pulsed CO_2 laser is used to produce a localized concentration of the radical of interest through photolysis of a suitable precursor molecule. Radicals produced in the region of the CO_2 laser focus are subsequently detected by their LIF, which can be excited by use of a dye laser tuned to resonance with a strong absorption feature of the radical. Radical removal rates may be obtained simply by varying the time delay between the CO_2 photolysis laser and the probing dye laser, and thereby measuring the LIF signal intensity (proportional to the concentration of excited state radicals, which itself is directly proportional to the concentration of ground electronic state radicals produced in the IRMPD process) as a function of the delay time. In order to establish a second-order rate constant for radical reaction with a particular diluent, it is necessary to carry out pseudo first order depletion rate measurements of this kind for a variety of diluent pressures, and thence derive the slope of the graph obtained when these measured rates are plotted against the respective diluent concentrations.

A number of potential complications can be envisaged in the method as outlined so far. These might include relaxation into and/or out of the particular rovibrational level(s) probed by the dye laser, the effects of quenching by the precursor molecule itself, as well as by any other fragments produced in the IRMPD process, and also the possibility of diffusive loss of the radical from the probed region; these effects could significantly influence the measured first-order decay rates and thus the overall experimentally determined reaction rate coefficient. In order to minimize these obvious sources of error a typical experiment of this kind utilizes a number of premixed equilibrated gas samples each containing the lowest (constant) pressure of precursor radical source compatible with an adequate fluorescence signal-to-noise ratio, and diluted by a sufficient (variable) quantity of reactant to ensure pseudo first order kinetics. In most of the studies it has been possible to prevent diffusive radical loss by the further addition of a relatively high pressure of inert diluent (He or Ar). Clearly, however, such a procedure is precluded in kinetic measurements of electronically excited radicals, such as $CH_2(\tilde{a}^1A_1)$, which are efficiently quenched by the rare gases.[195] Addition of inert diluent provides a further advantage in that the internal energy distributions within the radical fragments will be rapidly thermalized through collision, and as a result possible interpretive problems associated with the effects of intrafragment relaxation will be minimized. Fortunately internal relaxation, even in the absence of a relatively high pressure of buffer gas, is unlikely to constitute a major complication since the IRMPD process is by its very nature unlikely to produce highly excited radical fragments (see earlier). This is an important feature in favour of the IRMPD process as a source of radicals for kinetic study; in comparison, u.v. single and two-photon excitation methods such as are provided by the increasingly commonplace rare-gas halide excimer lasers often give rise to highly excited fragments with a markedly non-thermal internal energy distribution.

Table 4 lists those radicals, produced through IRMPD, which have so far proved amenable to LIF detection. Although LIF constitutes a very sensitive diagnostic probe, it is readily apparent from Table 4 that its real value lies in the detection of *small* radicals whose spectroscopy is, in general, already well characterized. Notable exceptions to this generalization however are the methylene and substituted halogenomethylene radicals. Stephenson and co-workers[196,197] have used the IRMPD of CF_2HCl and C_2F_3Cl as convenient sources of ground state CF_2 and CFCl radicals respectively and have produced a far more extensive account of the spectroscopy of both the $CF_2(\tilde{A}^1B_1—\tilde{X}^1A_1)$ and the $CFCl(\tilde{A}^1A''—\tilde{X}^1A')$ systems than was hitherto available. In addition, use of the pulsed method of radical production provided by i.r. laser photolysis in conjunction with time-resolved LIF detection techniques has allowed measurement of the pure radiative ('zero-pressure') lifetimes of several vibronic levels of the excited singlet states of CF_2[196] (all of which have τ_{rad} 61 $\pm$ 3 ns[197]) and CFCl[196] (for which τ_{rad} ~700 $\pm$ 10 ns[196]) as well as an estimate of the rate of vibrational relaxation within the electronically excited $CF_2(\tilde{A}^1B_1)$ state through non-reactive collision with He, Ne, N_2, and SF_6,[198] all of which are inefficient at promoting electronic relaxation to the ground singlet state. The IRMPD of normal and perdeuteriated acetic anhydride has been found to provide a quantitative source of the respective methylene radicals CH_2 and CD_2 in

[195] M. N. R. Ashfold, G. Hancock, G. W. Ketley, and J. P. Minshull-Beech, *J. Photochem.*, 1980, **12**, 75; M. N. R. Ashfold, M. A. Fullstone, G. Hancock, and G. W. Ketley, *Chem. Phys.*, in the press.

[196] S. E. Bialkowski, D. S. King, and J. C. Stephenson, *J. Chem. Phys.*, 1979, **71**, 4010.

[197] D. S. King, P. K. Schenck, and J. C. Stephenson, *J. Mol. Spectrosc.*, 1979, **78**, 1.

[198] D. L. Akins, D. S. King, and J. C. Stephenson, *Chem. Phys. Lett.*, 1979, **65**, 257.

their lowest singlet electronic state $(\tilde{a}^1A_1)$. Through analysis of the wave-length-resolved LIF emission from the $CH_2(\tilde{b}^1B_1)(0,14,0)$ vibronic level Feldmann *et al.*[199] have been able to determine vibrational and rotational constants for a number of vibrationally excited levels of the $CH_2(\tilde{a}^1A_1)$ state. Recent work carried out in our own laboratory has resulted in the first measurement of a part of the fluorescence excitation spectrum of the undocumented $CD_2(\tilde{b}^1B_1-\tilde{a}^1A_1)$ system; a progression of bands have been observed which, by analogy with CH_2, can be attributed to excitation of the v_2 bending mode in the excited state.[195]

Table 4 *Radicals produced through i.r. multiple photon dissociation (IRMPD) and detected by laser-induced fluorescence*

Radical	Parent molecule (refs. in parentheses)
$NH_2(\tilde{X}^2B_1)$	NH_3 (61), $MeNH_2$ (46, 200), CD_3NH_2 (46), N_2H_4 (200)
$OH(X^2\Pi)$	MeOH (201)
$CH(X^2\Pi)$	MeOH (201), MeCN (202), $MeNH_2$ (203), C_2H_4 (203), CH_2CO (203), EtOH (203)
$CH_2(\tilde{a}^1A_1)$	$(CH_3CO)_2O$ (195, 199, 204), CH_2CO (204), C_2H_4 (204), $CH_2(COOMe)_2$ (195)
$CD_2(\tilde{a}^1A_1)$	$(CD_3CO)_2O$ (195)
$C_2(X^1\Sigma^+)$	C_2H_3CN (205), C_2H_4 (205)
$C_2(a^3\Pi_u)$	C_2H_4 (206), C_2H_3CN (205, 207), EtOH (207), $EtNH_2$ (207), CH_2CO (204), MeCN (208), EtCN (208)
$CN(X^2\Sigma^+)$	MeCN (202), EtCN (208), C_2H_3CN (205)
$CF_2(\tilde{X}^1A_1)$	CF_2Cl_2 (91, 99), CF_2Br_2 (91, 99), CF_2HCl (99)
$CFCl(\tilde{X}^1A')$	C_2F_3Cl (196)
$CHF(\tilde{X}^1A')$	CH_2FCl (209)
$C_3(\tilde{X}^1\Sigma^+)$	C_3H_4 (205, 211)
$C_2O(\tilde{X}^3\Sigma^-)$	C_3O_2 (210)

Results of studies involving the use of pulsed i.r. lasers (or pulsed u.v. excimer lasers[212-214]) to generate radicals for kinetic measurements of the kind described in this section are summarized in Table 5. Clearly the use of LIF to monitor the

[199] D. Feldmann, K. Meier, R. Schmiedl, and K. H. Welge, *Chem. Phys. Lett.*, 1978, **60**, 30.
[200] S. V. Filseth, J. Danon, D. Feldmann, J. D. Campbell, and K. H. Welge, *Chem. Phys. Lett.*, 1979, **63**, 615.
[201] S. E. Bialkowski and W. A. Guillory, *J. Chem. Phys.*, 1978, **68**, 3339.
[202] M. L. Lesiecki and W. A. Guillory, *J. Chem. Phys.*, 1978, **69**, 4572.
[203] I. Messing, C. M. Sadowski, and S. V. Filseth, *Chem. Phys. Lett.*, 1979, **66**, 95.
[204] S. V. Filseth, J. Danon, D. Feldmann, J. D. Campbell, and K. H. Welge, *Chem. Phys. Lett.*, 1979, **66**, 329.
[205] H. Reisler, M. Mangir, and C. Wittig, *Chem. Phys.*, 1980, **47**, 49.
[206] N. V. Chekalin, V. S. Dolzhikov, V. S. Letokhov, V. N. Lokhman, and A. N. Shibanov, *Appl. Phys.*, 1977, **12**, 191; N. V. Chekalin, V. S. Letokhov, V. N. Lokhman, and A. N. Shibanov, *Chem. Phys.*, 1979, **36**, 415.
[207] J. D. Campbell, M. H. Yu, and C. Wittig, *Appl. Phys. Lett.*, 1978, **32**, 413.
[208] S. V. Filseth, J. Danon, J. D. Campbell, D. Feldmann, and K. H. Welge, unpublished results reported in ref. 211.
[209] M. N. R. Ashfold, F. Castano, G. Hancock, and G. W. Ketley, *Chem. Phys. Lett.*, 1980, **73**, 421.
[210] M. N. R. Ashfold, G. Hancock, and J. P. Minshull-Beech, unpublished results.
[211] M. L. Lesiecki, K. W. Hicks, A. Orenstein, and W. A. Guillory, *Chem. Phys. Lett.*, 1980, **71**, 72.
[212] J. E. Butler, L. P. Goss, M. C. Lin, and J. W. Hudgens, *Chem. Phys. Lett.*, 1979, **63**, 104.
[213] V. M. Donnelly and L. Pasternack, *Chem. Phys.*, 1979, **39**, 427.
[214] L. Pasternack and J. R. McDonald, *Chem. Phys.*, 1979, **43**, 173.

Table 5 *Measured removal rates for radicals produced through i.r. multiple photon dissociation (IRMPD), in units of* 10^{-11} cm^3 mol^{-1} s^{-1}

Reactant	Radical					
	$C_2(X^1\Sigma^+)$[a]	$C_2(a^3\Pi_u)$[a]	$C_3(X^1\Sigma_g^+)$[b]	$CH(X^2\Pi)$	$CH_2(\tilde{a}^1A_1)$[c]	$CD_2(\tilde{a}^1A_1)$[c]
He	—	—	—	—	0.31 ± 0.03	0.25 ± 0.04
Ne	—	—	—	—	0.42 ± 0.06	0.37 ± 0.04
Ar	—	$<2 \times 10^{-4}$[e]	—	—	0.60 ± 0.05	0.49 ± 0.06
Kr	—	0.02 ± 0.002	—	—	0.70 ± 0.05	0.93 ± 0.05
Xe	—	0.45 ± 0.04	—	—	1.6 ± 0.2	1.5 ± 0.3
N	—	—	—	2.1 ± 0.4[f]	—	—
O	0.14 ± 0.02	—	—	9.5 ± 1.1[f]	—	—
H$_2$	$<3 \times 10^{-3}$	$<3 \times 10^{-3}$	$<3 \times 10^{-3}$	—	13 ± 2	—
N$_2$	—	$<3 \times 10^{-3}$	$<3 \times 10^{-3}$	—	0.85 ± 0.05	—
O$_2$	0.3 ± 0.1[d]	—	2×10^{-3}	3.3 ± 0.4[g]	3.0 ± 0.4	—
CO	21 ± 3	7.5 ± 0.3[h]	$<3 \times 10^{-3}$	—	5.6 ± 0.5	—
NO	$<3 \times 10^{-3}$	$<3 \times 10^{-3}$	—	—	—	—
CO$_2$	$<3 \times 10^{-3}$	$<3 \times 10^{-3}$	—	—	—	—
H$_2$O	$<3 \times 10^{-3}$	$<3 \times 10^{-3}$	—	—	—	—
CF$_4$	1.7 ± 0.2	$<3 \times 10^{-3}$	$<3 \times 10^{-3}$	—	7.3 ± 0.5	—
CH$_4$	43 ± 4	—	—	—	—	—
C$_2$H$_2$	—	17 ± 2	—	—	—	—
C$_2$H$_4$	—	—	—	—	—	—
C$_2$H$_6$	—	—	$<3 \times 10^{-3}$	—	—	—
C$_2$H$_3$CN	44 ± 3	5.7 ± 0.4	—	—	—	—
C$_2$H$_3$F	20 ± 2	7.6 ± 0.5	—	—	—	—
C$_2$H$_3$Cl	49 ± 4	12 ± 1	—	—	—	—
C$_2$HCl$_3$	24 ± 2	3.8 ± 0.2	—	—	—	—
C$_2$Cl$_4$	26 ± 2	0.06 ± 0.005	—	—	—	—
C$_3$H$_4$	47 ± 4	26 ± 2	—	—	—	—
C$_3$H$_8$	33 ± 2	17 ± 1	—	—	—	—
C$_6$H$_6$	52 ± 4	7.6 ± 0.4	—	—	—	—

[a] From ref. 215 unless otherwise stated; [b] ref. 205; [c] ref. 195; [d] ref. 217; [e] ref. 216; [f] ref. 218; [g] ref. 203; [h] ref. 219

concentration of the reactant radical as a function of time only provides information regarding the *total* removal rate; in order to estimate whether or not a particular reaction contributes to the overall removal process, it is necessary to monitor the product build up as well. In a number of the reactions listed in Table 5 this has been possible because the reaction exothermicity is such that electronically excited products are formed. For example, the exothermicity of the reaction between C_2 radicals and O_2 is such that a number of excited electronic states of CO are accessible and, indeed, red emission from the $a'\,{}^3\Sigma^+$, $d^3\Delta_i$, and $e^3\Sigma^-$ excited triplet states[205] and vacuum u.v. emission from the $A^1\Pi$ state of CO have been observed.[217] Evidence that these emissions do indeed arise through the primary reaction of C_2 radicals with O_2 [and not, for example, through secondary reactions involving $C(^1S_0)$ or excited C_2O radicals] has been provided by the fact that the decay rate for each of the emissions shows the same dependence on O_2 pressure as the measured rate coefficients for removal of $C_2(X^1\Sigma^+)$[205] and $C_2(a^3\Pi_u)$.[217] In a subsequent detailed investigation of this particular system Wittig and co-workers[216] added hydrocarbon scavengers that reacted preferentially with $C_2(X^1\Sigma^+)$, and showed that O_2 is very efficient at inducing intersystem crossing between the singlet and triplet state of C_2. Kinetic analysis revealed that the rate coefficient for the triplet $\rightarrow$ singlet ISC brought about through collision with O_2 is roughly three times that of the reverse singlet $\rightarrow$ triplet ISC process and an order of magnitude faster than the rate of removal through reaction with O_2. Clearly, therefore, measurement of separate $C_2(X^1\Sigma^+)$ and $C_2(a^3\Pi_u)$ reaction rates with O_2 in these circumstances is impossible, and the previously measured rate constants for both $C_2(X^1\Sigma^+)$[205, 213] and $C_2(a^3\Pi_u)$[217] removal by O_2 have been re-interpreted as the rate constant for removal of *equilibrated* C_2 molecules. As a consequence of this rapid equilibration it is not yet clear whether both the singlet and triplet state of C_2 each react with O_2 to give all of the various observed excited states of CO or whether, for example, the combined requirements of spin and energy conservation rule out the formation of $CO(A^1\Pi)$ in the reaction of ground-state $C_2(X^1\Sigma^+)$ radicals with O_2. Some evidence for spin conservation is provided by the fact that the observed red emission from excited triplet states of CO has been shown to arise mainly from the reaction of $C_2(X^1\Sigma^+)$ with O_2.[216]

Application of these techniques has also been of considerable help in elucidating the mechanistic details of the reaction between C_2 radicals and NO. Time-resolved analysis of the fluorescence from electronically excited CN radicals (in both their $A^2\Pi$ and $B^2\Sigma^+$ states) produced in this exothermic reaction has revealed that these products are formed with the same rate coefficient as that determined through LIF measurements of the total $C_2(a^3\Pi_u)$ removal rate.[219] Since the reaction of ground-state $C_2(X^1\Sigma^+)$ radicals (which were inevitably also produced in the IRMPD radical generation process) with NO has been found[215] to proceed at a substantially faster rate (see Table 5), it has been concluded that

[215] H. Reisler, M. S. Mangir, and C. Wittig, *J. Chem. Phys.*, in the press.
[216] M. S. Mangir, H. Reisler, and C. Wittig, *J. Chem. Phys.*, 1980, **73**, 829.
[217] S. V. Filseth, G. Hancock, J. Fournier, and K. Meier, *Chem. Phys. Lett.*, 1979, **61**, 288.
[218] I. Messing, T. Carrington, S. V. Filseth, and C. M. Sadowski, *Chem. Phys. Lett.*, 1980, **74**, 56; to be published.
[219] H. Reisler, M. Mangir, and C. Wittig, *J. Chem. Phys.*, 1979, **71**, 2109.

electronically excited CN fragments arise predominantly from reaction of $C_2(a^3\Pi_u)$ with NO; this product state specificity has been rationalized using adiabatic state correlation arguments.[220] Spectral analysis of the emission from the electronically excited CN products has yielded vibrational energy distributions corresponding to Boltzmann temperatures of 13 000 K (for CN $A^2\Pi$) and 10 500 K (for $B^2\Sigma^+$), suggesting that the reaction mechanism forming CN is *via* a relatively long lived C_2NO complex.[219] As a final example, Messing *et al.*[203] have observed that the chemiluminescence from electronically excited $OH(A^2\Sigma^+)$ produced in the reaction of $CH(X^2\Pi)$ with O_2 follows the same decay rate as that obtained from LIF measurements of the total $CH(X^2\Pi)$ removal rate.

Thus it is apparent that IRMPD constitutes a new and convenient source of reactive fragments in concentrations that have hitherto been unavailable. As a result it may be confidently expected that, when combined with the ever improving sensitivity of LIF detection techniques, pulsed CO_2 lasers (complemented by the new generation of u.v. excimer lasers) will achieve a yet more important status in the field of photochemistry in general, and kinetics and spectroscopy in particular. As a final illustration of the power of this technique, Reisler *et al.*[205] have recently used LIF not only to detect ground state $CN(X^2\Sigma^+)$ radicals produced in the IRMPD of C_2H_3CN, but also to monitor the very much lower concentration of $NCO(X^2\Pi_i)$ radicals formed in their subsequent reaction with O_2.

Finally, two preliminary reports of complementary experimental methods should be mentioned in concluding this section on the applications of IRMPD to kinetic measurements. In the first study, which is based on an adaptation of the very low-pressure pyrolysis (VLPϕ) technique,[221] the IRMPD of CF_3I has been used to generate trifluoromethyl radicals whose subsequent reactions with Br_2, NOCl, NO_2, and O_3 have been investigated.[222] While the interpretation and analysis of kinetic data derived through use of this VLPϕ method may be complicated by secondary and wall reactions, it would appear that, when carefully applied, the technique provides a potential route for the study of reactions involving larger radical fragments that for one reason or another may not be suited to LIF detection methods. Of far wider potential applicability is the very direct method introduced by Gutman and co-workers,[223] which employs time-resolved photo-ionization mass spectrometry (PIMS) as a sensitive and selective means of radical detection. The power of this dynamic mass spectrometric technique has been illustrated in a study of the reaction between allyl radicals, C_3H_5, and NO_2.[223] Pulsed samples of the reactant radicals were formed along the length of a flow reactor through IRMPD of C_3H_5Br and their decay monitored in the presence of various added concentrations of NO_2 by photoionization mass spectrometric sampling of that fraction of the flowing gas mixture that effuses through a small orifice near the exit end of the reactor. This detection method is equally suited to monitoring product build-up; in this study[223] the formation rates of NO, C_3H_5O, and C_3H_4O products were also measured in order to provide a detailed picture of the overall reaction mechanism. Clearly this technique can be of considerable value in studies of the kinetics and mechanisms of numerous elementary reactions of polyatomic free radicals.

[220] H. Krause, *J. Chem. Phys.*, 1979, **70**, 3871.
[221] J. S. Chang, J. R. Barker, J. E. Davenport, and D. M. Golden, *Chem. Phys. Lett.*, 1979, **60**, 385.
[222] M. J. Rossi, J. R. Barker, and D. M. Golden, *J. Chem. Phys.*, 1979, **71**, 3722.
[223] I. R. Slagle, F. Yamada, and D. Gutman, *J. Am. Chem. Soc.*, in the press.

4

Ultraviolet Multiphoton Excitation: Formation and Kinetic Studies of Electronically Excited Atoms and Free Radicals

BY R. J. DONOVAN

1 Introduction

The recent commercial availability of high powered lasers operating in the u.v. region has opened up a challenging new area for photochemists. It is now possible to excite a molecule with two, three, or more photons and gain access to states that are forbidden for single-photon excitation. Thus, for example, with two-photon excitation it is possible to populate upper states having the same parity as the ground state (*i.e.*, $g \leftrightarrow g$ or $u \leftrightarrow u$), while for single-photon excitation the upper state must have the opposite parity ($g \leftrightarrow u$).[1,2] Furthermore, doubly excited states (states for which *two* electrons are excited) can be prepared by two-photon absorption. The formation of such states by single-photon absorption is strictly forbidden.

Another advantage of multiphoton excitation is that highly excited states, lying in the vacuum u.v. region ($\lambda < 200$ nm) can be readily populated without having to contend with the many problems normally encountered in the use of vacuum u.v. transmitting optical materials and high vacuum techniques. The highly excited states so formed may then undergo dissociation to yield excited photofragments. An interesting example of this type of process is the formation of electronically excited CH radicals[3] following the multiphoton excitation of CH_3I, *viz.*

$$CH_3I + 2h\nu(193 \text{ nm}) \rightarrow CH(A^2\Delta) + H_2 + I$$

Further examples of excited states that have been produced by multiphoton excitation are given in Table 1. It is clear from the wide range of species presented there, and considering the relatively short period in which the techniques for producing multiphoton excitation have been available, that a substantial research field lies ahead. By comparison with i.r. multiphoton excitation the field of u.v. multiphoton excitation is still in its infancy, but shows encouraging signs of being a prodigy.

[1] W. M. McClain, *Acc. Chem. Res.*, 1974, **7**, 129; W. M. McClain and R. A. Harris, in 'Excited States' vol. 3, ed. E. C. Lim, Academic Press, New York, 1977, p. 1.

[2] P. Lambropoulos, *Adv. At. Mol. Phys.*, 1976, **12**, 87.

[3] C. Fotakis, M. Martin, K. P. Lawley, and R. J. Donovan, *Chem. Phys. Lett.*, 1979, **67**, 1.

Table 1 *Excited states and ions* that have been produced by u.v. multiphoton excitation (see text for discussion and references)*

Excited states and ions	Parent molecule	Laser wavelength/nm
$S(3p^4, {}^1S_0)$	OCS	193 (ArF)
$Mn({}^6P_J, {}^6F_J, {}^8P_J)$	$Mn(CO)_5Br$, $Mn_2(CO)_{10}$	193 (ArF)
$Fe({}^3P, {}^3D, {}^3F, {}^3G, {}^5P, {}^5D, {}^5F, {}^5G, {}^7D)$	$Fe(CO)_5$	248 (KrF)
$Zn({}^1S_0, {}^3S_1, {}^1D_2, {}^3D_J)$	$ZnMe_2$	248 (KrF)
$I(6s^2P_j)$	CD_3I, I_2	193 (ArF)
$Hg(7^3D_3, 9^3D_3)$	HgX_2 (X = I, Br, Cl)	193 (ArF)
$Pb(7s^3P_J^0)$	$PbEt_4$	193 (ArF)
$H_2(E, F^1\Sigma_g^+)$	H_2	193 (ArF)
$CH(A^2\Delta, B^2\Sigma^-, C^2\Sigma^+)$ |	MeI, CD_3I, MeBr,	
$CD(A^2\Delta, B^2\Sigma^-, C^2\Sigma^+)$ |	MeOH, CH_2I_2	193 (ArF)
$NH(A^3\Pi, b^1\Sigma^+)$	NH_3	193 (ArF)
$PH(A^3\Pi)$	PH_3	193 (ArF)
$C_2(A^1\Pi_u, c^1\Pi_g, a^3\Pi_u, d^3\Pi_g)$	C_2H_2	193 (ArF)
$CN(H^2\Pi, F^2\Delta, D^2\Pi, B^2\Sigma)$	HCN, $(CN)_2$, ClCN	193 (ArF)
	PhCOCN	266 (Nd:YAG)
$CF(A^2\Sigma^+, B^2\Delta)$	CF_2Br_2	193 (ArF)
$Br_2[B^3\Pi(0_u^+)]$	CF_2Br_2	193 (ArF)
$I_2[B^3\Pi(0_u^+)]$	CH_2I_2	248 (KrF)
		193 (ArF)
$I_2[E^1\Sigma(0_g^+)]$	I_2	351 (Kr$^+$) +
		560—630 (Dye)
$CF_2(\tilde{A}^1B_1)$	CF_2Br_2	248 (KrF)
I^+	I_2	193 (ArF)
		360—380 (Dye)
I_2^+	I_2	193 (ArF)
		360—380 (Dye)
NO^+	NO	248 (KrF)
		450—455 (Dye)
$C_6H_5^+$	C_6H_6†	380—395 (Dye)

* A complete list of ionic species is given in Table 4.

† Numerous fragment ions, C_6—C_1, have been observed from excitation of C_6H_6; the yield of the lighter fragments is strongly dependent on the laser intensity.

In this Report we have concentrated on the use of multiphoton excitation as a means of producing atoms, ions, free radicals and excited state species (*i.e.*, to study *laser-induced chemistry*). We shall attempt, where possible, to identify the mechanisms by which such species are produced but will avoid discussion of the detailed aspects of multiphoton spectroscopy which is now an extensive field of research in its own right. We first discuss the experimental techniques and laser systems that are commonly used for u.v. multiphoton excitation. After this we briefly discuss the fundamental aspects of the multiphoton excitation process, and then consider the individual systems that have so far been studied. In the final section multiphoton ionization is discussed.

2 Techniques and Laser Systems

Multiphoton excitation frequently results in the formation of excited state dissociation products and many studies have simply recorded the fluorescence from photofragments. A typical experimental arrangement is shown in Figure 1. This arrangement can also be used to study excited state kinetics and lifetimes. Additional diagnostic equipment has been employed, including atomic resonance fluorescence[4] and laser-induced fluorescence.[5-7] Multiphoton ionization has been studied by placing electrodes in the photolysis cell[8,9] and also by interfacing it with a mass spectrometer.[10-12]

The most popular laser systems to be used for multiphoton excitation (MPE) studies have been the so called rare-gas halide excimer* lasers. These operate on the bound-free continua discovered by Golde and Thrush,[13] and provide very high peak powers at a series of wavelengths throughout the u.v. and visible regions. Further discrete wavelengths can be produced using stimulated Raman techniques.[14] Peak powers of several megawatts, with total energies of *ca.* 1.0 J (pulse duration *ca.* 20 ns), can be achieved with simple bench-top transverse discharge systems. The output from such systems has a typical bandwidth of *ca.* 1 nm, but this can be narrowed to *ca.* 0.1 nm ($\simeq 20$ cm^{-1}) using an intracavity two-prism dispersive element.[15] Higher energy systems are also available but require expensive electron beam pumping equipment. Typical parameters for commercially available excimer

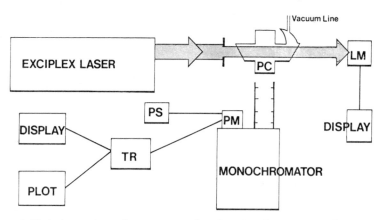

Figure 1 *Typical experimental arrangement for observing fluorescence and measuring excited-state lifetimes following multiphoton excitation; PC = photolysis cell, PM = photomultiplier, TR = transient recorder, LM = laser energy meter, PS = power supply*

* Many of the species involved are in fact *exciplexes*.

[4] M. C. Addison, R. J. Donovan, and C. Fotakis, *Chem. Phys. Lett.*, 1980, **74**, 58.
[5] V. M. Donnelly, A. P. Baronavski, and J. R. McDonald, *Chem. Phys.*, 1979, **43**, 271.
[6] J. J. Tiee, F. B. Wampler, and W. W. Rice, *J. Chem. Phys.*, 1980, **72**, 2925.
[7] F. B. Wampler, J. J. Tiee, W. W. Rice, and R. C. Oldenborg, *J. Chem. Phys.*, 1979, **71**, 3926.
[8] R. J. Donovan, C. Fotakis, and M. Martin, in *Proc. 4th Natl. Quantum Electron. Conf.*, ed. B. Wherrett, Wiley, New York, 1980, p. 173.
[9] J. H. Brophy and C. T. Rettner, *Chem. Phys. Lett.*, 1979, **67**, 351.
[10] S. Rockwood, J. P. Reilly, K. Hohla, and K. L. Kompa, *Opt. Commun.*, 1979, **28**, 175.
[11] L. Zandee and R. B. Bernstein, *J. Chem. Phys.*, 1979, **71**, 1359.
[12] T. G. Dietz, M. A. Duncan, M. G. Liverman, and R. E. Smalley, *Chem. Phys. Lett.*, 1980, **70**, 246.
[13] M. F. Golde and B. A. Thrush, *Chem. Phys. Lett.*, 1974, **29**, 486.
[14] T. R. Loree, R. C. Sze, and D. L. Barker, *Appl. Phys. Lett.*, 1977, **31**, 37.
[15] T. R. Loree, K. B. Butterfield, and D. L. Barker, *Appl. Phys. Lett.*, 1978, **32**, 171.

lasers are listed in Table 2. A detailed review of excimer lasers has been given by Shaw.[16]

High power output can also be achieved at fixed frequencies using the doubled and higher order harmonics, generated in non-linear crystals pumped by Nd:YAG and other commercially available solid state lasers.

The most versatile systems are undoubtedly frequency-doubled dye lasers which can be tuned almost continuously throughout the u.v. region; however, the available energy is much lower than for the systems already discussed. This limits their current applications for laser-induced chemistry and the main applications have been to spectroscopic studies where tightly focussed beams and small excitation

Table 2 *Typical parameters for commercially available excimer lasers**

Laser species	Wavelength/nm	Pulse energy/mJ	Peak power/MW
F_2	157	12	*ca.* 1.5
ArF	193	125	*ca.* 8
KrCl	222	60	*ca.* 3.5
KrF	248	10^3	*ca.* 10
XeBr	282	17	*ca.* 2
XeCl	308	90	*ca.* 5
N_2	337	9	*ca.* 1
XeF	351	90	*ca.* 5
N_2^+	427	2.5	*ca.* 0.5
F	713	2.5	*ca.* 0.4

* Other wavelengths available with electron beam pumped systems are discussed by Shaw.[16]

volumes are more acceptable. Dye lasers should also prove useful in the study of excited atom kinetics.

3 Fundamental Aspects of Multiphoton Excitation

Multiphoton excitation is a nonlinear process[2] and is only observed for high light intensities (*ca.* 10^6 W cm^{-2}). Consequently it is only with the advent of high power pulsed laser systems that this area has received any detailed attention, although the foundations were laid in 1931 by Göppert-Mayer.[17]

The simplest type of multiphoton process is two-photon absorption and we can distinguish two broadly different types of interaction.

Resonant Two-photon Excitation.—In this the photon is resonantly absorbed to produce a real intermediate state (A^*), having a well defined lifetime, typically in the range 10^{-6} to 10^{-9} s.† The second photon is then absorbed by the excited state, A^*, and carries the molecule to the final state, A^{**} (see Figure 2a). We shall also refer to this type of process as *sequential* excitation.

---

† This range refers to bound intermediate states, repulsive states have lifetimes of $\simeq 10^{-13}$ s.

[16] M. J. Shaw, *Prog. Quantum Electron.*, 1979, **6**, 3.
[17] M. Göppert-Mayer, *Ann. Phys.*, 1931, **9**, 273.

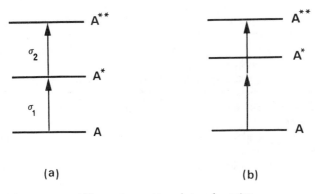

Figure 2 (a) *Resonant and* (b) *non-resonant two-photon absorption*

Non-resonant Two-photon Excitation.—In this no resonant intermediate states are available and a *virtual* intermediate state is created (*i.e.*, the molecule is dressed by a photon). This virtual state has an effective lifetime of *ca.* 10^{-15} s (the photon fly-by time) and is a mixture of states derived from those real states which lie close in energy and have the appropriate symmetry. It is only in the presence of a second photon that a *real* state (A^{**}) is produced (see Figure 2b). This type of process will also be referred to as *simultaneous* two-photon excitation.

Of the two processes, resonant two-photon absorption has the highest probability, although the two types of process merge as the frequency of the exciting radiation is tuned to approach the real intermediate state (A^{*}).

The rate at which the final state, A^{**}, is produced in a *resonant* (sequential) two-photon process is readily calculated using a conventional rate equation approach, provided the cross-sections (σ_1 and σ_2) and the rates of all loss processes from A^{*} are known.

To calculate the excitation rate for a *non-resonant* (simultaneous) two-photon process we need to know the two-photon cross-section or the *absorptivity*[18] for the transition to A^{**}. The absorptivity (δ), a measure of the strength or allowedness of the two photon process, is defined by

$$\frac{-dF}{dx} = N\delta F^2$$

where F is the photon flux density (photon cm^{-2} s^{-1}), N is the sample density (molecule cm^{-3}), and x is the distance (cm) along the beam propagation direction. The units of absorptivity are thus, cm^4 s $photon^{-1}$ $molecule^{-1}$. It is proportional to the square of the two-quantum matrix element M_{fg} connecting initial and final states:

$$M_{fg}(\omega) = 2 \sum_k \frac{\langle f| \hat{\varepsilon} \cdot \tilde{\mu} |k\rangle \langle k| \hat{\varepsilon} \cdot \tilde{\mu} |g\rangle}{E_{kg} - \hbar\omega}$$

[18] R. J. M. Anderson, G. R. Holtom, and W. M. McClain, *J. Chem. Phys.*, 1979, **70**, 4310.

where $\tilde{\mu}$ is the electric dipole operator, g, k, and f are the ground, intermediate, and final states, respectively, and $\hat{\varepsilon}$ is a unit vector denoting the polarization of the wave (note that here we consider both photons to have the same frequency, ω, although this need not necessarily be the case).

Typical values[18] of the absorptivity for allowed transitions are *ca.* 10^{-48} cm⁴ s photon⁻¹ molecule⁻¹; thus for a gaseous sample with a density of *ca.* 10^{17} molecule cm⁻³ and a modest photon flux of 10^{23} photons cm⁻² s⁻¹ we can expect to excite *ca.* 10^{14} molecule cm⁻³.

An alternative method for expressing the probability of a two-photon transition is in terms of a two-photon cross-section (σ).[19,20] It is common for this cross-section to be defined as an intensity-dependent parameter (σ/I) so that it is equivalent to the product δF. Typical values[20] for σ/I are given as *ca.* 10^{-31} cm⁴ W⁻¹.

Returning to the more general multiphoton process of order n, we might expect the excitation rate to be proportional to the nth power of the flux. However, the observed rate is frequently found to differ from this simple relationship. This is particularly common when resonant intermediate states are involved as one or more of the transitions may become saturated and thus rate limiting. Another factor which influences the excitation rate is the mode structure of the laser beam.[2] For single-mode operation there are no intensity fluctuations and the instantaneous value of the flux (F) is equal to the average flux $(\bar{F})$. However, for multimode operation F can be very much larger than $\bar{F}$, due to mode coupling, and this bunching of photons can lead to much *higher* excitation rates. In the limit of a very large number of modes (approaching a thermal light source) the rate of an nth order process is increased by $n!$ It is therefore an advantage in some applications, where excitation rate is the main requirement, to have multimode laser output. Rare-gas halide lasers are well known to give multimode output and this should be kept in mind for the discussions which follow.

For a spectroscopist the polarization properties of the two-photon absorption process are particularly interesting as the symmetry of a transition can be directly and uniquely determined from them.[1,2]

A bibliography of multiphoton work published between 1970 and 1978 is available.[21]

4 Direct Multiphoton Excitation of Atoms and Small Molecules

In this section we consider atoms and diatomic molecules which have been *directly* excited by multiphoton pumping. In the following section we shall consider excited state species formed indirectly by photofragmentation of larger molecules.

Atoms.—A number of atoms have been directly pumped to excited states using multiphoton techniques.[21] Particular attention has been given to alkali-metal atoms

[19] W. K. Bischel, P. J. Kelley, and C. K. Rhodes, *Phys. Rev. A*, 1976, **13**, 1817.
[20] C. K. Rhodes, in 'High-Power Lasers and Applications,' ed. K. L. Kompa and H. Walther, Springer-Verlag, Berlin, 1978.
[21] J. H. Eberly, J. W. Gallagher, and E. C. Beaty, 'Multiphoton Bibliography,' 1977 and 1978, available (free of charge) from Joint-Institute for Laboratory Astrophysics, Information Centre, University of Colorado, Boulder, Colorado 80309, USA.

(Li, Na, K, Rb, and Cs)[22-31] which are readily excited with dye lasers. The inert gas xenon has been excited[32] by use of a KrF laser (248 nm). Few kinetic studies have as yet been carried out but it has been clearly demonstrated that multiphoton excitation provides a useful and selective method for producing excited atoms.

Molecules.—Several diatomic molecules have been directly excited by multiphoton pumping, including H_2,[33] CO,[34] NO,[35-38] and I_2.[36d] However, kinetic studies have only been carried out with H_2[33] and NO.[37-40]

Hydrogen. Using the focussed output from an ArF laser (193 nm) Kligler and Rhodes[33] have excited the *double* minimum $E,F(^1\Sigma_g^+)$ state of H_2. Selective excitation of the $v' = 2$, $J' = 0, 1, 2$ levels occurs and excited state densities of *ca.* 10^{11} cm^{-3} were achieved with pulse energies of 15 mJ (15 ns duration). The two-photon 'cross section' was estimated to be, $\sigma/I = 10^{-32}$ cm^4 W^{-1}.

Radiative decay to the $B(^1\Sigma_u^+)$ state, by single-photon emission in the near-i.r., was observed and the radiative lifetime determined as 100 ± 20 ns. Quenching of the $E,F(^1\Sigma_g^+)$ state by H_2 and He was also studied and rate constants determined as $k_{H_2} = (2.1 \times 0.4) \times 10^{-9}$ and $k_{He} = (0.8 \pm 0.4) \times 10^{-9}$ cm^3 molecule^{-1} s^{-1}. Quenching was attributed to population transfer to the $C^1\Pi_u$ state which is nearly degenerate with the E,F state in the region of the inner minimum;[33]

$$H_2(E,F^1\Sigma_g^+) + H_2(X^1\Sigma_g^+) \rightarrow H_2(C^1\Pi_u) + H_2(X^1\Sigma_g^+)$$

Irradiation of D_2 under identical conditions did not produce any detectable excitation as there are no transitions from the ground state within the two-photon bandwidth of the laser. Excitation of HD is also expected to be negligible for the same reason.

This work[33] clearly opens the way for further detailed studies of both chemical and physical processes involving excited states of H_2, which previously have received very little attention. Selective excitation of H_2 in the presence of D_2 and HD can be achieved and provides a method for separating these isotopic species.

[22] S. A. Lee, J. Helmcke, J. L. Hall, and B. P. Stoicheff, *Opt. Lett.*, 1978, **3**, 141.

[23] J. Kowalski, R. Neumann, H. Sukr, K. Winkler, and G. Zu Putlitz, *Z. Phys.*, 1978, **A287**, 247.

[24] B. R. Marx and L. Allen, *J. Phys. B*, 1978, **11**, 3023.

[25] V. G. Arkhipkin, A. K. Popov, and V. P. Timofeev, *Appl. Phys.*, 1978, **16**, 209.

[26] J. N. Eckstein, A. I. Ferguson, and T. W. Hansch, *Phys. Rev. Lett.*, 1978, **40**, 847.

[27] W. Hartig, *Appl. Phys.*, 1978, **15**, 427.

[28] C. A. Von Dijk, P. J. T. Zeegers, G. Nienhuis, and C. T. Alkemade, *J. Quant. Spectrosc. Radiat. Transfer*, 1978, **20**, 55.

[29] M. E. Movsesyan, Z. O. Ninoyan, G. S. Sarkisyan, S. O. Sapondzhyan, and V. O. Chaltikyan, *Opt. Spectrosc. (USSR)*, 1977, **43**, 486.

[30] E. Campani, G. Degan, G. Govini, and E. Polacco, *Opt. Commun.*, 1978, **24**, 203.

[31] M. Y. Mirza and W. W. Duley, *J. Phys. B*, 1978, **11**, 1917.

[32] D. J. Kligler, D. Pritchard, W. K. Bischel, and C. K. Rhodes, *J. Appl. Phys.*, 1978, **49**, 2219.

[33] D. J. Kligler and C. K. Rhodes, *Phys. Rev. Lett.*, 1978, **40**, 309.

[34] R. A. Bernheim, C. Kittnell, and D. K. Viers, *J. Chem. Phys.*, 1978, **69**, 1308.

Nitric Oxide. Some of the earliest two-photon spectroscopic studies were carried out on nitric oxide.[35,36] Since then a number of investigators[37-40] have employed two-photon excitation to study quenching of the A, C, and D states of NO. Quenching by a wide range of collision partners, including NO, O_2, N_2, SO_2, $CHCl_3$, CCl_4, and $MeNO_2$, has been reported.[39] A mechanism involving charge transfer (NO^+Q^-) from the excited NO to the quenching molecule was suggested[39] to account for the observed rate data.

Multiphoton ionization of NO has been studied by several groups and will be discussed in Section 6.

5 Photofragmentation Studies

Most of the excited state species listed in Table 1 have been produced by photofragmentation following multiphoton excitation of polyatomic molecules. The wide variety of excited state species that have been observed suggests that this will be a very useful technique and should greatly facilitate the study of excited state chemistry. The individual systems studied to date will now be considered under headings giving the parent molecule which undergoes multiphoton excitation and from which the excited photofragments are derived.

Carbonyl Sulphide.—Two-photon excitation of OCS at 193 nm (ArF laser), results in dissociation to yield the second excited state of the sulphur atom, $S(3^1S_0)$,

$$OCS + 2h\nu \ (193 \ nm) \rightarrow CO + S(3^1S_0)$$

Excited sulphur atoms were detected[41] by collision-induced emission with Xe, which gives rise to a broad fluorescence band in the region 700—790 nm, from $XeS(2^1\Sigma^+ \rightarrow 1^1\Sigma^+)$. The $S(3^1S_0)$ densities produced in these experiments, using 7 mJ pulses (15 ns duration), were $\simeq 3 \times 10^9$ cm^{-3}. The two photon absorption 'cross-section' was estimated as $\simeq 2 \times 10^{-32}$ cm^4 W^{-1}.

It is interesting that the dissociative state of OCS, which is populated by two-photon absorption, lies 1.6 eV above the ionization limit (11.2 eV), and thus photofragmentation is seen to compete effectively ($\geqslant 10\%$) with photoionization (photoionization followed by dissociative ion recombination, as a mechanism for forming $S(3^1S_o)$, was considered but shown to be insignificant).[41]

Excitation of the OCS ν_2 (bending) mode using two CO_2 lasers prior to excitation with the ArF laser, was shown to enhance the yield of $S(3^1S_0)$ by approximately an order of magnitude. This was attributed to an improvement in the inter-mediate-state resonance for two-photon absorption when OCS is vibrationally excited.[41]

[35] R. G. Bray, R. M. Hochstrasser, and J. E. Wessel, *Chem. Phys. Lett.*, 1974, **27**, 167.
[36] (a) P. A. Freedman, *Can. J. Phys.*, 1977, **55**, 1387; (b) P. M. Johnson, M. R. Berman, and D. Zakheim, *J. Chem. Phys.*, 1975, **62**, 2500; (c) K. K. Innes, B. P. Stoicheff, and S. C. Wallace, *Appl. Phys. Lett.*, 1976, **29**, 715; (d) D. L. Rousseau and P. F. Williams, *Phys. Rev. Lett.*, 1974, **33**, 1368.
[37] H. Zacharias, J. B. Halpern, and K. H. Welge, *Chem. Phys. Lett.*, 1976, **43**, 41.
[38] M. Asscher and Y. Haas, *Chem. Phys. Lett.*, 1978, **59**, 231.
[39] M. Asscher and Y. Haas, *J. Chem. Phys.*, 1979, **71**, 2724.
[40] G. F. Nutt, S. C. Haydon, and A. I. McIntosh, *Chem. Phys. Lett.*, 1979, **62**, 402.
[41] D. J. Kligler, H. Plummer, W. K. Bischel, and C. K. Rhodes, *J. Chem. Phys.*, 1978, **69**, 4652.

Mercury Halides.—Photolysis of HgX_2 (X = Br or I) at 193 nm (ArF laser) leads to the formation of electronically excited $HgX(B^2\Sigma)$ by a *single*-photon process.[42, 43]

$$HgX_2 + h\nu \rightarrow HgX(B^2\Sigma) + X$$

Laser action has been reported from $HgX(B^2\Sigma)$ formed in this way.[42, 43]

Photolysis of $HgCl_2$ at 193 nm leads to the formation of ground state $HgCl(X^2\Sigma)$, but emission from excited Hg atoms is also observed[44] and must result from multiphoton absorption. The energetics for two-photon excitation of HgCl are illustrated in Figure 3. Single-photon excitation of $HgCl(X^2\Sigma)$ provides sufficient energy for $Hg(6^3P_1)$ formation and this state is indeed observed (it could, however,

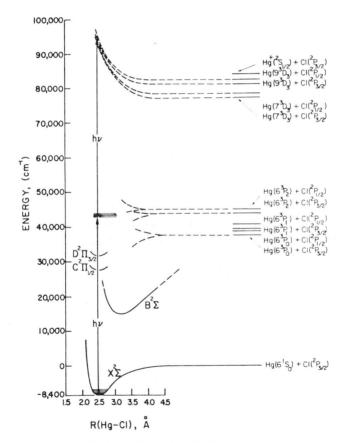

Figure 3 *Energetics for multiphoton dissociation of* HgCl
(Reproduced by permission from *Chem. Phys. Lett.*, 1978, **58**, 108)

[42] E. J. Schimitschek, J. E. Celto, and J. A. Trias, *Appl. Phys. Lett.*, 1977, **31**, 608.

[43] E. J. Schimitschek and J. E. Celto, *Opt. Lett.*, 1978, **2**, 64.

[44] T. A. Cool, J. A. McGarvey, and A. C. Erlandson, *Chem. Phys. Lett.*, 1978, **58**, 108.

be formed following cascade processes from higher states). Absorption of a second photon carries the HgCl* molecule to an energy just above the asymptote for Hg^+ formation. However, in this region we can expect to find a high density of molecular states correlating with the high density of excited Hg atomic states, and the observation of $Hg(9^3D_3)$ is not too surprising. Dissociation to neutral fragments thus competes effectively with photoionization of HgCl and Hg^+ formation.

Both resonant and near resonant two-photon processes, *via* bound and repulsive intermediate states of $HgCl_2$, can be expected to contribute to the formation of $Hg(9^3D_3)$, but secondary excitation of HgCl may also be important. Further work on the spectroscopy of HgCl in the 200 nm region will be required to resolve this point.

Both $HgBr_2$ and HgI_2 give rise to highly excited Hg atoms in addition to $HgX(B^2\Sigma)$ formation.[44] The mechanism is thought to be broadly similar to that discussed above for $HgCl_2$, but a contribution from secondary excitation of $HgX(B^2\Sigma)$ is also possible.

Ammonia and Phosphine.—Both NH_3 and PH_3 absorb strongly ($\tilde{A} \leftarrow \tilde{X}$ systems) in the region of the ArF laser line (193 nm) and both single- and two-photon excitation processes have been observed with these molecules. The $\tilde{A}$ states of both NH_3 and PH_3 are strongly predissociated and earlier conventional photochemical studies have shown that hydrogen atoms are formed with unit quantum efficiency. The NH_2 and PH_2 fragments are formed predominantly in their ground states, but laser studies have revealed that $NH_2(\tilde{A}^2A_1)$ is formed with 2.5% yield,[5] and $PH_2(\tilde{A}^2A_1)$ with 1.4% yield,[45] following 193 nm (ArF laser) photolysis of the respective parent molecules.

Of more direct interest in the present context is the observation of electronically excited NH and PH radicals formed by two-photon processes. Two excited states of NH, the $A^3\Pi$ and $b^1\Sigma^+$ states, are formed[5] from NH_3, but only one state, the $PH(A^3\Pi)$ state, was observed[45] from PH_3. The energetic requirements in the photolysis of PH_3 are illustrated in Figure 4.

From the available data it is not possible to distinguish between a simultaneous two-photon (non-resonant) and a sequential two-photon (resonant) excitation mechanism. However, we note that the formation of the $A^3\Pi$ states of NH and PH is likely to be accompanied by the formation of two hydrogen atoms, rather than H_2, as the latter process would be spin forbidden [this restriction does not apply for the formation of $NH(b^1\Sigma^+)$].

Water and Alcohols.—Photofragment fluorescence has been observed[46] following excitation of H_2O, MeOH, and EtOH with an ArF laser (193 nm). With H_2O, fluorescence was only observed at relatively high pressures (*ca.* 1.3 kN m^{-2}) and was attributed to the $OH(D^2\Sigma \rightarrow A^2\Sigma)$ transition. Fluorescence from $OH(A^2\Sigma)$ was not observed and it was suggested that this state is quenched at the pressures used. The fact that fluorescence does not occur at low pressures was taken to indicate[46] that emission arises from collisions involving two excited H_2O molecules. This appears to be an interesting system and deserves further study.

[45] C. L. Sam and J. T. Yardley, *J. Chem. Phys.*, 1978, **69**, 4621.
[46] W. M. Jackson, J. B. Halpern, and C.-S. Lin, *Chem. Phys. Lett.*, 1978, **55**, 254.

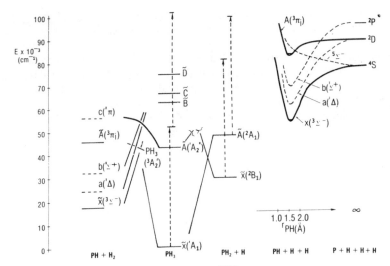

Figure 4 *Energetics for multiphoton dissociation of* PH_3
(Reproduced by permission from *J. Chem. Phys.*, 1978, **69**, 4621)

Both excited state CH and OH fluorescence is observed[46] with MeOH and EtOH. The systems identified were $CH(A^2\Delta \rightarrow X^2\Pi)$, $CH(B^2\Sigma \rightarrow X^2\Pi)$ and $OH(D^2\Sigma \rightarrow A^2\Sigma)$. Weak $OH(A^2\Sigma \rightarrow X^2\Sigma)$ emission was also observed with EtOH.

Cyanides.—Strong $CN(B^2\Sigma \rightarrow X^2\Sigma)$ fluorescence has been observed following laser excitation of ClCN,[46] $(CN)_2$,[46,47] and PhCOCN.[48] In all cases more than one photon is required on energetic grounds to produce $CN(B^2\Sigma)$. Weaker emission, assigned as $CN(D^2\Pi \rightarrow X^2\Sigma)$ and $CN(D^2\Pi \rightarrow A^2\Pi)$ is also observed from ClCN and $(CN)_2$.

Excitation of $(CN)_2$ was carried out with an ArF laser (193 nm) and the suggested mechanism involves *sequential* two-photon absorption *via* the bound $\tilde{B}(^1\Delta)$ state of $(CN)_2$. Predissociation from this intermediate state is expected from previous spectroscopic studies, and the quantum yield for $CN(B^2\Sigma)$ formation is found to be low (*ca.* 0.1%).[47]

The spectrum of ClCN at 193 nm shows no discrete structure but excitation with an ArF laser leads to $CN(B^2\Sigma \rightarrow X^2\Sigma)$ fluorescence. The quantum yield for $CN(B^2\Sigma)$ formation, from ClCN, was not reported[46] but must certainly be less than from $(CN)_2$. It seems likely that the excitation mechanism for ClCN involves simultaneous two-photon absorption, enhanced by the presence of continuum intermediate states.

Excitation of PhCOCN with a frequency-quadrupled Nd:YAG laser (266 nm) also gives rise to intense $CN(B^2\Sigma \rightarrow X^2\Sigma)$ fluorescence and this has been used to study the quenching of the $CN(B^2\Sigma)$ state by CH_4. A rate constant of $(1.5 \pm 0.5) \times 10^{-10}$ cm^3 molecule^{-1} s^{-1} for the quenching of both $v' = 0$ and $v' = 1$ levels of $CN(B^2\Sigma)$, by CH_4, was reported.[48]

[47] W. M. Jackson and J. B. Halpern, *J. Chem. Phys.*, 1979, **70**, 2373.
[48] J. B. Lurie and M. A. El-Sayed, *Chem. Phys. Lett.*, 1980, **70**, 251.

Excitation of HCN at 193 nm (ArF laser) produces weak HCN($\tilde{A}, {}^1A'' \rightarrow \tilde{X}{}^1\Sigma^+$) fluorescence.[49] No photofragment fluorescence was reported.

Halogenomethanes.—Intense   CH($A^2\Delta \rightarrow X^2\Pi$)   and   CH($B^2\Sigma^- \rightarrow X^2\Pi$) fluorescence is observed following excitation of MeBr,[46,50] MeI,[3] and CH_2I_2[8] at 193 nm (ArF laser). The mechanism favoured for the methyl halides involves simultaneous two-photon, *two-electron* excitation.[3] It was suggested that an electron on the halogen atom is excited to a C—X antibonding orbital, simultaneously with the excitation of a C—H antibonding orbital on CH_3. The molecule is thus pumped into a super-excited state from which it rapidly dissociates, *viz.*

$$CH_3X + 2h\nu \rightarrow CH_3^*X^* \rightarrow CH^* + H_2 + X$$

It has been pointed out[3] that the effects of predissociation must be taken into account when assessing the branching ratios for excited state formation. Thus a pronounced isotope effect is observed with the CH/CD($B^2\Sigma^- \rightarrow X^2\Pi$) fluorescence

Table 3 *Photofragment fluorescence from* MeI, CD_3I, *and* MeBr *following multiphoton excitation at* 193 nm *and* 248 nm

Molecule	Excitation wavelength	
	193 nm (ArF) *Observed fluorescence (intensity)*	248 nm (KrF) *Observed fluorescence*
MeI	CH($A^2\Delta \rightarrow X^2\Pi$) (1.0)	Only chemiluminescence,
	CH($B^2\Sigma^- \rightarrow X^2\Pi$) (0.07)	$I_2[B^3\Pi(0_u^+) \rightarrow X^1\Sigma(0_g^+)]$ observed
CD_3I	CD($A^2\Delta \rightarrow X^2\Pi$) (1.7)	Only chemiluminescence,
	CD($B^2\Sigma^- \rightarrow X^2\Pi$) (0.31)	$I_2[B^3\Pi(0_u^+) \rightarrow X^1\Sigma(0_g^+)]$ observed
MeBr	CH($A^2\Delta \rightarrow X^2\Pi$) (1.1)	None
	CH($B^2\Sigma^- \rightarrow X^2\Pi$) (0.06)	

from MeI and CD_3I (see Table 3). This is due to stronger predissociation of the CH($B^2\Sigma^-$) state, compared with CD($B^2\Sigma^-$), which reduces the fluorescence yield.

Weak fluorescence from CH($C^2\Sigma^+$) and I($5p^46s^2P_{\frac{1}{2}}$) was also observed from MeI and CH_2I_2 following excitation with an ArF laser (193 nm).[3,51]

Excitation of Me_3I at 248 nm (KrF laser) is dominated by *single* photon transitions to the repulsive $\tilde{A}$ state which dissociates to yield CH_3 and I($5^2P_{\frac{1}{2}}$). Chemiluminescence is observed following the sequence:[3]

$$I(5^2P_{\frac{1}{2}}) + Q \rightarrow I(5^2P_{\frac{3}{2}}) + Q^{\ddagger}$$

$$I(5^2P_{\frac{3}{2}}) + I(5^2P_{\frac{3}{2}}) + M \rightarrow I_2[B^3\Pi(0_u^+)] + M$$

$$I_2[B^3\Pi(0_u^+)] \rightarrow I_2[X^1\Sigma(0_g^+)] + h\nu(500\text{--}650 \text{ nm})$$

[49] A. P. Baronavski, *Chem. Phys. Lett.*, 1979, **61**, 532.
[50] A. P. Baronavski and J. R. McDonald, *Chem. Phys. Lett.*, 1978, **56**, 369.
[51] C. Fotakis, M. Martin, and R. J. Donovan, *J. Chem. Soc., Chem. Commun.*, 1979, 813.

No evidence for multiphoton excitation at 248 nm was found, although energetically ionization and formation of $CH(A^2\Delta)$ are both possible with two photons (see Figure 5). Any mechanism involving the steeply repulsive $\bar{A}$ state would require that a substantial kinetic energy be released along the C—I co-ordinate and this would effectively reduce the potential energy available to the products below the level required to produce $CH(A^2\Delta)$ or ions (for a *two*-photon process).

Multiphoton excitation of CH_2I_2 appears to be more complicated than for MeI. The fluorescence systems already discussed, for excitation at 193 nm, also appear following excitation of CH_2I_2 at 248 nm (KrF laser),[8] together with the $I_2[B^3\Pi(0_u^+) \to X^1\Sigma(0_g^+)]$ system. Substantial ion yields were also observed[8] following excitation at both 193 and 248 nm. The mechanism proposed involves two-photon absorption by the CH_2I radical, *viz.*

$$CH_2I_2 + h\nu \to CH_2I + I$$

$$CH_2I + h\nu \to CH_2I^*$$

$$CH_2I^* + h\nu \to CH^* + I + H$$

$$\to \text{ions}$$

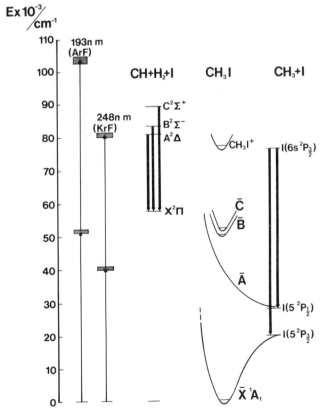

Figure 5 *Energetics for multiphoton dissociation of* CH_3I

Excitation of CF_2Br_2 at 248 nm (KrF) gives rise to an intense blue luminescence[7,52] (observable by eye) which extends well into the u.v. region (250—460 nm). This has been identified as being due to $CF_2(\tilde{A} \rightarrow \tilde{X})$ fluorescence and a mechanism involving formation of CF_2 in the primary step,

$$CF_2Br_2 + h\nu(248\ nm) \rightarrow CF_2(\tilde{X}^1A_1) + Br_2$$

followed by secondary excitation of CF_2 at 248 nm,

$$CF_2(\tilde{X}^1A_1, v = 0) + h\nu(248\ nm) \rightarrow CF_2(\tilde{A}^1B_1, v_2 = 6)$$

has been suggested.[7,52] Emission from excited bromine atoms was also observed and attributed to sequential excitation of Br_2 formed in the primary step.

Excitation of CF_2Br_2 at 193 nm (ArF) results[7,52] in emision from $CF(A^2\Sigma^+ \rightarrow X^2\Pi)$, $CF(B^2\Delta \rightarrow X^2\Pi)$, $CF_2(\tilde{A}^1B_1 \rightarrow \tilde{X}^1A_1)$, and $Br_2[B^3\Pi(0_u^+) \rightarrow X^1\Sigma(0_g^+)]$. Ground state CF_2 is also produced and has been identified by laser-induced fluorescence using a separate dye-laser system. It was suggested that formation of $CF(A^2\Sigma^+)$ and $CF(B^2\Delta)$ involves a three-photon process.

Photofragment fluorescence from CBr_4 (KrF laser excitation at 248 nm) is even more extensive[52] than from CF_2Br_2. Intense luminescence is observed in three spectral regions, 280—360, 430—520, and 615—864 nm. The emission in the latter region results from excited $(5p)$ bromine atoms, while the short wavelength emission is associated with $Br_2(D' \rightarrow A')$ fluorescence. The emission between 430—520 nm was identified as being due to the C_2 Swan bands and must result from secondary radical reactions.

Excitation of $CHBr_3$ at 193 nm (ArF laser) has been used[53] as a source of $CH(X^2\Pi)$ radicals for kinetic studies. No detailed investigation of the multiphoton process, which gives rise to this radical and probably other species, has yet been made.

The dominant species produced when $CFCl_3$ and CCl_4 are excited with an ArF laser (193 nm), are ground state $CFCl$ and CCl_2 radicals (detected by laser-induced fluorescence).[54] However, weak fluorescence from excited states of $CFCl$ and CCl_2 has also been detected[54] and will increase in importance with laser intensity.

From the foregoing studies on polyhalogenomethanes it would appear that a common excitation mechanism can be written. The primary step leads to two sets of products,

$$CX_4 + h\nu \rightarrow CX_3 + X$$
$$\rightarrow CX_2 + X_2(X_2^*)$$

the branching ratio depending on the particular molecule excited. Secondary excitation of these fragments then leads to the observed excited states:

$$CX_3 + h\nu \rightarrow CX_2^* + X$$
$$\rightarrow CX_2 + X^*$$
$$\rightarrow CX^* + X_2(X_2^*)$$

[52] C. L. Sam and J. T. Yardley, *Chem. Phys. Lett.*, 1979, **61**, 509.
[53] J. E. Butler, L. P. Goss, M. C. Lin, and J. W. Hudgens, *Chem. Phys. Lett.*, 1979, **63**, 104.
[54] J. J. Tiee, F. B. Wampler, and W. W. Rice, *Chem. Phys. Lett.*, 1979, **65**, 425.

and

$$CX_2 + h\nu \rightarrow CX_2^*$$

$$CX_2^* \rightarrow CX_2 + h\nu$$

$$\rightarrow CX^* + X(X^*)$$

In some cases, at high laser intensities, two-photon excitation of CX_3 may occur (*cf.* CH_2I).

The u.v. multiphoton excitation of polyhalogenomethanes clearly provides a convenient route to a number of novel ground and excited state species.

Metal Alkyls and Carbonyls.—Strong atomic fluorescence has been observed when metal alkyls ($ZnMe_2$ and $PbEt_4$) and metal carbonyls [$Fe(CO)_5$, $Cr(CO)_6$, $Mn_2(CO)_{10}$, and $Mn(CO)_5Br$] are irradiated with an ArF or KrF laser,[55] and this promises to be a useful method for generating sizable concentrations of excited metal atoms. Laser action has been observed when iron pentacarbonyl was pumped with a KrF laser,[56] but this was attributed to an excited *molecular* species.

It is interesting to note that only a limited range of excited states, and within these only specific J states, of the metal atom are populated in the primary photochemical step. Addition of a buffer gas enhances the overall emission intensity and causes the appearance of new lines. Atomic fluorescence is also observed when $Cr(CO)_6$ is irradiated with a nitrogen laser (337 nm), but it is weaker than with KrF or ArF lasers.[57]

The mechanism suggested[55] to account for these observations involves sequential absorption by the photofragments, *viz.*

$$ML_n + h\nu \rightarrow ML_{n-x} + x\,L$$

$$ML_{n-x} + h\nu \rightarrow M^* + (n-x)L$$

where M is a metal atom and L a ligand. More than two photons are required in some cases but the general mechanism of sequential stripping still applies.

Acetylene and Ethylene.—Strong $C_2(A^1\Pi_u \rightarrow X^1\Sigma_g^+)$ Phillips band emission (690—1100 nm) is observed when acetylene is irradiated with a high power ArF laser (193 nm).[58] Weaker emissions are also observed from $CH(A^2\Delta)$, $C_2(d^3\Pi_g)$, and $C_2(C^1\Pi_g)$. Earlier work[46] suggesting that CH is formed in the $B^2\Sigma^-$, $C^2\Sigma^+$, and $E^2\Pi$ states has been shown to be erroneous and these states resulted from the presence of acetone as an impurity in the acetylene.

The energetics for C_2, CH, and C_2H formation from acetylene are shown in Figure 6. It was suggested that formation of $CH(A^2\Delta)$ results from sequential absorption by acetylene,

$$C_2H_2 + h\nu \,(193\ nm) \rightarrow C_2H_2^*$$

$$C_2H_2^* + h\nu \,(193\ nm) \rightarrow CH(A^2\Delta) + CH(X^2\Pi)$$

[55] Z. Karny, R. Naaman, and R. N. Zare, *Chem. Phys. Lett.*, 1978, **59**, 33.
[56] D. W. Trainor and S. A. Mani, 'Optical Conversion Processes,' (Final Report), Avco Everett Research Laboratory Inc., Jan., 1978.
[57] C. Fotakis, W. H. Breckenridge, and R. J. Donovan, to be published.
[58] J. R. McDonald, A. P. Baronavski, and V. M. Donnelly, *Chem. Phys.*, 1978, **33**, 161.

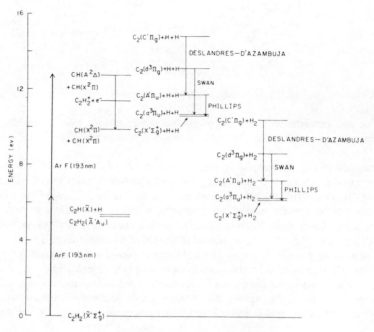

Figure 6 *Energetics for* C_2, CH, *and* C_2H *formation from acetylene*
(Reproduced by permission from *Chem. Phys.*, 1978, **33**, 161)

as there is only just sufficient energy to produce this state in a two-photon process (the formation of CH($A^2\Delta$) was found to vary with the square of the laser energy). The life-time of $C_2H_2^*$ was estimated[58] as $\geqslant 10^{-12}$ s which is sufficient to account for the observed yields (*ca.* 1%) of CH($A^2\Delta$).

The mechanism for formation of $C_2(A^1\Pi_u)$, which is the major product, was less clearly defined. The mechanism favoured by McDonald *et al.*[58] involves fragmentation of the initially formed $C_2H_2^*$, to yield C_2H, which is then excited by a further photon,

$$C_2H(\tilde{X}^2\Sigma) + h\nu \,(193\;\text{nm}) \rightarrow C_2(A^1\Pi_u) + H$$

The radiative lifetime of $C_2(A^1\Pi_u)$ was measured as $10.1 \pm 1.2\;\mu s$, and agrees well with the values obtained from theory.

Two-photon excitation of C_2H_2 at 193 nm has been used as a source of $C_2(a^3\Pi_u)$ radicals[59] and the kinetics of this species with CH_4, C_2H_2, C_2H_4, C_2H_6, and O_2 have been studied [i.r. multiphoton excitation has also been used to produce $C_2(a^3\Pi_u)$, see Chapter 3]. It was found that $C_2(a^3\Pi_u)$ is *less* reactive than the ground state, $C_2(X^1\Sigma_g^+)$ which accords with the electronic orbital correlations. Ground state $C_2(X^1\Sigma_g^+)$ is formed by two-photon (193 nm) excitation of hexafluorobut-2-yne.[60]

[59] V. M. Donnelly and L. Pasternack, *Chem. Phys.*, 1979, **39**, 427.
[60] L. Pasternack and J. R. McDonald, *Chem. Phys.*, 1979, **43**, 173.

Excitation of ethylene with the unfocussed output of an ArF laser leads to $CH(A^2\Delta \rightarrow X^2\Pi)$ and $CH(C^2\Sigma \rightarrow X^2\Pi)$ fluorescence. A mechanism involving sequential two-photon excitation has been suggested[46] on the basis of a preliminary study. Further detailed study of this system should prove rewarding.

6 Multiphoton Ionization

Multiphoton ionization (MPI) provides a highly sensitive means for observing two-photon and higher order excitation processes owing to the fact that very low levels of ionization, corresponding to the formation of only a few charges per second, are readily detectable. It has therefore been used widely in multiphoton spectroscopy[37, 61-67] where the ion yield provides a characteristic signature of resonant intermediate states (see later). Studies have been reported using static (bulk) samples[61a-c] and molecular beams,[61d, 68] both with[10, 68] and without[12] mass analysis of the ions produced.

A number of authors have emphasized the potential importance of MPI in mass spectrometry.[10-12, 68, 69] The ionization process is highly selective and one can ionize, and thus identify, trace amounts of a particular chemical species in the presence of a large excess of other species (*i.e.*, in very dilute systems). This essentially adds a second dimension[69] to mass spectrometry as one can identify molecules not only by their mass but, in addition, by their characteristic optical (MPI) spectrum. Further dimensions are added by the fact that the fragmentation pattern is both wavelength and laser-power dependent. That the efficiency of ionization (*ca.* 100%) is also much higher[12, 70] than in conventional ion sources is due to the high light intensities available with pulsed lasers. This increased efficiency should be of particular value to molecular beam studies[71] (note that MPI can also be quantum state selective). The probability of MPI increases rapidly with laser frequency, owing to the increase in density of electronic states, however the selectivity will decline.

The earliest work on MPI was concerned primarily with atomic species and with the physics of the ionization process.[2, 72] The first application of MPI to the study of molecular spectra was made by Johnson[36b, 61] and it has since been widely adopted.[62-67] The high sensitivity of the technique allows the study of species in

---

[61] (*a*) P. M. Johnson, *J. Chem. Phys.*, 1975, **62**, 4562; (*b*) 1976, **64**, 4143; (*c*) 1976, **64**, 4638; (*d*) D. Zakheim and P. M. Johnson, *ibid.*, 1978, **68**, 3644.

[62] G. Petty-Sil, C. Tai, and F. W. Dalby, *Phys. Rev. Lett.*, 1975, **34**, 1207; F. W. Dalby, G. Petty-Sil, M. H. Pryce, and C. Tai, *Can. J. Phys.*, 1977, **55**, 1033; C. Tai and F. W. Dalby, *ibid.*, 1978, **56**, 183.

[63] D. H. Parker, S. J. Sheng, and M. A. El-Sayed, *J. Chem. Phys.*, 1976, **65**, 5534; J. O. Berg, D. H. Parker and M. A. El-Sayed, *ibid.*, 1978, **68**, 5661; D. H. Parker and P. Avouris, *Chem. Phys Lett.*, 1978, **53**, 515.

[64] G. C. Nieman and S. D. Colson, *J. Chem. Phys.*, 1978, **68**, 5656.

[65] M. B. Robin and N. A. Kuebler, *J. Chem. Phys.*, 1978, **69**, 806.

[66] K. K. Lehmann, J. Smolanek, and L. Goodman, *J. Chem. Phys.*, 1978, **69**, 1569.

[67] (*a*) A. D. Williamson and R. N. Compton, *Chem. Phys. Lett.*, 1979, **62**, 295; (*b*) A. D. Williamson, R. N. Compton, and J. H. D. Eland, *J. Chem. Phys.*, 1979, **70**, 590.

[68] (*a*) L. Zandee, R. B. Bernstein, and D. A. Lichtin, *J. Chem. Phys.*, 1978, **69**, 3427; (*b*) L. Zandee and R. B. Bernstein, *ibid.*, 1979, **70**, 2574.

[69] U. Boesl, H. J. Neusser, and E. W. Schlag, *J. Chem. Phys.*, 1980, **72**, 4327.

[70] M. A. Duncan, T. G. Dietz, and R. E. Smalley, *Chem. Phys.*, 1979, **44**, 415.

[71] D. L. Feldman, R. K. Lengel, and R. N. Zare, *Chem. Phys. Lett.*, 1977, **52**, 413.

[72] J. S. Bakos, *Adv. Electron. Electron Phys.*, 1974, **36**, 57.

Table 4 *Ions observed following multiphoton excitation*

Parent species	Ion	Refs.
H(D)	$H^+(D^+)$	a
Ar	Ar^+	b, c
Kr	Kr^+	c
Xe	Xe^+	c
Li	Li^+	23 [d]
Na	Na^+	e, f
K	K^+	g
Rb	Rb^+	h
Cs	Cs^+	i
Ca	Ca^+	21 [j]
Sr	Sr^+	k, l
Ba	Ba^+	l
I	I^+	m
Hg	HG^+	n
U	U^+	o, p
H_2	H_2^+	10
N_2	N_2^+	b
NO	NO^+	10, 11
Li_2	Li_2^+ (isotope selective)	q
Na_2	Na^+, Na_2^+	r, s
K_2	K_2^+	r
Rb_2	Rb^+, Rb_2^+	t
Cs_2	Cs^+, Cs^-, Cs_2^+	t, u
RbCs	$Rb^+ Cs^+, Cs^-, RbCs^+$	t
BaCl	$BaCl^+$	71
I_2	I^+, I_2^+	11, 67a, 68a
H_2O	H_2O^+	v
D_2O	D_2O^+	w
NO_2	NO_2^+	x
NH_3	NH_3^+	y
H_2CO	H_2CO^+	x
C_6H_6	$C_6H_6^+$, ion fragments*	10, 11, 68b, 69
$C_6H_5NH_2$	$C_6H_5NH_2^+$	9, 12
$Fe(CO)_5$	Fe^+	70
$Cr(CO)_6$	Cr^+	70
$Mo(CO)_6$	Mo^+	70

* A range of ion fragments from $C_5H_n^+$ down to C^+ are observed, the yield being very sensitive to the laser intensity.

[a] G. C. Bjorklund, C. P. Ausschnitt, R. R. Freeman, and R. H. Storz, *Appl. Phys. Lett.*, 1978, **33**, 54. [b] M. J. Hollis, *Opt. Commun.*, 1978, **25**, 395. [c] L. A. Lompre, G. Mainfray, C. Manus, and J. Thebault, *Phys. Rev.*, 1977, **15**, 1604. [d] M. Shimazu, V. Takubo, and Y. Maeda, *Jpn. J. Appl. Phys.*, 1977, **16**, 1275. [e] R. M. Measures, *J. Appl. Phys.*, 1977, **48**, 2673. [f] P. Agostini, A. T. Georges, S. E. Wheatley, P. Lambropoulos, and M. D. Levenson, *J. Phys. B*, 1978, **11**, 1733. [g] K. J. Nygaard, R. J. Corbin, and J. D. Jones, *Phys. Rev. A*, 1978, **17**, 1543. [h] N. R. Isenor, in 'Multiphoton Processes', ed. J. H. Eberly and P. Lambropoulos, Wiley, New York, 1978, p. 179. [i] E. I. Toader, C. B. Collins, and B. W. Johnson, *Phys. Rev. A*, 1977, **16**, 1490. [j] V. V. Suran, *Ukr. Khim. Zh.* (*Russ. Ed.*), 1977, **22**, 2055. [k] I. S. Aleksakhin, I. P. Zapesochnyi, and V. V. Suran, *JETP Lett.*, 1977, **26**, 11. [l] R. J. Fonk, D. H. Tracy, D. C. Wright, and F. S. Tomkins, *Phys. Rev. Lett.*, 1978, **40**, 1366. [m] C. Tai and F. W. Dalby, *Can. J. Phys.*, 1978, **56**, 183. [n] C. Tai and F. W. Dalby, *Can. J. Phys.*, 1977, **55**, 434. [o] T. Mochizuki, M. Morikawa, and C. Yamanaka, *Appl. Phys. Lett*, 1978, **32**, 212. [p] H. D. V. Bohm, W. Michaelis, and C. Weitkamp, *Opt. Commun.*, 1978, **26**, 177. [q] E. W. Rothe, B. P. Mathur, and G. P. Reck, *Chem. Phys. Lett.*, 1978, **53**, 74; B. P. Mathur, E. W. Rothe, G. P. Reck, and A. J. Lightman, *ibid.*, 1978, **56**, 336. [r] A. Herrmann, S. Leutwyler, E. Schumacher, and L. Wöste, *Chem. Phys. Lett.*, 1977, **52**, 418. [s] D. E. Nitz, P. B. Hogan, L. D. Shearer, and S. J. Smith, *J. Phys. B*, 1979, **12**, L103. [t] E. H. A. Granneman, M. Klewer, K. J. Nygaard, and M. J. Van der Wiel, *J. Phys. B*, 1976, **9**, 865. [u] M. Klewer, M. J. M. Beerlage, J. Los, and M. J. Van der Wiel, *J. Phys. B*, 1977, **10**, 2809.

supersonic expanded jets, which simplifies their spectra due to rotational and vibrational cooling.[61c]

As emphasized earlier the efficiency of excitation depends on the presence of resonant intermediate states. In MPI spectroscopy it is these intermediate states which one wishes to identify. Problems can arise from overlapping intermediate state spectra and processes of different photon order but, with care, these can generally be sorted out. Several excitation mechanisms are possible, with resonant states occurring at energies equivalent to one, two, or more photons. The transition to the final ionization continuum can also involve multiphoton steps. A systematic way of describing these different mechanisms is needed and Figure 7 indicates a possible method. Processes involving resonant intermediate states are referred to as *resonance-enhanced* multiphoton ionization.

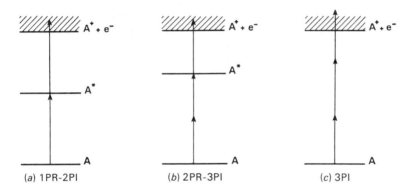

Figure 7 *Suggested nomenclature for multiphoton ionization processes. The number of photons, n, required to reach the resonant state, A^*, is written first (nPR = n-photon resonant) followed by the number of photons, m, required to go from A to the ionization continuum (mPI = m-photon ionization). Examples shown are, (a) one-photon resonant two-photon ionization, 1PR–2PI; (b) two-photon resonant three-photon ionization, 2PR–3PI; (c) three-photon non-resonant (simultaneous absorption) photoionization, 3PI*

Extensive ion-fragmentation is often observed with MPI, the extent depending sensitively on the laser intensity.[11, 68] The mechanism by which fragmentation occurs has been studied by Boesl *et al.*,[69] who suggest that, in benzene at least, the excitation does *not* proceed up an auto-ionization ladder of the neutral molecule, but by climbing two independent ladders, one up to the ionization potential of the neutral molecule, the second up the ladder of the parent molecular ions produced.[69] Two lasers operating at different wavelengths, and with the pulse from one delayed relative to the other, were used for this work.[69] In addition to providing valuable information on the ion fragmentation process this technique offers a new basis for ion spectroscopy.

[v] S. L. Chin, *Phys. Lett. A*, 1977, **61**, 311. [w] S. L. Chin and D. Faubert, *Appl. Phys. Lett.*, 1978, **32**, 303. [x] V. S. Antonov, I. N. Knyazev, V. S. Letokhov, and V. G. Movshev, *Sov. Phys. JETP Engl. transl.*, 1977, **46**, 697. [y] G. C. Nieman and S. D. Colson, *J. Chem. Phys.*, 1978, **68**, 5656.

   It is clear from this that we are now able to produce ions and to select particular fragments for the study of their gas phase kinetics. We can expect MPI techniques to broaden greatly the conditions under which ion–molecule reactions can be studied. Table 4 lists the ions which have been produced by MPI techniques.

7 Summary

The literature now contains abundant evidence to show that u.v. multiphoton excitation provides an efficient and convenient way of producing a wide range of excited state atoms and free radicals. Ionic species can also be produced under appropriate conditions and this allows a powerful new approach to mass spectrometry and a sensitive and general method of detection for multiphoton spectroscopy. Selective excitation of one molecular species in the presence of a large excess of other species can be readily achieved and permits kinetic studies to be made under well defined collision conditions. Access to electronically excited states which are forbidden for single-photon excitation and which lie in the vacuum u.v. energy range can also be achieved by multiphoton excitation. The rapid pace of development of high power and tunable u.v. laser systems can be expected to accelerate further research in these areas.

Acknowledgement. I am greatly indebted to Drs. C. Fotakis, M. Martin, and A. G. A. Rae for their guidance and helpful comments during the preparation of this report.

5

Gas Phase Reactions of Hydroxyl Radicals

BY D. L. BAULCH AND I. M. CAMPBELL

1 Introduction

Since the substantial reviews of Drysdale and Lloyd[1] and Wilson[2] some 400 measurements of OH radical reaction rate constants in the gas phase have been reported, which represents an increase in the available data by approximately a factor of four in the last eight years. This dramatic increase has been due to two main factors, firstly improvements in experimental methods for obtaining OH radical reaction rate data and secondly the appreciation of the key role which OH radicals play in the chemistry of the atmosphere.

The most notable advances in techniques have been the application of resonance fluorescence methods for OH radical detection and the great improvements in the collection and processing of transient signals which have given a new lease of life to such techniques as flash photolysis. Because of the impetus provided by atmospheric pollution, data have been obtained for the reactions of OH radicals with a wide range of organic compounds, but in general only at ambient temperatures. This is also true of reactions with inorganic species where most effort has also centred on species of importance in atmospheric chemistry, e.g., CO, NO, NO_2, ClO, HO_2, COS, H_2S. Thus, despite the known importance of OH radicals in combustion systems there is a paucity of data available at elevated temperatures. Another area of importance, which has received little attention, is the reactions of OH with organic radicals, where both rate and mechanistic data are very sparse. On the other hand a number of reactions with atoms (O, S, N) and inorganic radicals (OH, HO_2, SO, ClO) have received attention.

One of the few reactions which have been extensively studied over the whole temperature range from 300 to >2000 K is reaction (1), which is of practical

$$CO + OH \rightarrow CO_2 + H \tag{1}$$

importance for combustion and atmospheric pollution but in which much of the interest has been generated by the demonstration that its rate data conforms to a curved Arrhenius plot. Such curvature was first suggested by Dryer et al.[3] and explained on the basis of relatively simple theoretical arguments. Although there are

[1] D. D. Drysdale and A. C. Lloyd, *Oxidation Combust. Rev.*, ed. C. F. H. Tipper, 1970, **4**, 157.
[2] W. E. Wilson, jun., *J. Phys. Chem. Ref. Data*, 1972, **1**, 535.
[3] F. Dryer, D. Naegeli, and I. Glassman, *Combust. Flame*, 1971, **17**, 270

aspects which still need clarification, the recent experimental results are now numerous and precise enough to confirm the curvature, and represent the first clear demonstration of this effect for a gas phase atom transfer reaction. These findings have stimulated the search for the same feature in other OH reactions, with some success.

The literature surveyed in this Report covers the period 1972 to October 1979. During its preparation a comprehensive review by Atkinson *et al.*[4] has described reactions of OH radicals with organic compounds, particularly with regard to atmospheric pollution. For that reason, we have dealt in more detail with inorganic species and those aspects of the reactions with organic compounds, such as high temperature data, which Atkinson *et al.* have not covered. However, for completeness, the principal features of the reactions with organic compounds have been presented.

The rate data for OH reactions of importance in hydrocarbon oxidation has most recently been surveyed by Walker.[5]

2 Experimental Techniques

The methods which have been used to study OH reactions are, in general, standard gas phase kinetic techniques the fundamentals of which are well understood. We therefore merely outline recent developments and their contribution to OH reaction studies, but discuss in more detail two features common to all the methods, namely the generation of OH radicals and the ways in which the reaction progress may be followed.

Rate Measurements.—While most of the available techniques for gas phase rate measurements have been applied to OH radical reactions the recent development of fluorescence methods for measuring small OH concentrations has led to discharge flow and flash photolysis becoming the pre-eminent techniques for absolute rate measurements at low temperature.

One of the previous major limitations of flow discharge studies has been that imposed by the need to use detectable radical concentrations (10^{12}—10^{13} radicals cm^{-3} for most detection methods) and low pressures. With such restrictions it is difficult to achieve a sufficiently high concentration ratio of substrate to OH to be sure that the kinetics of removal of OH are truly pseudo-first order. The alternative has been to work under conditions where reactions of the OH with the products of the initial attack of OH on the substrate are important. Measurement of the reaction stoicheiometry then becomes necessary to deduce the rate constant. In many cases this has proved difficult and has been a major source of error in flow discharge measurements. Using lower OH concentrations measurable by fluorescence detection methods allows the use of sufficiently high substrate concentrations to ensure first-order conditions and elimination of much of the ambiguity associated with interpretation of earlier flow discharge results.

[4] R. Atkinson, K. R. Darnall, A. C. Lloyd, A. M. Winer, and J. N. Pitts, jun., *Adv. Photochem.* 1979, 11, 375.
[5] R. W. Walker, in 'Gas Kinetics and Energy Transfer', ed. P. G. Ashmore and R. J. Donovan (Specialist Periodical Reports), The Chemical Society, London, 1977, Vol. 2, p. 296; 1975, Vol. 1, p. 161.

In the case of flash photolysis the advantages of more sensitive detection methods lie not so much in any possible increase in relative concentration of substrate to radical, but rather in the lower absolute concentrations which may be used. Unlike discharge flow methods flash photolysis is not limited to low pressure and hence relatively high concentrations of substrate to radical have not been difficult to achieve. But less sensitive detection required higher radical concentrations promoting the possible occurrence of radical–radical reactions and reactions of the radical with reaction products. The lower radical concentrations involved with resonance fluorescence detection mitigate both these problems, as does the use of repetitive flashing with signal averaging and use of flowing rather than static systems.

With the very rapid developments in signal processing in recent years it is becoming the rule rather than the exception in flash photolysis work to use a large number of regularly repeated flashes ($\leqslant 100$ J flash^{-1}) rather than one large flash (*ca.* 1000 J flash^{-1}). The absorption of fluorescence signal from each of the low-energy flashes is separated into channels corresponding to a certain time period after the flash and stored in a multichannel analyser. Data from many flashes are accumulated 'building up' the kind of signal which would have been obtained from the large single flash and thus it is limited in its precision only by the number of flashes and their reproducibility. By also flowing the gases through the photolysis cell, thus sweeping out reaction products, the problems of the radicals reacting with themselves, other radicals, or reaction products can largely be eliminated.

Shock tube measurements remain the only significant source of absolute rate data for temperatures > 1000 K. Between the usual operating temperatures of flash photolysis and flow discharge (< 500 K) and the lower temperature end of the shock tube range (> 1000 K) there is a paucity of data for OH reactions as is the case for most other species. That it is possible to bridge the gap by extending to higher temperatures the conventional low temperature techniques has been demonstrated in a few instances. For example, Zellner and Steinert[6] have studied the reaction of OH with CH_4 by flash photolysis–resonance absorption up to 900 K using a conventional furnace to heat the gases, and Westenberg and de Haas[7] have used a discharge flow system up to 915 K in studying the reaction of OH with CO.

Relative rate constant measurements continue to make important contributions. Usually the technique involves generation of the OH in a static system which, after a fixed time, is analysed for products or for the extent of removal of reactants. Only rarely are rate constant ratios deduced from monitoring the OH concentration. Product ratios can often be measured very precisely and the final accuracy of the derived rate constant may largely be limited by the accuracy of the reference rate constant. In a number of instances these are sufficiently well known for error limits to be at least as narrow as those for the existing absolute data. These measurements are also important because they often involve monitoring the yields of reaction products. They therefore complement the absolute measurements which almost invariably follow the rate of disappearance of OH.

[6] R. Zellner and W. Steinert, *Int. J. Chem. Kinet.*, 1976, **68**, 397.
[7] A. A. Westenberg and N. de Haas, *J. Chem. Phys.*, 1973, **58**, 4061.

Relative rate determinations have largely been limited to temperatures close to ambient. A notable exception is the work of Baldwin *et al.*,[8] whose technique, based on addition of compounds to the hydrogen–oxygen combustion system, yields rate constants relative to that for reaction (2) at *ca.* 750 K, a region not usually covered by other methods.

$$OH + H_2 \rightarrow H_2O + H \tag{2}$$

Further details of relative rate methods are given later.

Mechanistic Studies.—In contrast to the recent rapid increase in output of rate data, mechanistic studies remain few. With simple molecules there is often only one likely pathway for reaction but it needs very little increase in molecular complexity to increase the number of possible reaction channels. Since in practical situations, such as air pollution and combustion systems, the OH reaction may initiate a number and variety of subsequent steps, identification of the initial channels and their branching ratios is of considerable importance.

One recent development is the use of photo-ionization mass spectrometry to detect products from crossed molecular beams of OH radicals and organic species.[9] Such studies give some insight into the initial stages of the reaction but are limited to very low pressures which may lead to results not applicable at more normal pressures. Niki and his co-workers,[10] in studying HONO–NO–air–organic compound mixtures to simulate smog production, have used long-path Fourier transform i.r. spectroscopy to identify products and some intermediates *in situ*. This technique is obviously of considerable value and promise.

These two methods apart, the main technique for mechanistic studies remains that of product separation and gas chromatographic analysis. Such work is often exacting and time consuming and, despite its importance, it is not difficult to understand why it has been neglected.

Generation of OH Radicals.—Hydroxyl radicals may be produced directly or indirectly by photolysis, *via* electrical discharge or by thermal means. In the first of these a molecule absorbs light, either to produce OH in the primary act or to yield an intermediate which subsequently reacts with another molecule to give OH. The advantages of such sources lie in their specificity and the fact that the primary act is largely independent of conditions, so that reactions of the OH can be studied over a wide range of pressures and temperatures. Their main limitation is the problems arising from absorption of the photolysing light by molecules other than the source molecules.

An electrical discharge in a gas at low pressures in a flow tube is a convenient source for a variety of atoms but is not a convenient direct source of OH. Discharge sources therefore rely upon production of a species which may be reacted

[8] R. R. Baldwin and R. W. Walker, *J. Chem. Soc., Faraday Trans. 1*, 1979, **75**, 140, and refs. therein.

[9] J. R. Kanofsky, D. Lucas, F. Pruss, and D. Gutman, *J. Phys. Chem.*, 1974, **78**, 311; I. R. Slagle, J. R. Gilbert, R. E. Graham, and D. Gutman, *Int. J. Chem. Kinet. Symp. 1*, 1975, 317; T. M. Sloane, *Chem. Phys. Lett.*, 1978, **54**, 269.

[10] H. Niki, P. D. Maker, C. M. Savage, and L. P. Breitenbach, *J. Phys. Chem.*, 1978, **82**, 132, and refs. therein.

downstream in the gas flow with other compound to yield OH. Thus the method has the advantage that the production of OH is separated from the reaction region, but the technique is limited to low pressures and a small pressure range.

Thermal sources too depend mostly upon decomposition of a molecule to produce a reactive species which subsequently yields OH by its reaction with another molecule. The method suffers from lack of independent control of the conditions of OH production and subsequent reaction but can sometimes be used to study reactions which would be impossible using other sources.

Another indirect source which has been used is the irradiation of humidified argon with an electron beam.[11] The argon absorbs the energy and efficiently transfers it to the water, producing dissociation into $H(^2S)$ and $OH(X^2\Pi)$ as the only long-term (> 100 ns) effect.

Photolytic Methods. The molecules which have been used to produce OH by direct photodissociation are H_2O, H_2O_2, HNO_2, and HNO_3, and their usage is illustrated by the following examples. Water vapour has an energy threshold for dissociation into ground state products, $H(^2S)$ and $OH(X^2\Pi)$, which corresponds to a wavelength of *ca.* 240 nm. However, strong absorption only sets in below 185 nm and the dissociation is usually effected by vacuum–u.v. light between this upper wavelength and the LiF cut-off at *ca.* 105 nm. Water vapour has been the commonest source of OH in flash photolysis systems, to which its use is largely restricted. Although hydrogen peroxide vapour absorbs strongly at longer wavelengths than water vapour, with an upper wavelength limit of *ca.* 310 nm and so is dissociated conveniently into $OH(X^2\Pi)$ radicals using the strong mercury emission at *ca.* 254 nm, H_2O_2 is susceptible to attack by these radicals in reaction (3) and this additional complexity has precluded its wide use. Similar problems arise

$$OH + H_2O_2 \rightarrow H_2O + HO_2 \qquad (3)$$

with photodissociation of nitric acid (upper wavelength threshold *ca.* 330 nm) where a unity quantum yield for production of $OH(X^2\Pi)$ has been shown to apply down to 200 nm wavelength[12] but the rate constant of the subsequent reactions (4) is around

$$OH + HNO_3 \rightarrow H_2O + NO_3 \qquad (4)$$

an order of magnitude less than that of reaction (3),[13,14] thus reducing the likelihood of reaction (4) competing with the reaction under study. Compared to H_2O_2 and HNO_3, H_2O has the additional advantage of thermal stability. For example H_2O has been used as the source of OH in a combined flash photolysis–shock tube system[15] at temperatures above 1000 K whereas HNO_3 is subject to thermal decomposition above 400 K.[16] H_2O_2 is less thermally stable than HNO_3 owing to heterogeneous decomposition, although it has been used[17] as a homogeneous

[11] S. Gordon and W. A. Mulac, *Internat. J. Chem. Kinet. Symp. 1*, 1975, 289.

[12] H. S. Johnston, S.-G. Chang, and G. Whitten, *J. Phys. Chem.* 1974, **78**, 1.

[13] D. Gray, G. Lissi, and J. Heicklen, *J. Phys. Chem.* 1972, **76**, 1919.

[14] R. Zellner and I. W. M. Smith, *Chem. Phys. Lett.*, 1974, **26**, 72.

[15] J. Ernst, H. Gg. Wagner, and R. Zellner, *Ber. Bunsenges. Phys. Chem.*, 1978, **82**, 409.

[16] C. Anastasi and I. W. M. Smith, *J. Chem. Soc., Faraday Trans. 2*, 1978, **74**, 1056.

[17] D. Booth, D. J. Hucknall, and R. J. Sampson, *Combust. Flame*, 1973, **21**, 261.

thermal source of OH radicals at temperatures above 613 K. Nitrous acid (HNO_2) is less thermally stable than H_2O_2 or HNO_3, but its use as a direct photo-dissociation source of OH is promoted by the fact that it provides the only known direct means of generating OH cleanly using radiation of wavelengths above 300 nm (threshold wavelength *ca*. 380 nm). The quantum yield for the process represented as:

$$HNO_2 + h\nu \rightarrow OH + NO$$

is close to unity and at ambient temperatures the rate constant of the second process (5) has been estimated[18] as $k_5 = (1.3 \pm 0.1) \times 10^9$ dm^3 mol^{-1} s^{-1}, about three times

$$OH + HNO_2 \rightarrow H_2O + NO_2 \qquad (5)$$

larger than k_3. At room temperature HNO_2 vapour can be prepared at around 90 % purity in an equilibrium mixture with its decomposition products H_2O, NO, and NO_2,[18] and in this form has been used as a photolysis source of OH in steady state but not flash photolysis systems.

Indirect photolysis sources rely upon the generation of the lowest excited state of atomic oxygen, $O(^1D)$, as a reactive intermediate from parent molecules N_2O, NO_2, O_3, and O_2. Nitrous oxide is photodissociated by u.v. radiation of wavelength less than *ca*. 260 nm to yield ground state N_2 and $O(^1D)$ while NO_2 yields $O(^1D)$ and ground state NO at wavelengths less than *ca*. 240 nm. The upper wavelength threshold for $O(^1D)$ production from ozone photodissociation is *ca*. 313 nm while O_2 yields $O(^1D) + O(^3P)$ in the Schumann–Runge continuum with a threshold at *ca*. 175 nm. $O(^1D)$ will react rapidly with many hydrogen-containing substrates to yield $OH(X^2\Pi)$, including water vapour:

$$O(^1D) + H_2O \rightarrow 2OH(X^2\Pi)$$

Other substrates such as H_2, CH_3OH, and NH_3 have also been used. Examples of the various combinations are: O_2–NH_3,[19] NO_2–H_2O,[20] N_2O–H_2,[21,22] N_2O–CH_3OH,[23] and O_3–H_2.[24]

The apparently most complex photolytic source of OH radicals is the 'synthetic photochemical smog' system, where mixtures of NO_x and small amounts of a hydrocarbon, *e.g.*, olefin, in air are irradiated with simulated solar radiation. At the outset, such systems are daunting when it is considered that at least 200 elementary reactions can be involved for just one hydrocarbon species added. However, the reactions with significant rates have kinetic parameters which are established with sufficient certainty for computer modelling to match experimental concentration *vs.* time profiles for species such as NO, NO_2, CO, and the hydrocarbon, and it has

[18] R. A. Cox, *J. Photochem.*, 1974, **3**, 175 and 291.
[19] M. J. Kurylo, *Chem. Phys. Lett.*, 1973, **23**, 467.
[20] R. Simonaitis and J. Heicklen, *Int. J. Chem. Kinet.*, 1972, **4**, 529.
[21] I. W. M. Smith and R. Zellner, *J. Chem. Soc., Faraday Trans. 2*,1973, **69**, 1617.
[22] R. Overend and G. Paraskevopoulos, *J. Phys. Chem.*, 1978, **82**, 1329.
[23] T. L. Osif, R. Simonaitis, and J. Heicklen, *J. Photochem.*, 1975, **4**, 233.
[24] A. R. Ravishankara, G. Smith, R. T. Watson, and D. D. Davis, *J. Phys. Chem.*, 1977, **81**, 2220.

been established[25-32] that, in the period before significant concentrations of O_3 are formed, the consumption rate of a hydrocarbon simply reflects the rate of its elementary bimolecular reaction with OH. The major sources of OH in these systems are the photodissociation of nitrous acid formed by the equilibrium:

$$NO + NO_2 + H_2O \rightleftharpoons 2HNO_2$$

and the reaction of HO_2 radicals with NO.

The OH concentrations can be calculated from the observed rate of disappearance of a hydrocarbon, e.g., isobutene,[30] by use of the absolute value of the OH + isobutene rate constant established by other methods.

Electrical Discharge Sources. The most common source in this category is the discharge flow system reaction (6) of NO_2 with hydrogen atoms produced by a discharge in an inert gas (He, Ar) carrier containing small amounts ($\leqslant 1\%$) of H_2.

$$H + NO_2 \rightarrow NO + OH \qquad (6)$$

The rate constant for this reaction has recently been revised upwards by a factor of more than two, compared with the previously accepted value; in three separate studies[33-35] values of $k_6 = (7.7 \pm 1.5) \times 10^{10}$, $(7.95 \pm 1.08) \times 10^{10}$, and $(6.93 \pm 1.33) \times 10^{10}$ dm^3 mol^{-1} s^{-1} at 298 K were obtained, with an averaged activation energy[34] of 3.4 ± 0.5 kJ mol^{-1}. The revision is obviously helpful to the use of reaction (6) as an OH source.

Reaction (6) has an exothermicity of 117 kJ mol^{-1} which is sufficient to allow formation of $OH(X^2\Pi)(v = 0, 1, 2, \text{and } 3)$. Since there are several known cases of reactions where the rate constants are sensitive to v, it is important to consider the possibility of survival of the $OH(v \geqslant 1)$ into the reaction zone. The results of studies of reaction (6) under conditions minimizing $OH(v)$ loss processes[36-38] show that at least 25% of the exothermicity is used to populate $v > 0$. However Spencer and Glass[37] have shown that H atoms and NO rapidly relax v to zero, while Potter *et al.*[39] have found that vibrational relaxation takes place on every collision with the walls, so that for the usual flow time between the point of generation of OH by reaction (6) and the point of addition of the substrate, $OH(X^2\Pi)(v = 0)$ should be the only significant species. However, it has also been shown[40] that if the mixing of

[25] G. J. Doyle, A. C. Lloyd, K. R. Darnall, A. M. Winer, and J. N. Pitts, *Environ, Sci. Technol.*, 1975, **9**, 237.

[26] A. C. Lloyd, K. R. Darnall, A. M. Winer, and J. N. Pitts, *J. Phys. Chem.*, 1976, **80**, 789.

[27] A. M. Winer A. C. Lloyd, K. R. Darnall, and J. N. Pitts, *J. Phys. Chem.*, 1976, **80**, 1635.

[28] A. C. Lloyd, K. R. Darnall, A. M. Winer, and J. N. Pitts, *Chem. Phys. Lett.*, 1976, **42**, 205.

[29] C. H. Wu, S. M. Japar, and H. Niki, *Environ. Sci. Health-Environ, Sci. Eng.*, 1976, **A11**, 191.

[30] W. R. L. Carter, A. C. Lloyd, J. L. Sprung, and J. N. Pitts, *Int. J. Kinet.*, 1979, **11**, 45.

[31] K. R. Darnall, A. M. Winer, A. C. Lloyd, and J. N. Pitts, *Chem. Phys. Lett.*, 1976, **44**, 415.

[32] K. R. Darnall, R. Atkinson, and J. N. Pitts, *J. Phys. Chem.*, 1978, **82**, 1581.

[33] H. Gg. Wagner, H. Welzbacher, and R. Zellner, *Ber. Bunsenges. Phys. Chem.*, 1976, **80**, 1023.

[34] M. A. A. Clyne and P. B. Monkhouse, *J. Chem., Soc., Faraday Trans. 2*, 1977, **73**, 298.

[35] P. P. Bemand and M. A. A. Clyne, *J. Chem. Soc., Faraday Trans. 2*, 1977, **73**, 394.

[36] J. A. Silver, W. L. Dimpfl, J. H. Brophy, and J. L. Kinsey, *J. Chem. Phys.*, 1976, **65**, 1811.

[37] J. E. Spencer and G. P. Glass, *Chem. Phys.*, 1976, **15**, 35.

[38] J. C. Polanyi and J. J. Sloan, *Int. J. Chem. Kinet. Symp. 1*, 1975, 51.

[39] A. E. Potter, R. N. Coltharp, and S. D. Worley, *J. Chem. Phys.*, 1971, **54**, 992.

[40] G. K. Smith and E. R. Fisher, *J. Phys. Chem.*, 1978, **82**, 2139.

H and NO_2 is poor, then significant concentrations of $v = 0$ can appear downstream due to the prolongation of the production process.

Mackenzie *et al.*[41] found that rates of disappearance of OH downstream from the point of mixing of H and NO_2 demanded a substantial wall destruction term and this increased by a factor of *ca.* 3.5 on crossing the end point from conditions of excess H to excess NO_2. This confirmed earlier observations[42–44] of significant wall effects. Baked phosphoric acid surfaces in flowtubes have been used to minimize (but not eliminate) OH heterogeneous destruction[45,46] and it is also worth mentioning that Spencer and Glass[37] found that halogenocarbon wax coatings inhibited wall relaxation of $OH(v = 1)$.

Another reaction which has been used as a discharge–flow system source of OH occurs when water vapour is added to fluorine atoms in an inert carrier:

$$F + H_2O \rightarrow OH + HF$$

The rate constant of this reaction is *ca.* 7×10^9 dm^3 mol^{-1} s^{-1} at 300 K.[47] Since the OH contains an original substrate bond, vibrational excitation is unlikely.

The ready generation of vibrationally excited OH in discharge flow systems may be used to advantage for studying the kinetics of specific v levels. Spencer and Glass[48] have shown that when HBr is added to flowing $O(^3P)$ atoms, reaction (7)

$$O + HBr \rightarrow OH + Br \tag{7}$$

$(k_7 = 2.7 \times 10^7$ dm^3 mol^{-1} $s^{-1})$ produces $OH(X^2\Pi)(v = 1)$ in more than 97 % yield and $OH(X^2\Pi)(v = 2)$ formation is precluded by a corresponding endothermicity of 23 kJ mol^{-1}. Another source of $OH(v = 1)$ has been developed by Light,[49] where it has been shown that reaction between $O(^3P)$ and H_2 in the first excited vibrational level (generated by flowing H_2 over a heated tungsten filament) proceeds largely according to the equation

$$O(^3P) + H_2(v = 1) \rightarrow OH(v = 1) + H$$

and for sufficient concentrations of $H_2(v = 1)$ at *ca.* 300 K, the main source of $OH(v = 0)$ in the system is physical quenching of $OH(v = 1)$ rather than direct reactive production.[50] Another system of interest in this connection is the laser-enhanced reaction of HCl with $O(^3P)$ atoms.[51] Here vibrationally excited HCl was generated *in situ* by excitation with the TEA HCl chemical laser and the $OH(v)$ product populations from the reaction

$$O(^3P) + HCl(v) \rightarrow OH(v) + Cl$$

[41] A. McKenzie, M. F. R. Mulcahy, and J. R. Steven, *J. Chem. Soc., Faraday Trans. 1*, 1974, **70**, 549.
[42] J. E. Breen and G. P. Glass, *J. Chem. Phys.*, 1970, **52**, 1082.
[43] M. F. R. Mulcahy and R. H. Smith, *J. Chem. Phys.*, 1971, **54**, 5215.
[44] A. McKenzie, M. F. R. Mulcahy, and J. R. Steven, *J. Chem. Phys.*, 1973, **59**, 3244.
[45] J. G. Anderson, J. J. Margitan, and F. Kaufman, *J. Chem. Phys.*, 1974, **60**, 3310.
[46] J. S. Chang and F. Kaufman, *J. Phys. Chem.*, 1978, **82**, 1683.
[47] W. Hack, A. W. Preuss, and H. Gg. Wagner, *Ber. Bunsenges. Phys. Chem.*, 1978, **82**, 1167.
[48] J. E. Spencer and G. P. Glass, *Int. J. Chem. Kinet.*, 1977, **11**, 97.
[49] G. C. Light, *J. Chem. Phys.*, 1978, **68**, 2831.
[50] G. C. Light and J. H. Matsumoto, *Chem. Phys. Lett.*, 1978, **58**, 578.
[51] J. E. Butler, J. W. Hudgens, M. C. Lin, and G. K. Smith, *Chem. Phys. Lett.*, 1978, **58**, 216.

were probed by laser-induced fluorescence. The reaction with HCl($v = 0$) to yield OH($v = 0$) is slightly endothermic and the activation energy barrier is less than the energy of an HCl vibrational quantum. On energetic grounds OH($v = 1$) can only be formed by reaction of HCl($v = 2$). The laser enhancement resulted in *ca.* 30% of the OH being produced in $v = 1$ and indicated that the rate constant for the reaction with HCl($v = 2$) and OH($v = 1$) was over 2×10^4 times faster than that with HCl($v = 0$) and OH($v = 0$) at ambient temperature.

Higher degrees of vibrational excitation of OH are produced in other reactions. The familiar reaction (8) of H atoms with O_3 is 322 kJ mol^{-1} exothermic and hence

$$H + O_3 \rightarrow OH + O_2 \qquad (8)$$

yields OH with $v \leqslant 9$. Arrested relaxation experiments[38] indicate that nearly all of this energy channels into the OH with 90 and 3% in vibration and rotation respectively. Two recent determinations[34,52] of the rate constant k_8 at 300 K gave $(1.06 \pm 0.13) \times 10^{10}$ and $(1.7 \pm 0.1) \times 10^{10}$ dm^3 mol^{-1} s^{-1}. This system lends itself to the determination of reaction rate constants for OH($v = 9$) by using the decrease of intensity of the OH infrared emission [*e.g.*, the (9—7) band[39] at 2.15 μm] down the flow tube to measure the decay rate. The mean radiative lifetime[39] of OH($v = 9$) is 0.064 s, and under normal conditions collisional quenching by molecules like O_2 and Ar is negligible,[39] but the species is quenched on every collision with the wall of the tube.[39] The addition of O_3 led to a substantial increase in the decay rate which was ascribed to the reaction of O_3 with OH($v = 9$) proceeding about an order of magnitude faster at ambient temperatures than with OH($v = 0$).

Excitation of OH($v = 10$) has been detected in the reaction system of O(^3P) with molecules such as ethylene and formaldehyde[53] and has been ascribed to the reaction:

$$O(^3P) + HCO \rightarrow OH(v \leqslant 10) + CO$$

This is supported by the identification of the observed rotational population of $v = 10$ as being the primary distribution and which leads to a value for the heat of formation of the HCO radical of $44 \cdot 8 \pm 4 \cdot 0$ kJ mol^{-1}, in good agreement with literature values, and to a dissociation energy of $D(\text{H--CO}) = 57.8 \pm 4.0$ kJ mol^{-1}.

Thermal Sources. Several sources of hydroxyl radicals from reactions not involving the use of a discharge or radiation have been developed. Work in this laboratory has shown that at ambient temperatures the reactions in gas phase mixtures of H_2O_2, NO_2, and CO involve a linear chain cycle where OH is the carrier.[54,55] The initiating reaction (9) is heterogeneous and is promoted by a thin

$$H_2O_2 + NO_2 \rightarrow OH + HNO_3 \qquad (9)$$

coating of boric acid on the walls. With a substantial excess of CO and of NO_2 over H_2O_2 the subsequent propagation steps are as given in equations (1) and (6).

[52] J. H. Lee, J. V. Michael, V. A. Payne, and L. J. Stief, *J. Chem. Phys.*, 1978, **69**, 350.

[53] K. H. Becker, H. Lipmann, and U. Schurath, *Ber. Bunsenges. Phys. Chem.*, 1977, **81**, 567.

[54] I. M. Campbell, B. J. Handy, and R. M. Kirby, *J. Chem. Soc., Faraday Trans. 1*, 1975, **71**, 867.

[55] I. M. Campbell and P. E. Parkinson, *J. Chem. Soc., Faraday Trans. 1*, 1979, **75**, 2048.

Termination is mainly effected (under excess of NO_2 conditions) by reaction (10).

$$OH + NO_2 (+M) \rightarrow HNO_3 (+M) \tag{10}$$

The yield of CO_2 is the measured parameter and its decrease on addition of a substrate S which induces the additional termination reaction

$$OH + X \rightarrow \text{Non-propagating products} \tag{X}$$

yields relative values of k_X which can be calibrated using an absolute rate constant for one particular X; n-butane has been used in this role. It has been shown that unity stoicheiometry between OH released into the gas phase and H_2O_2 reacted can be obtained for some surfaces,[55] but it is usually less; however, the method only depends on the preservation of a constant stoicheiometry during a series of experiments with different X species.

The OH radical is an important intermediate in the H_2–O_2 reaction and the effects of addition of particularly hydrocarbon species on explosion limits or the slow reaction depends in part upon the rate of reaction of OH with the added gas. Baldwin and his co-workers[8] have exploited such systems, generally at temperatures in the vicinity of 770 K, to obtain rate constants k_X for a variety of hydrocarbon substrates relative to that for reaction (2) for which much absolute kinetic data are available. This method, involving measurements of the rate of change of [X] and the total pressure of the slow reaction system, and the matching of these reaction profiles to that generated by a computer program based on the established full mechanism, provides access to kinetic parameters in a temperature region which is difficult to study by other means.

Finally in this survey of OH generation methods we come to the high temperature regions, above 1000 K, in which flames and shock-heated gas mixtures are the experimental systems used.

In hydrogen–hydrocarbon–oxygen flames the chain propagating species H, O, OH are generated at supra-equilibrium levels in the flame front and may be used as a source of these species to react with added substrates. However there have been no recent developments in the use of such sources and the reader is referred to previous reviews for further information.[1,2]

Mention has already been made of the use of photolytic generation of OH combined with shock heating of the reaction mixture.[15] The only other OH source used in recent shock tube work has been hydrogen peroxide.[56] To prevent rapid heterogeneous decomposition of the peroxide it was found necessary to coat the shock tube walls with an epoxy resin (Araldite) and to 'age' the walls by repeatedly exposing them to peroxide vapour beforehand. With this procedure reproducible OH concentrations could be obtained. The rate of disappearance of OH with various additives yielded relative rate constants at temperatures close to 1300 K.

Detection Methods.—The principal techniques used recently to detect and measure OH concentrations have been mass spectrometry, electron spin resonance, light absorption, resonance fluorescence, and laser magnetic resonance. The kinetics of OH reactions may also be followed by detection of products, by these methods, or

---

[56] J. N. Bradley, W. D. Capey, R. W. Fair, and D. K. Pritchard, *Int. J. Chem. Kinet.*, 1976, **8**, 549.

Table 1

Technique	$[OH]/mol\ dm^{-3}$
Mass spectrometry	10^{-9}
E.s.r.	10^{-9}
Light absorption (*ca.* 308 nm)	10^{-8}
Resonance fluorescence	10^{-10}
Laser magnetic resonance	10^{-11}

by chemical analysis. Typical OH concentrations which may be used with these techniques are given in Table 1, although considerably lower limits are possible, particularly for the two resonance methods.

The advantages of the last two listed methods are obvious but the usual practice with these two techniques of working with very low concentrations of OH to substrate in rate constant determinations also has its dangers. When investigating relatively slow reactions the presence of small quantities of highly reactive impurities or reaction products may lead to error. For example it has proved difficult to study the reaction of OH with COS by flash photolysis for this reason (see later). In such cases it would be desirable to have rate constant measurements where the substrate concentration was followed in a large excess of OH. Such measurements are possible, but difficult, and there are few examples.

Mass Spectrometry. Mass spectrometry is the only technique which, in principle, offers routinely the opportunity of simultaneous monitoring of all, or several, of the reactants and reaction products. A number of the earlier absolute values of OH radical reaction rate constants were obtained using mass spectrometry,[57] particularly by Niki and his co-workers[58-60] in their work on OH reactions with a number of organic compounds. A discharge flow system was used and the concentration of the organic substrate monitored in an excess of OH. In the case of the reactions with ethene and propene, mass peaks corresponding to the OH–olefin adduct were observed, demonstrating the value of the technique in mechanistic studies.

In studying reactions of OH in crossed molecular beams with alkenes, alkynes, and aromatic compounds, both Sloane *et al.*[61,62] and Gutman *et al.*[63,64] have used quadrupole mass spectrometry to detect both stable and transient species. In order to obtain the necessary selectivity for detecting low radical concentrations in the presence of large amounts of background and beam gases, Gutman *et al.*[63,64] used photoionization by inert gas atomic resonance lamps.

However, despite its value in such work, mass spectrometry has the disadvantage of complexity of equipment and not particularly high sensitivity and for these reasons in rate constant measurements it has largely been superseded by optical methods.

[57] J. T. Herron and R. D. Penzhorn, *J. Phys. Chem.*, 1969, **73**, 191.
[58] E. D. Morris, jun., D. H. Stedman, and H. Niki, *J. Am. Chem. Soc.*, 1971, **93**, 3570.
[59] E. D. Morris, jun., and H. Niki, *J. Phys. Chem.*, 1971, **75**, 3640.
[60] E. D. Morris, jun., and H. Niki, *J. Chem. Phys.*, 1971, **55**, 1991.
[61] T. M. Sloane, *Chem. Phys. Letters*, 1978, **54**, 269.
[62] T. M. Sloane and R. J. Brudzinski, *J. Am. Chem. Soc.*, 1979, **101**, 1495.
[63] J. R. Kanofsky, D. Lucas, F. Pruss, and D. Gutmann, *J. Phys. Chem.*, 1974, **78**, 311.
[64] I. R. Slagle, J. R. Gilbert, R. E. Graham, and D. Gutman, *Int. J. Chem. Kinet., Symp. 1*, 1975, 317.

Magnetic Resonance Methods. The use of electron spin resonance (e.s.r.) to monitor radicals has a number of advantages. The e.s.r. spectrum is specific to the particular radical, the technique detects only radicals and hence is insensitive to interference from large quantities of background gases, and measurements are made external to the reaction system. Furthermore the absolute concentration of the OH can be established by calibration with NO, which undergoes a similar type of e.s.r. transition.[65]

E.s.r. has been widely and successfully used by a number of groups, *e.g.* those of Westenberg[7,66] and Wagner.[67] In appropriate cases more than one atom and/or radical can be readily monitored since there is little probability of overlap of e.s.r. spectra. Thus in a recent study of reaction (11) both [S] and [OH] were followed as

$$S + OH \rightarrow SO + H \qquad (11)$$

a function of reaction time.[68] A more striking example is the work of Spencer and Glass[48] on the reaction between $O(^3P)$ and HBr in which absolute concentrations of the species $OH(v = 0)$, $H(^2S)$, $O(^3P)$, and $Br(^2P_{3/2})$ could be monitored. In this instance concentrations of OH as low as 3×10^{-11} mol dm^{-3} were detectable.

Because of the relatively high concentrations of radicals required, the use of e.s.r. is limited to flow systems where it also has the disadvantages of poor time resolution, due to the length of the cavity required (*ca.* 2 cm), and the necessity to use a narrow flowtube through the cavity enhancing the pressure drop down the tube and wall reactions. It is also unable to detect complex radicals and, of course, the equipment is elaborate and expensive.

A number of these disadvantages are overcome in the laser magnetic resonance technique.[69] In this method i.r. radiation is used to produce transitions between rotational levels in paramagnetic molecules. The rotational energy levels are split by application of a magnetic field which may be varied to bring the energy splitting into exact coincidence with the energy quanta from a laser operating in the far i.r. trained on the absorption cell. Achievement of the resonance condition results in absorption of the laser radiation. The power absorbed by the radicals is usually only a small fraction (often <0.1 %) of the total laser power and to measure such small changes it is necessary to modulate the magnetic field and use phase-sensitive detection of the changes in the laser power.

In flow systems, with which the technique is used, the detection volume is defined by the dimensions of the laser beam, resulting in good spatial, and hence time, resolution. The technique is extremely sensitive, *e.g.* for OH radicals concentrations as low as *ca.* 10^{-13} mol dm^{-3} are detectable, and although higher concentrations are usually used, they are low enough readily to achieve pseudo-first-order conditions in kinetics measurements. It is also applicable to larger radicals such as HCO and HO_2 which cannot be handled by e.s.r.

[65] A. A. Westenberg, *J. Chem. Phys.*, 1965, **43**, 1544.

[66] A. A. Westenberg and N. de Haas, *J. Chem. Phys.*, 1965, **43**, 1550.

[67] J. N. Bradley, W. Hack, R. Hoyermann, and H. Gg. Wagner, *J. Chem. Soc., Faraday Trans. 1*, 1973, **69**, 1889; W. Hack, G. Mex, and H. G. Wagner, *Ber. Bunsenges. Phys. Chem.*, 1977, **81**, 677.

[68] J. L. Jourdain, G. Le Bras, and M. Combourieu, *Int. J. Chem. Kinet.*, 1979, **11**, 569.

[69] C. J. Howard and K. M. Evenson, *J. Chem. Phys.*, 1974, **61**, 1943, and refs. therein.

Light Absorption. Absorption of u.v. light in the $OH(X^2\Pi \rightarrow A^3\Sigma^+)$ band system has been used to monitor OH concentrations in a variety of experimental systems. In much of the earlier work[70-72] photographic plates were used to detect absorption by individual rotational lines in the band system, using light from a flash lamp, which was effectively a continuum in the spectral region of interest. This flash lamp–photographic plate combination has been replaced in more recent work by photomultiplier detection combined with a light source employing a microwave discharge through an inert gas–water vapour mixture, isolating rotational lines by means of a monochromator. Other light sources which have been used include a discharge in an inert gas containing a trace of bismuth vapour ($\lambda = 301.7$ nm),[73] a flash lamp containing water vapour,[74] and high-pressure mercury[75] or xenon[76] arcs.

With a discharge lamp–photomultiplier arrangement a pathlength of up to several dm is required to measure with any accuracy radical concentrations in the 10^{-11}—10^{-12} mol dm^{-3} range. When used with discharge-flow systems, where the pathlength is limited by the diameter of the flowtube, this is achieved by multiple pass optical systems.[77,78] In flash photolysis experiments the length of the photolysis cell is usually adequate for a single pass to suffice.[79] In shock tube and flame work both single[56] and multiple pass[15] systems are used.

To improve the signal-to-noise ratio the bandwidth of the monochromator is sometimes set so wide that absorption by several rotational transitions becomes significant. Under these circumstances light absorption will, in general, not follow the normal Beer–Lambert relationship. If the contributions from more than one line are small the Beer–Lambert Law may hold up to a reasonable extent of absorption. For example, Morley and Smith[79] used an optical system centred on the 308.15 nm line but receiving contributions from three neighbouring lines and their satellites: considerations of the thermal population and transition probabilities indicated that deviations from Beer's Law were $<6\%$ at 30% absorption. In contrast in the work of Ernst *et al.*[15] a bandwidth of 10 nm was used involving many rotational lines. In that case it was necessary to use a modified absorption law of the form $\ln(I_0/I) = \varepsilon_{\text{eff}}([OH]1)^\gamma$ and to establish values of the constants ε_{eff} and γ, experimentally.[15] Bradley *et al.*[56] using a much smaller band width (0.62 nm) used $\ln(I_0/I) = (\varepsilon_{\text{eff}}[OH])^\gamma$ to deal with the problem.

In experiments which are capable of rapid and reproducible repetition, *e.g.*, repetitive flashing in a flash photolysis system, the problem of poor signal-to-noise ratio may be overcome by repeating the experiment many times and accumulating the output in digital form.[80]

[70] N. R. Greiner, *J. Chem. Phys.*, 1966, **45**, 99.

[71] N. R. Greiner, *J. Chem. Phys.*, 1967, **46**, 2795.

[72] N. R. Greiner, *J. Phys. Chem.*, 1968, **72**, 406.

[73] W. C. Gardiner, K. Morinaga, D. L. Ripley, and T. Takeyama, *J. Chem. Phys.*, 1968, **48**, 1665.

[74] G. L. Schott, *J. Chem. Phys.*, 1960, **32**, 710.

[75] H. A. Olschewski, J. Troe, and H. Gg. Wagner, Proc. 11th Symp. (Internat.) Combust., 1967, p. 155.

[76] T. Singh and R. F. Sawyer, Proc. 13th Symp. (Internat.) Combust., 1971, p. 403.

[77] F. Kaufman and F. P. Del Greco, *J. Chem. Phys.*, 1961, **35**, 1895; *Discuss. Faraday Soc.*, 1962, **33**, 128; D. M. Golden, F. P. Del Greco, and F. Kaufman, *J. Chem. Phys.*, 1963, **39**, 3034.

[78] A. Pastrana and R. W. Carr, jun., *Int. J. Chem. Kinet.*, 1974, **6**, 587.

[79] C. Morley and I. M. W. Smith, *J. Chem. Soc., Faraday Trans. 2*, 1972, **68**, 1016.

[80] R. P. Overend, G. Paraskevopoulos, and R. J. Cvetanovic, *Can. J. Chem.*, 1975, **53**, 3374.

An interesting recent development is the first application[81] to OH radical kinetics of molecular modulation spectroscopy of the type developed by Hunziker and used by Atkinson and Cvetanovic[82] to study $O(^3P)$ kinetics. In this work the OH concentration was monitored by absorption measurements. The system studied was mercury photosensitized production of $O(^3P)$ from N_2O in the presence of propane resulting in the sequence of reactions (12)—(14).

$$Hg(^3P) + N_2O \rightarrow Hg + N_2 + O(^3P) \tag{12}$$

$$O(^3P) + C_3H_8 \rightarrow OH + C_3H_7 \tag{13}$$

$$OH + C_3H_8 \rightarrow H_2O + C_3H_7 \tag{14}$$

The light source used to excite the mercury was a radiofrequency excited low-pressure mercury lamp ($\lambda = 253.7$ nm), part of the intensity of which was sinusoidally modulated at audiofrequencies in the range 1—126 kHz. The total intensity of absorption I at any time t is given by $I = a + b \sin \omega t$, where a is the constant fraction (unmodulated) of the light absorbed, b is the amplitude of the modulated fraction, and ω is the angular modulation frequency. The excited mercury is quenched very rapidly by reaction (12) so that its concentration growth and decay remains in phase with the intensity of the exciting light. Thus the modulated 253.7 nm radiation transmitted through the cell acts as a reference signal for the rise and fall of $O(^3P)$ production in the system.

The OH radicals present were detected in absorption (307.5—309.5 nm) by passage of the light from a 500 W xenon arc through the cell and into a monochromator and photomultiplier. If reactions (13) and (14) were very fast the intensity of OH absorption would also be in phase with that of the exciting light. However the modulation frequency chosen is such that its period is comparable with the timescale on which the OH concentration is changing due to the occurrence of reactions (13) and (14). In these circumstances the absorption intensity is no longer in phase with the exciting light but shifted by a phase angle ϕ (in radians) which is simply related to the reaction rate and the modulation frequency by

$$\phi = \tan^{-1}(-\omega/k_{13}[C_3H_8]) + \tan^{-1}(-\omega/k_{14}[C_3H_8])$$

Both reactions (13) and (14) contribute to the phase shift, but because their rate constants are sufficiently different (a factor of 30) it is possible to separate the two contributions by measuring ϕ as a function of $\omega/[C_3H_8]$ and analysing the non-linear curve so obtained by a curve fitting procedure. The values of k_{13} and k_{14} derived are in good agreement with literature values.

This is the first application of this method to OH reactions. It is elegant and offers a method of photolytic generation of OH radicals in circumstances where direct photolytic generation might be difficult.

[81] A. B. Harker and C. S. Burton, *Int. J. Chem. Kinet.*, 1975, **7**, 907.
[82] R. Atkinson and R. J. Cvetanovic, *J. Chem. Phys.*, 1971, **55**, 659, and refs. therein.

Resonance Fluorescence. Resonance fluorescence was first used to detect OH radicals in kinetic studies by Stuhl and Niki.[83] It was rapidly adopted by a number of groups for both flash photolysis[19,84,85] and discharge flow studies.[86] The technique involves irradiation of the reaction zone by light in the 306 nm region which is absorbed by any OH present ($A^2\Sigma^+$; $v' = 0 \leftarrow X^2\Pi$; $v'' = 0$). Resonance fluorescence from the excited OH is detected at right angles to the analysing beam. The usual light source is a microwave discharge through water vapour alone or mixed with an inert gas which gives rise to emissions, the strongest of which is the (0,0) band of the OH($A^2\Sigma^+ \rightarrow X^2\Pi$) system. Typically the rotational temperature in the source is 500—700 K but the overlap with the absorption spectrum of the 'cooler' OH in the reaction system is substantial with the bandwidth of the filters (2—20 nm) usually used to isolate the radiation.

The major advantages of the technique are simplicity and sensitivity. Concentrations $< ca.$ 10^{-11} mol dm^{-3} can readily be measured and the response is linear over a wide concentration range. Consequently it is now the major method for monitoring OH concentrations in flash photolysis and discharge flow studies.

Fourier Transform Infrared Analysis. Product analysis by i.r. absorption is, in principle, attractive because of the excellent diagnostic properties of spectroscopy in this wavelength region. However, it has been little applied to kinetics because of its relatively low sensitivity and the difficulty of gathering spectral information on a timescale comparable with reaction times normally encountered in OH studies.

Recent studies by Niki and co-workers demonstrate that i.r. methods can be successfully applied.[10] They investigated reactions initiated by the photolysis of HONO to produce OH radicals in air containing small quantities (*ca.* 50 p.p.m.) of organic compounds. Problems of low sensitivity were overcome by using a long-path absorption cell (1 m length, 40 passes) and, with Fourier transform methods, sufficiently rapid scanning could be achieved to monitor the changes in concentration of several of the species present, which were changing on a timescale of minutes. The principles of Fourier transform i.r. techniques may be found in the book by Griffiths[87] and for recent applications of fast scanning see the article by Durana and Mantz.[88]

Niki *et al.*[10] used this method to study OH reactions with aldehydes, deducing relative rate constants by following rates of disappearance of reactants, but in other systems they studied, a variety of reaction products was identified at the p.p.m. level and their concentration changes followed with time.[89] The ability to monitor a range of species and observe their evolution *in situ* makes this technique attractive for handling these HONO–air–substrate systems, which are normally analysed by means of gas chromatography.

[83] F. Stuhl and H. Niki, *J. Chem. Phys.*, 1972, **57**, 3671.
[84] D. D. Davis, S. Fischer, and R. Schiff, *J. Chem. Phys.*, 1974, **61**, 2213.
[85] R. Atkinson, D. A. Hansen, and J. N. Pitts, jun., *J. Chem. Phys.*, 1975, **62**, 3284.
[86] J. G. Anderson and F. Kaufman, *Chem. Phys. Lett.*, 1972, **16**, 375.
[87] P. R. Griffiths, 'Chemical Infra-red Fourier Transform Spectroscopy', Wiley, New York, 1975.
[88] J. F. Durana and A. W. Mantz, 'Fourier Transform Infra-red Spectroscopy, Applications to Chemical Systems', ed. J. R., Ferraro and L. J. Basile, vol. 2 Academic Press, New York, 1979, p. 1.
[89] H. Niki, P. D. Maker, C. M. Savage, and L. P. Breitenbach, *Chem. Phys. Lett.*, 1979, **61**, 100 and refs. therein.

3 Reactions with Inorganic Species

Bimolecular Reactions with Atoms.—The rate constant for the rapid reaction (15) with $O(^3P)$ atoms is now well established; recent studies in discharge flow systems,

$$O + OH \rightarrow O_2 + H \qquad (15)$$

have yielded rate constants at ambient temperatures of $k_{15} = (2.65 \pm 0.52) \times 10^{10}$ dm^3 mol^{-1} s^{-1},[90] and $(2.59 \pm 0.78) \times 10^{10}$ dm^3 mol^{-1} s^{-1}.[42] The former value was deduced from a measurement of $k_{15}/k_1 = 260 \pm 20$, and k_1 is so well defined in the temperature range 290—500 K, where it shows no significant temperature variation, that the value of k_{15} deduced from k_{15}/k_1 and k_1 (with combined error limits) is more precisely defined than the latter absolute measurement. This is some encouragement to those using methods which can only yield relative rate constant values. Since k_{15} has a value more than 10% of the collision frequency, it is unlikely to have a significant or even measurable temperature coefficient in the temperature ranges usually involved in kinetic studies.

Spencer and Glass[48] have measured $k'_{15} = (6.3 \pm 3.2) \times 10^{10}$ dm^3 mol^{-1} s^{-1} when $OH(v = 1)$ is the reactant; despite their claim that reaction (24) is enhanced in rate by a factor of 2 to 3 by vibrational excitation of OH, it is apparent that the error limits for k_{15} and k'_{15} encompass the possibility of no enhancement at all. However, on the basis of the mean k'_{15} being one half of the collision frequency, they consider that it is likely that reaction of $O(^3P)$ with $OH(v = 1)$ occurs at every collision of O with the oxygen end of the OH radical. Collision at the hydrogen end is postulated to result in vibrational deactivation by the atom exchange process

$$O_a + HO(v = 1) \rightarrow O_aH(v = 0) + O$$

which occurs with a rate constant of $(8.7 \pm 1.5) \times 10^{10}$ dm^3 mol^{-1} s^{-1},[48] overlapping with k'_{15} for $OH(v = 1)$ and again, around half of the collision frequency, supporting the postulate. It seems reasonable therefore to postulate that reaction (15) only occurs when the O atom collides with the oxygen end of $OH(v = 0)$ in which case the mean value of k'_{15} could be considered to approximate to the maximum pre-exponential factor of k_{15}. In this event k_{15} would have an upper limiting activation energy of *ca.* 2 kJ mol^{-1}. Alternatively Spencer and Glass[48] have considered that vibrational excitation of a reactant may only affect the pre-exponential factor, in which case the mean value of k'_{15} suggests a truly zero energy barrier for reaction (15).

In the case of the analogous reactions (11) and (16) of $S(^3P)$ and $N(^4S)$ atoms

$$N + OH \rightarrow NO + H \qquad (16)$$

with OH there are only single determinations available. Jourdain *et al.*[68] have determined $k_{11} = (4.0 \pm 0.8) \times 10^{10}$ dm^3 mol^{-1} s^{-1} at 298 K in a discharge flow study where the reactants were generated together in a discharge flow system by addition of H_2S and NO_2 to flowing hydrogen atoms in an inert carrier. E.s.r. was used to measure [S] and [OH] as a function of time; [OH] decays were pseudo-first-order since [S]/[OH] was established as 40—150. The value of k_{11} thus

[90] I. M. Campbell and B. J. Handy, *Chem. Phys. Lett.*, 1977, **47**, 475.

appears to be slightly larger than that for the corresponding O atom reaction (k_{15}). The study also showed an increased heterogeneous decay of OH on the boric-acid coated walls when S was present compared with when it was absent. Since the measured [S] in the gas phase was significantly lower than that which should have been produced from the [H_2S] added to excess of hydrogen atoms, it was considered that S was adsorbed on the walls and induced the heterogeneous analogue of reaction (11). Such heterogeneous accelerations by wall-adsorbed species have been postulated before for H, NO, and NO_2.[41,42] For reaction (16) there is only the relative rate-constant determination[91] of $k_{16}/k_{15} = 1.4 \pm 0.1$ at 320 K which yields $k_{16} = (3.7 \pm 1.0) \times 10^{10}$ dm^3 mol^{-1} s^{-1} in combination with the most precise value of k_{15}. Both of these reactions are sufficiently important in combustion and atmospheric chemistry to merit further determinations of their rate constants.

Bimolecular Reactions with Diatomic Molecules.—Reaction (1) is of major

$$CO + OH \rightarrow CO_2 + H \qquad (1)$$

importance in the chemistry of the atmosphere and in combustion systems. In addition, their theoretical interest has caused the problems of the pressure and temperature dependence of the reaction to have been studied extensively in recent years.

There is no remaining doubt that the Arrhenius plot of k_1 at low pressures is strongly curved. This was first suggested by Dryer *et al.*,[3] who justified their suggestion in terms of a simple transition-state model proceeding through the activated complex HOCO.[±] Their calculation suggested that with an energy barrier close to zero, the differing temperature dependences of the vibrational partition functions of the reactants and activated complex would lead to k_1 having an increasing positive temperature dependence with rising temperature. The corresponding activation energy would be approximately zero for temperatures up to 600 K but would rise sharply to about 20 kJ mol^{-1} at temperatures around 1400 K. These calculations could be fitted quite well to the available rate constant data; in particular they fitted those measurements in the vicinity of 1000 K, which fell well below an Arrhenius plot based on the low- and high-temperature data. In two substantial review articles which followed[92,93] almost identical empirical expressions were derived for k_1 as a function of temperature.

$$\log (k_1/\text{dm}^3 \text{ mol}^{-1} \text{ s}^{-1}) = 7.83 + 3.94 \times 10^{-4} \, T/K$$

Smith and Zellner[21] and Smith[94] have pursued the transition-state theory approach showing that the bent HOCO configuration postulated by Dryer *et al.* generated a k_1 value which was too large and indicated isotope effects (replacement of OH by OD) that were not in agreement with earlier experimental results.[95] Better agreement with experiment was obtained with a model in which the reaction proceeded by way of two intermediates, the first effectively a 'collision complex'

[91] I. M. Campbell and B. A. Thrush, *Trans. Faraday Soc.*, 1968, **64**, 1265.
[92] D. L. Baulch and D. D. Drysdale, *Combust. Flame*, 1974, **23**, 215.
[93] W. Steinert and R. Zellner, Proc. 2nd European Symp. Combust., 1975, p. 31.
[94] I. W. M. Smith, *Chem. Phys. Lett.*, 1977, **49**, 112.
[95] A. A. Westenberg and W. E. Wilson, *J. Chem. Phys.*, 1966, **45**, 338.

identified with the bent configuration of HOCO, followed by passage through a transition state in which the three heavy atoms were collinear. The energy barriers to reversion of the 'collision complex' to $CO + OH$ and reaction to $H + CO_2$ were required to be almost equal and the true transition state of the overall reaction was identified with the maximum in the latter path. The 'collision complex' lay in a potential well on the reaction hypersurface reflecting its ability to survive at least a few vibrations and collisions with other molecules before its destruction. Smith[94] also considered that this collision complex could be stabilized by collisional deactivation to yield an HOCO species which was so stable to dissociation on thermochemical grounds that alternative mechanisms for its destruction were required. However, in linking the interpretation of the temperature dependence of k_1 to its pressure dependence (to be discussed later), Biermann *et al.*[96] suggest that although the collision complex can be de-energized by collisions, it cannot become immune to dissociation through subsequent collisional re-energization. From the observed decays of OH in the presence of CO in a reaction chamber in which OH was generated by a pulsed radiation source, these last-named workers found that a lifetime of longer than 1 ms for the stabilized HOCO species was unlikely, and interpreted this as indicating a binding energy for HO–CO of $\leqslant 61$ kJ mol^{-1}. This value is in conflict with calculations of the thermochemical properties of the HOCO radical by Gardiner *et al.*[97] and experimental evidence cited by Smith[55] which suggest binding energies nearer to 160 kJ mol^{-1}. Bierman *et al.*[96] have therefore suggested that the stabilized species may be the isomeric formate radical (HCO_2) which has a binding energy of 75 ± 21 kJ mol^{-1}, more compatible with their experimental requirement. However, the formation of HOCO rather than HCO_2 has been observed[98] in a matrix of CO at 14 K which may argue against the formation of HCO_2 in the gas phase experiments. Another possibility is that substantial de-energization of the HOCO collision complex may be sufficiently slow on the timescale concerned to keep most of it available for dissociation. Trainor and von Rosenberg[99] have found that there is no detectable excitation of the vibrational modes of the CO_2 product from the $CO + OH$ reaction and concluded that most of the exothermicity appeared as translational energy of the hydrogen atoms, in agreement with the predictions made by Smith and Zellner.[21]

The debate on the curvature of the Arrhenius plot for the $CO + OH$ reaction has been broadened by the findings of similar curvature in other reactions of OH radicals,[100] in particular reactions (2), (17), and (18). In no case can the low- and

$$OH + H_2 \rightarrow H_2O + H \qquad (2)$$

$$OH + CH_4 \rightarrow CH_4 + H_2O \qquad (17)$$

$$OH + OH \rightarrow H_2O + O \qquad (18)$$

high-temperature data be compatible with an Arrhenius form. An alternative general explanation for the curvature, applicable to all these reactions, has been

[96] H. W. Biermann, C. Zetsch, and F. Stuhl, *Ber Bunsenges. Phys. Chem.*, 1978, **82**, 633.
[97] W. C. Gardiner, D. B. Olsen, and J. N. White, *Chem. Phys. Lett.*, 1978, **53**, 134.
[98] D. E. Milligan and M. E. Jacox, *J. Chem. Phys.*, 1971, **54**, 927.
[99] D. W. Trainor and C. W. von Rosenberg, *Chem. Phys. Lett.*, 1974, **29**, 35.
[100] R. Zellner, *J. Phys. Chem.*, 1979, **83**, 18.

considered for CO + OH by Gardiner and co-workers.[101] They suggest that it could arise from the reactions of thermally equilibrated, vibrationally excited molecules.

It is frequently the case where vibrational energy is present in either OH or its co-reactant molecule that the rate constant is markedly increased compared to the reaction without vibrational excitation of reactants. As described earlier, sources of OH($v = 1$) have been developed and in the case of OH + HBr, Spencer and Glass[48] have found a rate constant accelerated by a factor of 9 for OH($v = 1$) as opposed to OH($v = 0$) at 295 K, which increase, they argue, does not arise from any difference in activation energy for the reactions of OH($v = 0$) and OH($v = 1$), but is due solely to differences in the pre-exponential factors. For thermally equilibrated OH at room temperature the fraction of OH in $v = 1$ is only 2×10^{-8} so that the faster reaction of OH($v = 1$) exerts negligible effect on the overall OH + HBr reaction rate. On the basis that at, say, 2000 K [OH($v = 1$)]/[OH($v = 0$)] would be *ca* 0.07, the overall rate would be enhanced by over 60% due to the higher reactivity of OH($v = 1$) and the activation energy between 2000 and 2500 K was estimated to be some 16 kJ mol^{-1} larger than that at 300 K, *i.e* the Arrhenius plot curves sharply upwards for temperatures in excess of 1000 K. When they studied the reaction of OH($v = 1$) with CO, Spencer and Glass[48] could not obtain sufficiently precise rate constants for the chemical reaction (its relative slowness is not well suited to the technique used) for them to be able to decide unequivocally whether or not the rate constant was enhanced sufficiently for $v = 1$ over $v = 0$ to explain the curved Arrhenius plot for CO reacting with thermally equilibrated OH. They estimated that an enhancement factor of 15 was required but could only derive an upper limiting value of $\lesssim 30$, which did not exclude the possibility. At the same time they derived an upper limiting rate constant for physical deactivation of OH($v = 1$) by CO to OH($v = 0$) which had an upper limit of only twice the value of the rate constant for the chemical reaction of OH($v = 0$) with CO. Since the mechanism for vibrational deactivation is likely to involve the formation of the collision complex HOCO, this result seems inconsistent with the Smith and Zellner model[21] where the transient HOCO species is required to dissociate more frequently to OH + CO than to H + CO$_2$. However, in accord with the foregoing postulate, it could also be taken as evidence for a rather short lifetime of the HOCO species, insufficient for the transfer of vibrational energy out of the OH bond before redissociation and, presumably, then also insufficient for extensive relaxation down the HOCO manifold, *i.e.*, the species remains liable to redissociation.

Other points relevant to the general debate on the relationship between vibrational excitation and curvature of Arrhenius plots are provided by reactions (2) and (17). As pointed out by Zellner[100] in these reactions vibrational excitation of OH might not be expected to lead to a dramatic enhancement of the rate constants, since to a large extent the energy would be in the 'wrong bond': vibrational excitation of the bond which is broken might be expected to be more effective. In this respect these cases differ from the CO + OH case. (However, see later discussion of the OH + HBr reaction where it appears that vibrational excitation of the OH significantly increases the rate constants.) Zellner *et al.*[102] have

[101] W. C. Gardiner, W. G. Mallard, M. McFarland, K. Morinaga, J. H. Owens, W. T. Rawlins, T. Takeyama, and B. F. Walker, Proc. 14th Symp. (Internat.) Combust., 1972, p. 61.
[102] R. Zellner, W. Steinert, and H. G. Wagner, cited in ref. 100.

found that OH + $H_2(v = 1)$ is at least 100 times faster than is OH + $H_2(v = 0)$ at 298 K, with an upper limit of 250, well within that of 1000 found by Light and Matsumoto.[50] This is, as expected, a much larger enhancement than the upper limit of *ca.* 2 formed by Spencer *et al.*[103] for the effect of vibrational excitation of OH in reaction with $H_2(v = 0)$. Whilst the temperature coefficient of OH + $H_2(v = 1)$ is not yet known, Zellner[61] has shown that the use of reasonable kinetic parameters can produce the curved Arrhenius plot for the overall OH + H_2 reaction, when the H_2 has an equilibrium vibrational distribution.

The vibrational excitation mechanism offers a more general explanation than do transition-state models but more definitive kinetic data are required on the relative reactivities of specific vibrational levels of species, and their temperature dependences, before the arguments can be resolved.

Returning to the CO + OH reaction, its rate constant at ambient temperatures at atmospheric pressure in air has been found to be enhanced by a factor of *ca.* 2 compared to the limiting low pressure value.[104] A similar enhancement was observed by use of H_2 and SF_6 as the bulk gases,[105] but none was observed when He or Ar were used within this pressure range.[11,71,83,84,106,107] With SF_6[106] and H_2[105] the variation of rate constant with increasing pressure was found to be a smooth function but Butler *et al.*[108] have found an anomalous inflected curve for the increase in O_2 + N_2 mixtures where the transition from the low to the high pressure value of the rate constant took place sharply in the pressure range 13—40 kPa. This appears to be associated with the observation by Biermann *et al.*[96] that the rate constant is independent of pressure in N_2 at pressures up to atmospheric when extensive precautions are taken to remove all traces of O_2 from the system. Upon addition of traces of O_2 at higher pressures the measured value of the rate constant increased to a limiting value, *e.g.* for an atmosphere of N_2, the pressure of O_2 required to achieve the limiting value was *ca.* 0.03 % of the total. Similar increases were produced by the presence of other (undefined) impurities.

It is pointed out that in the various studies where a pressure dependence was found, either O_2 was necessarily present (air, N_2 + O_2 mixtures), or was inadvertently present at an impurity level sufficient to be significant. A study of the kinetics at high pressures in say SF_6, where O_2 and other potentially reactive impurities have been excluded at the trace level, would be most valuable.

The mechanism advanced by Biermann *et al.*[96] takes the form shown in Scheme 1 where HOCO* represents the collision complex discussed earlier and $HOCO(HCO_2)$ is a de-energized form but subject to redissociation. Thus in the absence of O_2 (or at low pressure) the vertical progression of the mechanism as written is inhibited, and the 'low pressure' rate constant is obtained. In the presence of a species reactive towards $HOCO(HCO_2)$, such as O_2, the leakage downwards increases the overall rate of removal of OH. The analysis of the variation of the rate constant for OH removal as a function of $[N_2]$ and $[O_2]$ led to ratios $k_b/k_c = 4.10 \times 10^{-2}$ and $k_d/k_e \approx 3.5 \times 10^5$ dm^3 mol^{-1}. On the basis that k_c and k_d cannot be

[103] J. E. Spencer, H. Endo, and G. P. Glass, Proc. 16th Symp. (Internat.) Combust., 1977, p. 829.
[104] R. A. Cox, R. G. Derwent, and P. M. Holt, *J. Chem. Soc., Faraday Trans. 1*, 1976, **72**, 2031.
[105] B. K. T. Sie, R. Simonaitis, and J. Heicklen, *Int. J. Chem. Kinet.*, 1976, **8**, 85.
[106] R. A. Perry, R. Atkinson, and J. N. Pitts, jun., *J. Chem. Phys.*, 1977, **67**, 5577.
[107] R. Atkinson, R. A. Perry, and J. N. Pitts, jun., *Chem. Phys. Lett.*, 1976, **44**, 204.
[108] R. Butler, I. J. Solomon, and A. Snelson, *Chem. Phys. Lett.*, 1978, **54**, 19.

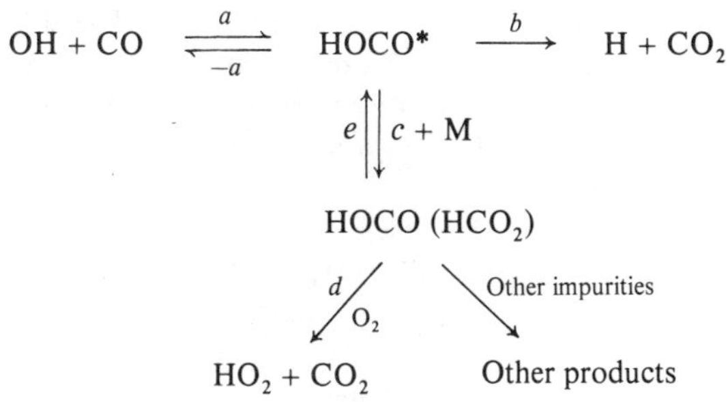

$$OH + CO \xrightleftharpoons[{-a}]{a} HOCO^* \xrightarrow{b} H + CO_2$$

$$e \left\|\, c + M\right.$$

$$HOCO \ (HCO_2)$$

$$\overset{d}{\diagup}\underset{O_2}{} \qquad \overset{\text{Other impurities}}{\diagdown}$$

$$HO_2 + CO_2 \qquad\qquad \text{Other products}$$

Scheme 1

larger than the collision frequency (6×10^{10} dm^3 mol^{-1} s^{-1}) upper limits of $k_b \leqslant 2.5 \times 10^9$ and $k_e < 2 \times 10^5$ s^{-1} were derived. However, while the basic features of this mechanism are similar to others suggested, and are probably correct, it has been assumed that reaction c is pressure dependent whilst its reverse, reaction e is not. An alternative pressure-independent path to compete with reaction d would be preferable.

An interesting analogue to the pressure-dependent component of the reaction of CO with OH has been found by Phillips[109] in his study of the reaction between OH and HCN in a discharge flow system at 373 K. The overall reaction is postulated to be

$$OH + HCN \rightarrow H_2O + CN$$

On following the decay of OH, the apparent bimolecular rate constant increased with total pressure of argon or helium carrier gas. At very low pressures the increase was directly proportional to pressure but fell away from direct proportionality as the pressure increased to the maximum value used, 2.5 kPa.

The reactions of OH with HCl, HBr, and HI have all been studied. For reaction (19) five separate measurements, three using flash photolysis, with detection of OH

$$OH + HCl \rightarrow H_2O + Cl \qquad\qquad (19)$$

by resonance fluorescence[24,110] or resonance absorption,[111] and two using discharge flow systems with e.s.r.[112] or resonance fluorescence detection,[113] agree with the value of $k_{19} = (4.0 \pm 0.4) \times 10^8$ dm^3 mol^{-1} s^{-1} obtained by Ravishankara *et al.*[24] Of the four studies in which temperature dependence was investigated over the range 200—500 K, three are in agreement on a small activation energy with values[24,110] of (3.92 ± 0.33) and (4.4 ± 0.2) kJ mol^{-1},[111] whereas the fourth found a much smaller value of (2.5 ± 0.1) kJ mol^{-1}.[112] There is no obvious explanation for the discrepancy except that it appears to be one of a number of instances where an anomalous temperature coefficient is obtained from a discharge flow study using

[109] L. F. Phillips, *Chem. Phys. Lett.*, 1978, **57**, 538.
[110] A. R. Ravishankara, P. H. Wine, and A. O. Langford, *Chem. Phys. Lett.* 1979, **63**, 479.
[111] I. W. M. Smith and R. Zellner, *J. Chem. Soc., Faraday Trans. 2*, 1974, **70**, 1045.
[112] W. Hack, G. Mex, and H. Gg. Wagner, *Ber Bunsenges. Phys. Chem.*, 1977, **81**, 677.
[113] M. S. Zahniser, F. Kaufman, and J. G. Anderson, *Chem. Phys. Lett.*, 1974, **27**, 507.

H + NO_2 as the OH source (*cf.* OH + H_2S). Smith and Zellner[111] obtained a rate constant expression of $(2.8 \pm {}^{1.4}_{0.9}) \times 10^9 \exp[-(780 \pm 40)K/T]$ dm^3 mol^{-1} s^{-1} for OH + DCl, indicating an activation energy for this reaction *ca.* 2 kJ mol^{-1} larger than that of reaction (19), but little of significance emerged from an attempt to interpret the isotope effect.

Two measurements of the rate constant of reaction (20), one using discharge-

$$OH + HBr \rightarrow H_2O + Br \tag{20}$$

flow–e.s.r.[114] and the other flash photolysis–absorptiometry[111] agree on a result of $k_{20} = (2.7 \pm 0.6) \times 10^9$ dm^3 mol^{-1} s^{-1} at ambient temperatures. However, Ravishankara *et al.*,[110] using flash photolysis–resonance fluorescence with the photolysis of O_3 in the presence of H_2O as the source of OH obtain, $k_{20} = (7.17 \pm 0.84) \times 10^9$ dm^3 mol^{-1} s^{-1}, independent of temperature in the range 249—416 K. In these experiments the pseudo-first-order decay of OH in the presence of excess of HBr was measured, and the important precaution was taken of measuring the HBr concentration *in situ* by u.v. absorptiometry. This proved necessary because of significant absorption of HBr on the glass walls of the reactor, which may explain the considerably larger value of k_{20} obtained in this work compared with those from other studies. Although the performance of the apparatus was tested by obtaining values of k_{19} in agreement with those from other studies, confirmatory independent measurements of k_{20} would be valuable.

As well as these studies on OH($v = 0$) there have also been some measurements of k_{20} for OH($v = 1$). Spencer and Glass[48] have obtained a rate constant value of $(2.7 \pm 0.8) \times 10^{10}$ dm^3 mol^{-1} s^{-1} for the reaction of OH($v = 1$) at 295 K, a significant enhancement whatever value is accepted for OH($v = 0$). This is a surprising result since the excitation is in the 'wrong' bond, *i.e.* not that which is broken in the reaction, and also an important result in view of the arguments outlined earlier on the relevance of vibrationally excited reactants to curved Arrhenius plots. It would be interesting to know the effect of vibrational excitation of HBr on the rate constant of reaction (20), which might be expected to be larger than that of OH, and there is also a need to determine a reliable temperature variation of k_{20} over a wider range before there can be any worth-while discussion of the possibility of curvature of the Arrhenius plot.

The only indication we have of the behaviour of k_{20} at high temperatures is the value of $k_{20} = 1.60 \times 10^{10}$ dm^3 mol^{-1} s^{-1} at 1925 K obtained from flame inhibition studies.[115] Compared to the low-temperature expression for k_{20}, obtained by Ravishankara *et al.*,[110] this represents enhancement of around a factor of 2 and may support a curved Arrhenius plot. The comparison is in reasonable agreement with the predictions made by Spencer and Glass[48] on the basis that k_{20} depends only on the enhanced reactivity of a thermal equilibrium population of OH($v = 1$).

The two determinations[111,114] of the rate constant of reaction (21) have been

$$OH + HI \rightarrow H_2O + I \tag{21}$$

made in parallel to those of HBr and agree within the wide error limits of $(5.4 \pm 2.4) \times 10^9$ dm^3 mol^{-1} s^{-1} at 298 K.

[114] G. A. Takacs and G. P. Glass, *J. Phys. Chem.*, 1973, **77**, 1948.
[115] W. E. Wilson, J. T. Donovan, and R. M. Fristrom, Proc. 12th Symp. (Internat.) Combust., 1969, p. 929.

Leu and Lin[116] have measured the pseudo-first-order rate of disappearance of OH in the presence of excess of Cl_2 or Cl_2O. For Cl_2 the rate constant at 298 K for the second-order process was obtained as $(3.3 \pm 0.2) \times 10^7$, while that for Cl_2O was $(3.9 \pm 0.3) \times 10^9$ dm^3 mol^{-1} s^{-1}, with no indication of the reaction channels.

Bimolecular Reactions with Inorganic Polyatomic Molecules.—The rate constant for reaction (22) of OH with O_3 now appears to be established with some precision.

$$OH + O_3 \rightarrow HO_2 + O_2 \tag{22}$$

Two recent studies[117,118] have used flash photolysis–resonance fluorescence methods to obtain results of $k_{22} = (1.09 \pm ^{0.21}_{0.17}) \times 10^9 \exp[-(930 \pm 50)K/T]$ over the temperature range 238—357 K and $(1.29 \pm 0.13) \times 10^9 \exp[-(969 \pm 40)K/T]$ dm^3 mol^{-1} s^{-1} over the temperature range 223—353 K. As pointed out by Chang and Kaufman,[46] the results from earlier discharge flow system studies where OH was produced by the H + NO_2 reaction, are affected by the recent upward revision,[119] by over an order of magnitude over previously accepted values, of the rate constant of reaction (23). Evidently in the presence of NO, if reaction (23) is

$$HO_2 + NO \rightarrow OH + NO_2 \tag{23}$$

sufficiently fast, the HO_2 product of reaction (22) will reform OH *via* (23) thus reducing the apparent decay rate. Taking this into account Chang and Kaufman[46] obtained $k_{22} \simeq 3.8 \times 10^7$ dm^3 mol^{-1} s^{-1} at 295 K as a best estimate, but rather lower than $k_{22} = (4.79 \pm 0.23) \times 10^7$ dm^3 mol^{-1} s^{-1} at 298 K measured by Ravishankara *et al.*[117] An interesting point made by these last-named workers was that care had to be taken to ensure that OH($v = 0$) was the only OH species produced in significant amounts on the reaction timescale. This was achieved by use of O(^1D) + H_2O as the source of OH, which is capable of generating OH($v \leqslant 3$) but the H_2O present induces very efficient vibrational deactivation. When, however, they attempted to use O(^1D) + H_2 as the source of OH, vibrational relaxation of OH($v > 0$) occurred on the same timescale as reaction (22) and the decay of OH($v = 0$) appeared to be significantly slower as a result. Addition of water vapour to the O_3–H_2 mixtures restored the faster observed decay of OH($v = 0$). There seems to be no obvious explanation for this anomalous effect but it is clear that care should be exercised when O(^1D) + H_2 is used as an hydroxyl radical source and only the OH($v = 0$) concentration is monitored.

There have been two studies in which the effects of vibrational excitation of the OH on the reaction rate have been measured. Streit and Johnston[120] generated OH($v \leqslant 9$) using the reaction of H atoms with excess of O_3 in a large (340 dm^3) steel tank, determining the ozone concentration by u.v. absorption, and the OH(v) populations from the vibration–rotation emission bands in the spectral range 550—850 nm. Losses of OH from the observation region were minimized by the use of low pumping speeds and the effects of the bimolecular disproportionation

[116] M. T. Leu and C. L. Lin, *Geophys, Res. Lett.*, 1979, **6**, 425.
[117] A. R. Ravishankara, P. H. Wine, and A. O. Langford, *J. Chem. Phys.*, 1979, **70**, 984.
[118] S. Fischer and D. D. Davis, cited in ref. 117.
[119] C. J. Howard and K. M. Evenson, *Geophys. Res. Lett.*, 1977, **4**, 437.
[120] G. E. Streit and H. W. Johnston, *J. Chem. Phys.*, 1976, **64**, 95.

reaction OH + OH and of self-absorption of emission were negligible with the low [OH] involved. It was found that the rotational temperature of the OH(v) was *ca.* 1500 K, independent of total pressure across the range 0.11—0.25 Pa. Coltharp *et al.*[121] carried out similar work but at a higher pressure (40 kPa). Other measurements[122] at those pressures have indicated rotational temperatures of *ca.* 550 K for the OH, which implies that rotational relaxation is probably significant in the experiments of Coltharp *et al.* Results from both studies, at ambient temperatures, are listed in Table 2. These values represent an enhancement by vibrational energy in OH compared to the mean value of $k_{22} = 4.8 \times 10^7$ dm^3 mol^{-1} s^{-1} at 298 K for OH($v = 0$) and presumably with a more equilibrated rotational temperature. Accepting the above sets of results, it is clear that the reaction rate is also enhanced by rotational excitation of OH, since Streit and Johnston's results are some 40% larger than those of Coltharp *et al.* determined at higher pressures. Hence even the second set cannot be compared properly with the above rate constant for OH($v = 0$), since they must still reflect some rotational enhancement. Moreover both sets show a discontinuity between $v = 5$ and $v = 6$, with a disproportionate increase evident in the rate constants between these two levels. It is

Table 2

v	2	3	4	5	6	7	8	9	Ref.
$10^{-9}k$/dm^3 mol^{-1} s^{-1}			2.2	2.7	4.3	5.1	5.4	6.6	120
$10^{-9}k$/dm^3 mol^{-1} s^{-1}	1.1	1.4	1.7	2.0	3.2	3.9	4.0	4.6	121

likely that this reflects the opening of a new reaction channel for $v \geqslant 6$. The basic reaction channel for all v is represented as (22) but other possible channels are

$$OH + O_3 \rightarrow HO_2 + O_2 \tag{22}$$

(22a) and (22b) which are endothermic for OH($v = 0$) by *ca.* 33 and *ca.* 107 kJ

$$OH + O_3 \rightarrow H + O_2 + O_2 \tag{22a}$$

$$OH + O_3 \rightarrow OH + O_2 + O \tag{22b}$$

mol^{-1} respectively. Streit and Johnston incorporated a factor into their kinetic analysis to take account of the production of secondary H atoms by reaction (22a), but the results yielded a very small value for this factor indicating an insignificant contribution. Reaction (22b) becomes feasible on energetic grounds for OH($v = 3$), but the foregoing tabulated results suggest that it makes its most significant impact only for OH($v \geqslant 6$), largely independent of the degree of rotational excitation. However, reaction (22b) could also lead to a secondary source of H atoms if it were followed by reaction (15) rather than (23). At ambient temperature k_{15} approaches the collision frequency $(2.7 \pm 10^{10}$ dm^3 mol^{-1} s^{-1}) while k_{23} is more than three orders of magnitude lower. Nevertheless, since Streit and Johnston specifically eliminated a secondary H atom source, it may be presumed that the [O$_3$]/[OH] ratio in their experiments was large enough to make reaction (23) much faster than reaction (15).

[121] R. N. Coltharp, S. D. Worley, and A. E. Potter, *Appl. Opt.*, 1971, **10**, 1786.
[122] R. E. Murphy, *J. Chem. Phys.*, 1971, **54**, 4852.

$$O + OH \rightarrow O_2 + H \qquad (15)$$

$$O + O_3 \rightarrow 2O_2 \qquad (23)$$

There have been comparatively few studies of the reaction of OH with H_2O_2 but the results obtained are in surprisingly good agreement. The reaction is taken to be (24). Hack *et al.*[123a] have used discharge flow–e.s.r. to obtain for the temperature

$$OH + H_2O_2 \rightarrow H_2O + HO_2 \qquad (24)$$

range 298—670 K the expression $k_{24} = (4.8 \pm 1.0) \times 10^9 \exp[-670 \pm 70)K/T]$, which gives $k_{24} = (5.1 \pm 1.1) \times 10^8$ dm^3 mol^{-1} s^{-1} at 298 K. Within the error limits this agrees with other 298 K values of $(4.1 \pm 0.8) \times 10^8$ obtained by flash photolysis–resonance fluorescence[123b] and $(3.6 \pm 0.6) \times 10^8$ dm^3 mol^{-1} s^{-1} obtained[124] by photolysis of H_2O_2 in the presence of CO. The latter value was based on measurement of CO_2 yields and hence gave k_{24} relative to the rate constant for the CO + OH reaction. For the latter a value of $(8.8 \pm 0.7) \times 10^7$ dm^3 mol^{-1} s^{-1} was used since the measurements were carried out at relatively low pressures (*ca.* 6.5 kPa).

The one other high temperature value is that of Atri *et al.*,[125] from relative yields of CO_2 produced from CO added to slowly reacting H_2–O_2–N_2 mixtures. The value at 773 K is *ca.* 1.5×10^9 dm^3 mol^{-1} s^{-1}, also agreeing well with the expression of Hack *et al.*

Three absolute determinations of the rate constant for reaction (25) have yielded

$$OH + NH_3 \rightarrow H_2O + NH_2 \qquad (25)$$

values of k_{25} at ambient temperature in substantial agreement. All used flash photolysis–resonance-fluorescence or absorptiometry. At ambient temperatures the values of $10^{-8} k_{25}$/dm^3 mol^{-1} s^{-1} are (0.98 ± 0.096),[126] (0.90 ± 0.24),[127] and 0.95.[128] These are also supported by a value of $k_{25} = (0.72 \pm 0.24) \times 10^8$ dm^3 mol^{-1} s^{-1} derived by Cox *et al.*[129] from the photolysis of nitrous acid in the presence of ammonia. However, two other values, one determined by flash photolysis–resonance fluorescence[19] and the other by a discharge flow–e.s.r. method[130] are respectively a factor of 4 lower and *ca.* 50% higher than the mean of the above values. The first method had been criticized by Smith and Zellner[128] on the grounds that photolysis of O_2 and subsequent reaction of $O(^1D)$ with NH_3 was used as the source of OH, in contrast to the H_2O photolysis used in the other flash photolysis studies. It is suggested that with the O_2 source, OH production by slower side reaction is possible on the reaction timescale, as indicated in an experiment by Stuhl.[127] The e.s.r. result may be yet another instance of the apparent unreliability of the discharge flow method for bimolecular reactions of OH. Two of the studies[126,128]

[123] (a) W. Hack, K. Hoyermann, and H. Gg. Wagner, *Int. J. Chem. Kinet. Symp. 1*, 1975, 329; (b) G. W. Harris and J. N. Pitts, jun., *J. Chem. Phys.*, 1970, **70**, 2881.

[124] J. F. Meagher and J. Heicklen, *J. Photochem.*, 1974/5, **3**, 455.

[125] G. M. Atri, R. R. Baldwin, D. Jackson, and R. W. Walker, *Combust. Flame*, 1977, **30**, 1.

[126] R. A. Perry, R. Atkinson, and J. N. Pitts, jun., *J. Chem. Phys.*, 1976, **64**, 3237.

[127] F. Stuhl, *J. Chem. Phys.*, 1973, **59**, 635.

[128] I. W. M. Smith and R. Zellner, *Int. J. Chem. Kinet., Symp. 1*, 1975, 341.

[129] R. A. Cox, R. G. Derwent, and P. M. Holt, *Chemosphere*, 1975, **4**, 201.

[130] W. Hack, K. Hoyermann, and H. Gg. Wagner, *Ber. Bunsenges. Phys. Chem.*, 1974, **78**, 386.

agree on both activation energy and pre-exponential factor, for the former quoting values of 6.7[126] and 7.3 kJ mol^{-1}.[128] The pre-exponential factors average 1.6×10^9 dm^3 mol^{-1} s^{-1}.

On the question of whether this reaction might have a curved Arrhenius plot (like OH + CH$_4$), an upper limiting value of $k_{25} \leqslant 1.0 \times 10^9$ dm^3 mol^{-1} s^{-1} has been derived at 1923 K by Dove and Nip[131] from a shock tube study of the kinetics in shocked H$_2$–O$_2$–inert gas mixtures containing a trace of ammonia. On extrapolation to 1923 K the two low temperature Arrhenius expressions (using the mean value pre-exponential factor) predict $k_{25} = 9.1 \times 10^8$ dm^3 mol^{-1} s^{-1} [128] and 1.1×10^9 dm^3 mol^{-1} s^{-1},[126] which appears to preclude significant curvature of this Arrhenius plot. Confirmation of both this conclusion and the apparently distinct curvature of the Arrhenius plot for OH + CH$_4$ will present a considerable puzzle as to what type of reaction dynamics can produce such a difference between two reactions with similar energetics and not widely different Arrhenius parameters in the low temperature range. Accurate determination of k_{25}/k_{17} at higher temperatures would be valuable in this connection.

Hydroxyl radicals react with hydrazine in a primary process which has been postulated on the basis of crossed beam–mass spectrometric experiments[132a] to be

$$OH + N_2H_4 \rightarrow NH_3 + NH_2O$$

In a discharge flow–e.s.r. study[130] it was found that 2.5 OH radicals were removed per N$_2$H$_4$ molecule, leading to N$_2$ and H$_2$O as the main products, which seems inconsistent with the postulated breaking of the N–N bond in the primary step. More likely therefore is a primary step abstracting hydrogen

$$OH + N_2H_4 \rightarrow H_2O + N_2H_3$$

The primary-step rate constant has been deduced by combining this stoicheiometry with the observed decay rate of OH in the presence of excess of N$_2$H$_4$ as $(1.3 \pm 0.3) \times 10^{10}$ dm^3 mol^{-1} s^{-1} at 298 K, but this value must be used with caution since as noted elsewhere k_{25} (for OH + NH$_3$) determined in the same work appeared to be too large compared with other measurements of k_{25}, and the value of $(3.7 \pm 0.6) \times 10^{10}$ dm^3 mol^{-1} s^{-1}, recently obtained by Harris *et al.*,[132b] using flash photolysis–resonance fluorescence, is to be preferred.

The reaction of OH radicals with nitric acid vapour is significant for stratospheric chemistry, where the substrate is a sink species for nitrogen oxides. Accordingly the main kinetic interest has been in the low temperature range. There is good

$$OH + HNO_3 \rightarrow H_2O + NO_3 \qquad (26)$$

agreement between the available rate constant data; k_{26} is not significantly temperature dependent in the range 240—470 K, a value of $k_{26} = (4.8 \pm 0.6) \times 10^7$ dm^3 mol^{-1} s^{-1} being obtained in a flash photolysis–absorptiometric study[128] and $k_{26} = (5.4 \pm 0.8) \times 10^7$ dm^3 mol^{-1} s^{-1} by discharge flow–resonance fluorescence.[133]

[131] J. E. Dove and W. S. Nip, *Can. J. Chem.*, 1974, **52**, 1171.
[132] (*a*) M. Gehring, K. Hoyermann, H. Gg. Wagner, and J. Wolfrum, *Z. Naturforsch., Teil A*, 1970, **25**, 675; (*b*) G. W. Harris, R. Atkinson, and J. N. Pitts, jun., *J. Phys. Chem.*, 1979, **83**, 2557.
[133] J. J. Margitan, F. Kaufman, and J. G. Anderson, *Int. J. Chem. Kinet., Symp. 1*, 1975, 281.

This is a case where flash photolysis and discharge flow studies have produced excellent agreement.

Another potentially important stratospheric species is chlorine nitrate ($ClONO_2$) and the rate constant for its reaction with OH has been given as $(7.17 \pm 0.06) \times 10^8$ $\exp[-(333 \pm 22)K/T]$ dm^3 mol^{-1} s^{-1} for 246—387 K on the basis of a discharge flow study[134] where pseudo-first-order decays of OH were measured by resonance fluorescence. A value of $(2.2 \pm 0.1) \times 10^8$ dm^3 mol^{-1} s^{-1} at 245 K obtained by flash photolysis–resonance fluorescence[135] is in reasonable agreement, but *ca.* 20% higher than the mean value calculated from the Arrhenius expression. There is no evidence on the reaction channel involved, with the analogue of reaction (26) (Cl transfer to OH):

$$OH + ClONO_2 \rightarrow HOCl + NO_3$$

or release of NO_2 by ClO transfer to OH:

$$OH + ClONO_2 \rightarrow HOClO + NO_2$$

both possibilities being endothermic and unlikely to have a substantial energy barrier in the reaction co-ordinate.[134] This rate constant is too small for effective competition of this reaction with photo-dissociation of $ClONO_2$ in the stratosphere.[134]

Two other nitrogen-containing species whose rate constants for reaction with OH have been measured are N_2O and C_2N_2. In both cases flash photolysis–resonance fluorescence was used by Atkinson *et al.*[136] At ambient temperatures upper limits of 1.2×10^5 with N_2O and 3×10^7 dm^3 mol^{-1} s^{-1} for C_2N_2 were obtained. The reaction of C_2N_2 is thought to be thermoneutral.

$$OH + C_2N_2 \rightarrow HOCN + CN$$

There have been several measurements of rates of reactions with the sulphur-containing triatomic species H_2S, CS_2, and COS because of their significance in atmospheric chemistry. Reaction (27) with H_2S is very rapid. Two

$$OH + H_2S \rightarrow H_2O + HS \tag{27}$$

studies in which the pseudo-first-order removal of OH in an excess of H_2S was followed (one using discharge flow–e.s.r.[137] and the other by flash photolysis–resonance fluorescence[126]) have produced values at 298 K of $k_{27} = (3.3 \pm 0.2) \times 10^9$ and $(3.16 \pm 0.32) \times 10^9$ dm^3 mol^{-1} s^{-1}, in evident good agreement. However, the results disagree at higher temperatures with Perry *et al.*[126] finding no significant temperature coefficient across the range 297—424 K, whereas Westenberg and de Haas[137] found an increase by a factor of 2.7 across the range 298—885 K, corresponding to an activation energy of 3.7 kJ mol^{-1}. The technique used by Perry *et al.*[126] has proved reliable for a wide variety of reactions, whereas the discharge flow technique suffers from some difficulties, particularly with regard to

[134] M. S. Zahniser, J. S. Chang, and F. Kaufman, *J. Chem. Phys.*, 1977, **67**, 997.
[135] A. R. Ravishankara, D. D. Davis, G. Smith, G. Tesi, and J. Spencer, *Geophys. Res. Lett.*, 1977, **4**, 7.
[136] R. Atkinson, R. A. Perry, and J. N. Pitts, jun., *Chem. Phys. Lett.*, 1976, *44*, 204; *Combust. Flame*, 1978, **31**, 213.
[137] A. A. Westenberg and N. de Haas, *J. Chem. Phys.*, 1973, **59**, 6685.

heterogeneous reactions. Unfortunately the position is not resolved by the earlier flash photolysis–resonance fluorescence study by Stuhl[138] which produced $k_{23} = (1.9 \pm 0.3) \times 10^9$ dm^3 mol^{-1} s^{-1} at 298 K, in disagreement with both results.

Atkinson *et al.*[139] have applied flash photolysis–resonance fluorescence to the OH + CS$_2$ and OH + COS reactions in the range 299—430 K. However, it was found that the OH decay rates at constant substrate concentration were dependent on the flash energy which suggested photodissociation of the substrate in the flash and reaction of OH with the photoproducts. Only upper limits could be given for reactions (28) and (29). Kurylo[140] has attempted to overcome these problems by

$$OH + OCS \rightarrow CO + SOH \text{ or } CO_2 + SH \tag{28}$$

$$OH + CS_2 \rightarrow CS + SOH \text{ or } OCS + SH \tag{29}$$

using low flash intensities, a flow system, and by restricting the photolysis wavelengths to $\lambda \geqslant 165$ nm, where substrate absorption coefficients compared to those of water vapour are less than those at $\lambda < 165$ nm. As a result, well-behaved decays of OH in the flash photolysis–resonance fluorescence system were observed, and at 296 K values of $k_{28} = (3.41 \pm 0.73) \times 10^7$ and $k_{29} = (1.11 \pm 0.20) \times 10^8$ dm^3 mol^{-1} s^{-1} were obtained. Points needing further clarification are the identification of the reaction products and whether the reactions proceed initially by addition rather than abstraction[109] (*cf.* OH + HCN). The pressure dependences if any, and the temperature coefficients, are also of importance in considering the involvement of OCS and CS$_2$ in stratospheric sulphur chemistry.

Bimolecular Reactions with Radicals.—Over the years there have been many kinetic studies of the disproportionation reaction (18). All the measurements have

$$OH + OH \rightarrow H_2O + O \tag{18}$$

been either at close to ambient temperatures or in the high temperature regime > 1000 K. For the latter region values of the rate constant were calculated by combination of the rate constant for the reverse reaction with the equilibrium constant. As pointed out by Zellner[100] these data are not adequate to decide whether the reaction has a curved Arrhenius plot, which is a distinct possibility if vibrational excitation has any influence on the rate constant (*cf.* CO + OH). Although no definitive study of this has been attempted a strong indication of the possibility has been furnished by Smith's measurement,[141] in a discharge flow system, of the ratio of rate constants k_{30}/k_{18}. Formaldehyde was added to a

$$OH + HCHO \rightarrow H_2O + CHO \tag{30}$$

concentration approximately equal to that of OH and the concentrations of products measured mass spectrometrically. A computer matching of concentration profiles was used to extract values of k_{18} using a calibration value of k_{30} at temperatures of 268, 298, and 334 K. If the activation energy E_{18} of reaction (18)

[138] F. Stuhl, *Ber Bunsenges. Phys. Chem.*, 1974, **78**, 230.
[139] R. Atkinson, R. A. Perry, and J. N. Pitts, jun., *Chem. Phys. Lett.*, 1978, **54**, 14.
[140] M. J. Kurylo, *Chem. Phys. Lett.*, 1978, **58**, 238.
[141] R. H. Smith, *Int. J. Kinet.*, 1978, **10**, 519.

was given the value of 4 kJ mol^{-1} required to produce a linear Arrhenius plot of k_{18}, then that, E_{30}, of reaction (30) came out as 6 ± 2 kJ mol^{-1}. On the other hand when E_{18} was made zero in this temperature range, then E_{30} was 3 ± 2 kJ mol^{-1}. Atkinson and Pitts[142] have used a flash photolysis–resonance fluorescence to obtain an absolute expression $k_{30} = 7.52 \times 10^9 \exp[-(88 \pm 150)K/T]$ dm^3 mol^{-1} s^{-1} over the temperature range 299—426 K, which is clearly inconsistent with $E_{30} = 6 \pm 2$ kJ mol^{-1}, and in combination with Smith's results strongly indicates $E_{18} \simeq 0$. This supports Zellner's proposal that k_{18} could be given by an empirical expression similar to that adopted for k_1 *viz* $k_{18} = \exp[20.2 + 1.5 \times 10^{-3} T/K]$ dm^3 mol^{-1} s^{-1}. The absolute measurements of k_{18} in the temperature range 250—400 K, which are in progress,[100] are awaited with interest.

The related reaction between OH and ClO radicals has two channels (31*a*) and (31*b*) open to it which, owing to the weaker bond in ClO compared with that in OH,

$$\text{OH} + \text{ClO} \rightarrow \text{HO}_2 + \text{Cl} \ (\Delta H^{\ominus}_{298} = -18 \text{ kJ mol}^{-1}) \tag{31a}$$

$$\text{OH} + \text{ClO} \rightarrow \text{HCl} + \text{O}_2 \ (\Delta H^{\ominus}_{298} = -234 \text{ kJ mol}^{-1}) \tag{31b}$$

are both exothermic. However, the channel (31*c*), directly analogous to reaction

$$\text{OH} + \text{ClO} \rightarrow \text{HOCl} + \text{O} \tag{31c}$$

(18), is endothermic and therefore unlikely. In a discharge flow study, Leu and Lin[116] have mixed OH produced from H + NO$_2$, with ClO radicals, generated by the quantitative titration of Cl atoms with O$_3$ or Cl$_2$O, and used resonance fluorescence to measure the decay rate of OH under pseudo-first-order conditions. The overall rate constant k_{31} obtained was $(5.5 \pm 0.8) \times 10^9$ dm^3 mol^{-1} s^{-1} at 298 K, with the absence of effects due to vibrationally excited reactions shown by the unchanged value of k_{31} on addition of SF$_6$, an efficient vibrational quencher. The formation of HO$_2$ as a product was demonstrated by addition of NO which induced the rapid conversion of HO$_2$ into OH through the reaction:

$$\text{HO}_2 + \text{NO} \rightarrow \text{NO}_2 + \text{OH}$$

These experiments indicated an HO$_2$ yield corresponding to reaction (31*a*) being at least 65% of the total reaction (31), the uncertainty arising from the difficulty of allowing quantitatively for side-reactions also occurring in this system. A future improvement in detection sensitivity in this system should allow measurement of the yield of HCl from channel (31*b*), an important measurement since this step represents chain termination for ozone destruction cycles in the stratosphere. This is also a reaction where it is clear that a channel of lower exothermicity involving simple atom transfer is favoured over a much more exothermic channel involving a four-centre transition state.

The important stratospheric reaction (32) between OH and HO$_2$ radicals (acting as a sink of HO$_x$ species) has attracted recent interest and measurement of this rate

$$\text{OH} + \text{HO}_2 \rightarrow \text{H}_2\text{O} + \text{O}_2 \tag{32}$$

[142] R. Atkinson and J. N. Pitts, jun., *J. Chem. Phys.*, 1978, **68**, 3581.

constant has been made possible by the development of laser magnetic resonance (l.m.r.) as the only existing method of measuring concentrations of HO_2 in discharge flow systems.[143] Burrows, Harris, and Thrush[144] have used the reaction of OH radicals with small additions of H_2O_2 vapour as a source of HO_2 *via* reaction (24), while $H + NO_2$ were used as the source of OH. The HO_2 radicals are

$$OH + H_2O_2 \rightarrow H_2O + HO_2 \qquad (24)$$

effectively maintained in a *quasi*-stationary state in this system since k_{24} is $\ll k_{32}$. The measured parameter was the ratio $[HO_2]/[OH]$ from the respective l.m.r. signals and the OH signal was calibrated using $H + NO_2$ to generate a known [OH]. The resultant $[HO_2]$ yields k_{32}/k_{24} and using $k_{24} = 4.8 \times 10^8$, these workers obtained $k_{32} = (3.1 \pm 1.0) \times 10^{10}$ dm^3 mol^{-1} s^{-1}. Hack *et al.*[47] have generated OH and HO_2 radicals together using the fast reactions

$$F + H_2O \rightarrow OH + HF$$

$$F + H_2O_2 \rightarrow HF + HO_2$$

with reaction (32) taking place to create a *quasi*-stationary-state HO_2 concentration. Absolute calibrations of the l.m.r. signals with respect to [OH] and $[HO_2]$ were made using $H + NO_2$ as calibration for [OH] and the fast interconversion reaction

$$HO_2 + NO \rightarrow OH + NO_2$$

to calibrate $[HO_2]$. From the measured [OH] and $[HO_2]$ and their decay profiles down the flowtube, $k_{32} = (1.8 \pm 0.6) \times 10^{10}$ dm^3 mol^{-1} s^{-1} at 293 K was obtained. These two results are indistinguishable within the stated error limits.

Chang and Kaufman[46] have added excess O_3 to OH in a discharge flow system, inducing the reaction

$$OH + O_3 \rightarrow HO_2 + O_2$$

which has a rate constant *ca.* 4×10^7 dm^3 mol^{-1} s^{-1} and is the only reaction forming HO_2 in this system; [OH] was measured by resonance fluorescence and this measurement with and without addition of excess of NO (to convert HO_2 into OH) just upstream of the detector yielded [OH] and $[OH] + [HO_2]$. This system is, however, complicated by the fact that twelve reactions have significant rates; a computer-matching procedure was therefore adopted for the concentration profiles and this depended upon several rate constant values derived from the literature. The best agreement was obtained for k_{32} in the range 1—2×10^{10} with an upper limit of 3×10^{10} dm^3 mol^{-1} s^{-1}. These conclusions are in satisfactory agreement with the absolute values.

Demore[145] has used the photolysis of H_2–N_2–O_2–CO mixtures at a total pressure of 1 atmosphere, with light of wavelength 184.9 nm, as a means of studying reaction (32). Measurement of the steady state concentration of ozone in the absence and presence of a small partial pressure of CO allowed evaluation of k_{32} as a ratio to a

[143] H. E. Radford, K. M. Evenson, and C. J. Howard, *J. Chem. Phys.*, 1974, **60**, 3178.
[144] J. P. Burrows, G. W. Harris, and B. A. Thrush, *Nature (London)*, 1977, **267**, 233.
[145] W. Demore, *J. Phys. Chem.*, 1979, **83**, 1113.

combination of other well established rate constants. At ambient temperature a value of k_{32} in excess of 6×10^{10} dm^3 mol^{-1} s^{-1} was obtained, which agreed with the order of magnitude of the value of k_{32} required to account for HO$_2$ decays in the flash photolysis of H$_2$O–O$_2$–Ar mixtures at atmospheric pressure,[146] the only other high pressure study. The fact that these lower limits are at least a factor of two larger than the lower pressure values seems to indicate that reaction (32) may have a pressure-dependent component, perhaps akin to that of reaction (1), and further experimental work over a pressure range would be worthwhile to check this possibility.

There have been two determinations of the rate constant at ambient temperatures of reaction (33) between SO and OH. Fair and Thrush[147] measured the ratio k_{15}/k_{33} = 0.40 ± 0.07 in a discharge flow study. In combination with the most precise value

$$SO + OH \rightarrow SO_2 + H \qquad (33)$$

of k_{15}, this yields $k_{33} = (6.6 \pm {}^{3.0}_{2.0}) \times 10^{10}$ dm^3 mol^{-1} s^{-1}. The absolute determination by Jourdain *et al.*,[68] using discharge flow–e.s.r., where SO was generated by the reaction of S atoms with O$_2$, yielded $k_{33} = (5.1 \pm 0.9) \times 10^{10}$ dm^3 mol^{-1} s^{-1}, in good agreement with the value of Fair and Thrush and substantially more precise.

Three-body Combination Reactions.—Because of their importance in atmospheric chemistry, the combination reactions of OH with NO, NO$_2$, and SO$_2$ are of major current interest. All show a transition from third- to second-order kinetics with increasing pressure across the pressure range of interest in the atmosphere.

Reaction (34) has been studied directly in its third-order regime (total pressure

$$OH + NO + M \rightarrow HONO + M \qquad (34)$$

below 7 kPa) using discharge flow techniques[45,69,148,152] and an entry to this regime can also be made using vacuum u.v. violet flash photolysis.[16,85,149,150] However, direct access to the limiting second-order region, which lies well above standard pressures for most gases, has not been achieved as yet. Although results for the limiting third-order rate constant k_{34} at ambient temperatures show disagreement, which cannot be accounted for within quoted error limits, the majority view[16,45,69,85] is that it lies in the range $(1.2—1.6) \times 10^{-11}$ dm^6 mol^{-2} s^{-1} for M = Ar, with N$_2$ being 1.5—2 times as efficient as a third body, and He slightly less efficient than Ar. With this level of uncertainty in k_{34}, which can be measured by direct means, it is hardly surprising that a much wider level of disagreement (values differing by an order of magnitude) exists for the value of the high-pressure second-order rate constant estimated by extrapolation of lower pressure results. The inaccuracy of the extrapolation is exacerbated by the different approaches adopted.

Overend *et al.*[150] found that for M = N$_2$, SF$_6$, CF$_4$, and H$_2$O, the combination of a simple energy transfer mechanism, represented by

[146] C. J. Hochanadel, J. A. Ghormley, and P. J. Ogren, *J. Chem. Phys.*, 1972, **56**, 4426.
[147] R. W. Fair and B. A. Thrush, *Trans. Faraday Soc.*, 1969, **65**, 1557.
[148] G. W. Harris and R. P. Wayne, *J. Chem. Soc., Faraday Trans. 1*, 1975, **71**, 610.
[149] F. Stuhl and H. Niki, *J. Chem. Phys.*, 1972, **57**, 3677.
[150] R. Overend, G. Paraskevopoulos, and C. Black, *J. Chem. Phys.*, 1976, **64**, 4149.

$$OH + NO \rightleftharpoons HONO^*$$

$$HONO^* + M \rightarrow HONO + M$$

with simple Lindemann theory for the decomposition of the transient complex (HONO*) produced agreement with their experimental results, which, it must be noted, are considerably lower than most of the values of k_{34} for M = He, Ar, and N_2 found in the other studies. On the assumption that M = H_2O possessed a collision efficiency of unity for de-energization of HONO* the other M had collisional efficiencies of 0.41 (CF_4), 0.37 (SF_6), 0.12 (N_2), 0.5 (Ar), and 0.02 (He), although the pressure variation of k_{34} with M = He could not be matched with either Lindemann or RRKM unimolecular rate theory. Anastasi and Smith[16] have performed the most extensive study of the kinetics of reaction (34), determining the rate constant in the temperature range 233—505 K for M = He, Ar, and N_2 and varying [M] over close to two orders of magnitude. They found that the pressure dependence of their experimental rate constants could be fitted by an equation based on an empirical procedure developed by Troe[151] in connection with unimolecular dissociation rate theory. The temperature dependence[16] of the third-order k_{34} for M = N_2 corresponded to a median value of the activation energy of −6.6 kJ mol⁻¹, which compared favourably with that (−7.1 kJ mol⁻¹) found[45] from a discharge flow study (M = He), and the k_{34} values at 296 K were in close agreement despite the very different techniques. An earlier discharge flow study[152] yielded an activation energy of −9.2 kJ mol⁻¹ for k_{34} (M = He), but the ambient temperature value appeared anomalously high. Even the sophisticated approach to evaluation of the high-pressure limiting rate constant adopted by Anastasi and Smith[16] seemed to yield little definite information on its temperature coefficient.

The approaches to the kinetics of reaction (10) present a similar pattern to those

$$OH + NO_2 + M \rightarrow HNO_3 + M \qquad (10)$$

to reaction (34). The low-pressure third-order rate constant k_{10} has been determined in discharge flow systems[45, 69, 86, 148, 152, 153] and in a flash photolysis study.[154] At ambient temperatures the agreement is good with a majority in favour of k_{10} (M = He) in the range (3.3—4.0) × 10¹¹ dm³ mol⁻¹ s⁻¹. The third-body efficiencies for He:Ar:N_2:CO_2 are roughly 1:1:2:4.[153, 154]

As for reaction (34), there has been no measurement of k_{10} at sufficiently high pressures for it to be in its limiting second-order region, but there have been two studies of the pressure dependence of k_{10} which allow extrapolation to both the third- and the second-order limits. Atkinson *et al.*[154] used flash photolysis–resonance fluorescence to measure k_{10} across the pressure range 3—86 kPa for M = Ar at 298 K. At the lowest pressures the kinetics are third-order and yield k_{10} in the middle of the range quoted (3.70 × 10¹¹ dm⁶ mol⁻² s⁻¹), with strong curvature towards the second-order limit evident at the highest pressure. However, extrapolation towards the limiting high pressure value of the second-order rate constant by the simple procedure adopted must leave some uncertainty. There is

[151] J. Troe, *Ber. Bunsenges. Phys. Chem.*, 1974, **78**, 478.
[152] A. A. Westenberg and N. de Haas, *J. Chem. Phys.*, 1972, **57**, 5375.
[153] K. Erler, D. Field, R. Zellner, and I. W. M. Smith, *Ber. Bunsenges. Phys. Chem.*, 1977, **81**, 22.
[154] R. Atkinson, R. A. Perry, and J. N. Pitts, jun. *J. Chem. Phys.*, 1976, **65**, 306.

evident scope for a study conducted at the apparently moderate pressures required to ensure approximation of this rate constant to its second-order limit. Anastasi and Smith,[155] using flash photolysis–resonance absorption, have measured values of the rate constant across the temperature range 222—550 K and pressure range (M = N_2) of 1.2—60 kPa. By use of a fitting and extrapolation procedure similar to that applied to reaction (34), they were able to extrapolate to obtain a third-order k_{10} value for M = N_2 of 9.4×10^{11} dm^6 mol^{-2} s^{-1} at 296 K, and to estimate third-body efficiencies relative to He of 1.2, 2.0, 2.9, and 7.4 for Ar, O_2, N_2, and SF_6. The temperature coefficient of k_{10} corresponded to an activation energy of -6.8 kJ mol^{-1} (M = He, N_2) in good agreement with the direct measurements of -9.2,[152] -7.5,[45] and -7.1 kJ mol^{-1}.[153] However, the value of the rate constant at the high-pressure limit, obtained by extrapolation, appeared to show a rather sharp decrease with increasing temperature compared with expectations based on RRKM theory, and more results are desirable to clarify this feature of the work.

There is even less certainty about the kinetics of reaction (35). The absolute study

$$OH + SO_2 + M \rightarrow HSO_3 + M \qquad (35)$$

conducted by Atkinson *et al.*[154] using flash photolysis and pressures of 3—86 kPa of Ar at 298 K provides the only definitive work. The third-order regime was approached (but not apparently attained) at the lowest pressures and a rate constant of $k_{35} = (5.9 \pm 1.2) \times 10^{10}$ dm^6 mol^{-2} s^{-1} for M = Ar may only represent a lower limit. Harris and Wayne[148] determined $(1.6 \pm 0.5) \times 10^{11}$ dm^6 mol^{-2} s^{-1} in a discharge flow study with M = Ar, but this is likely to be too high since their values of k_{10} and k_{34} in the same study appear to be too large compared to other values. Castleman and Tang[156] have used steady-state photolysis of water vapour in a static system with M = N_2 at 297 K where minor additions of SO_2 and CO induced competition between reaction (35) and reaction (1) for the OH produced: the yields

$$CO + OH \rightarrow CO_2 + H \qquad (1)$$

of CO_2 in the presence and absence of SO_2 allowed evaluation of the ratio k_{35}/k_1. Unfortunately no account seems to have been taken of the potential pressure dependence of k_1 and there are no indications that precautions were taken to ensure the absence of O_2 or other impurities which might have affected k_1. If it is presumed that reaction (1) showed its full pressure dependence, corrected values of k_{35} (M = N_2) can be calculated using the values of k_1 derived from Biermann, Zetsch, and Stuhl's work.[96] Since Castleman and Tang[156] applied the low-pressure value of k_1 across the entire pressure range (2.7—133 kPa), this procedure exerts little influence on the values of k_{35} at the lowest pressures but increases those at the highest pressures by a factor approaching 2. As a result, reaction (35) shows kinetics less close to the second-order limiting condition at the highest pressures than is indicated by their stated values of the rate constant. Further a plot of the reciprocal of the revised values of k_{35} *vs.* pressure shows a curvature for M = N_2 perhaps more in keeping with that exhibited by the results of Atkinson *et al.*[154] for M = Ar than the strongly linear plot given by Castleman and Tang. Of

[155] C. Anastasi and I. W. M. Smith, *J. Chem. Soc., Faraday Trans. 2*, 1976, **72**, 1459.
[156] A. W. Castleman and I. N. Tang, *J. Photochem.*, 1976/7, **6**, 349.

course there is always the possibility that in Castleman and Tang's system the impurity levels were low enough to allow correct application of a pressure-independent value of k_1. However, the doubt surrounding these results means that only the value of $k_{35} = 5.1 \times 10^{10}$ dm^6 mol^{-2} s^{-1} at the lowest pressures of N$_2$ can be taken as valid; this is close to the absolute value of k_{35} for the same pressure of Ar at 298 K determined by Atkinson *et al.*,[154] but even this is anomalous since N$_2$ is usually a more efficient third body than Ar [as for reactions (10) and (34)].

Trainor and von Rosenberg[157] have measured the rate constant of the three-body combination reaction (36) in a flash photolysis system using pressures up to 53 kPa,

$$OH + OH + N_2 \rightarrow H_2O_2 + N_2 \tag{36}$$

under which conditions reaction (36) was approximately equal in rate to that of the bimolecular disproportionation step

$$OH + OH \rightarrow H_2O + O$$

Their result [$k_{36} = (9.1 \pm 1.1) \times 10^{10}$ dm^6 mol^{-2} s^{-1}] at room temperature was in good agreement with the theoretical rate constant estimated from application of phase space theory.[158] However, no consideration was given to the possibility that reaction (36), [*cf.* reaction (34)] might have reached a transitional region between third- and second-order kinetics at the highest pressures used. Although the scatter of data points is substantial, some tendency in this direction can be discerned.

4 Reactions with Organic Species

The kinetics and mechanisms of the reactions of the hydroxyl radical with organic compounds in the gas phase have very recently been reviewed.[4]

Since the impetus for the article comes from the importance of such reactions in tropospheric chemistry the review is largely restricted to data at ambient temperatures. The proceedings[159] of a 'workshop' on 'Chemical Kinetic Data Needs for Modelling the Lower Troposphere,' covers much the same ground. Because of the existence of these surveys we concentrate here on work published more recently and the few high temperature studies, but for completeness we also outline the main features of hydroxyl radical reactions with different types of organic compound, referring the reader to the more specialized reviews for further details. The low-temperature rate data published before 1979 have been tabulated in the review by Atkinson *et al.*,[4] and that by Drysdale and Lloyd[1] tabulates most of the high temperature results up to 1970.

Alkanes.—The reactions of hydroxyl radicals are unambiguously hydrogen-atom abstraction and are substantially exothermic: *e.g.* for abstraction from CH$_4$, $\Delta H^\ominus$ (298 K) is -60.4 kJ mol^{-1}. Data are available for a sufficient number of alkanes for temperatures in the range 300—700 K to be able to relate the overall rate constant

[157] D. W. Trainor and C. W. von Rosenberg, *J. Chem. Phys.*, 1974, **61**, 1010.
[158] J. Keck and A. Kalelkar, *J. Chem. Phys.*, 1968, **49**, 3211.
[159] 'Chemical Kinetic Data Needs for Modelling the Lower Troposphere', ed. J. T. Herron, R. E. Huie, and J. A. Hodgeson, N.B.S. Special Publ. 557, 1979.

to the number and type of hydrogen atoms being abstracted (see later). However, only methane and ethane have been studied at temperatures > 1000 K and for the latter the data are meagre.

Aspects of reaction (17) of OH with CH_4 were discussed earlier in the context of

$$OH + CH_4 \rightarrow H_2O + CH_3 \qquad (17)$$

curved Arrhenius plots. It is also a reaction of importance in combustion and atmospheric chemistry and continues to receive regular attention from the experimentalists. At temperatures close to ambient there is remarkably good agreement between measurements of the rate constant by discharge flow–e.s.r.,[160] l.m.r.,[161] resonance fluorescence,[162] flash photolysis–absorptiometry[6, 80, 163] and –fluorescence,[84] and by photolysis and product analysis,[129, 164] giving values at 298 K in the range $(3.9—6.5) \times 10^6$ dm^3 mol^{-1} s^{-1}. In several of these studies measurements have been extended up to temperatures of *ca.* 500 K, and here too the agreement is good and in accord with other measurements in this region[11] and clearly shows that an extrapolation of these results to temperatures > 1000 K, on the basis of a linear Arrhenius plot, yields values for the rate constant very substantially smaller than the measured values at these high temperatures.

Zellner and Steinert[6] have made the most determined effort to bridge the gap between the low- and high-temperature values. By extending the normal working range of their flash photolysis–resonance absorption technique they have obtained data in the range 300—900 K. An Arrhenius plot of the data to which the expression $k_{17} = 3.45 \; T^{3.08} \exp(-1010 \; K/T)$ dm^3 mol^{-1} s^{-1} was fitted shows distinct curvature. The data points of Baldwin *et al.*[165] at 773 K, obtained from the addition of CH_4 to slowly reacting mixtures of H_2 and O_2, and the relative rate data of Baulch *et al.* (443—663 K),[166] are also in agreement with this expression, but its extrapolation to temperatures > 1000 K gives rate constant values which are higher than the most reliable values in that region and for this reason it has been recently reassessed by Ernst *et al.*[15]

At temperatures > 1000 K there are results from flame studies at 1285 K[167] and 1090—1857 K,[136] a conventional shock tube study with H_2O_2 as the OH source at 1300 K,[56] and a combined flash photolysis–shock heating experiment at 1300 K.[15] These agree well enough to define the rate constant in this region to within a factor of 2. The earlier values from flame studies[169] appear to be unacceptably high. The experiment in which flash photolysis of H_2O to produce OH was followed rapidly by shock heating of the system is novel and of considerable value since it allowed the relatively 'clean' production of high concentrations of OH radicals which is

[160] W. E. Wilson, jun., and A. A. Westenberg, Proc. 11th Symp. (Internat.) Combust., 1967, p. 1143.

[161] C. J. Howard and K. M. Evenson, *J. Chem. Phys.*, 1976, **64**, 197.

[162] J. T. Margitan, F. Kaufman, and J. G. Anderson, *Geophys. Res. Lett.*, 1974, **1**, 80.

[163] N. R. Greiner, *J. Chem. Phys.*, 1970, **53**, 1070.

[164] R. A. Cox, R. G. Derwent, P. M. Holt, and J. A. Kerr, *J. Chem. Soc., Faraday Trans, 1*, **72**, 2044.

[165] R. R. Baldwin, D. E. Hopkins, A. C. Norris, and R. W. Walker, *Combust. Flame*, 1970, **15**, 33.

[166] D. L. Baulch, D. Din, D. D. Drysdale, and D. J. Richardson, Proc. 1st European Symp. Combust., 1973, p. 29.

[167] G. Dixon-Lewis and A. Williams, Proc. 11th Symp. (Internat.) Combust., 1967, p. 951.

[168] J. Peeters and G. Mahnen, Proc. 14th Symp. (Internat.) Combust., 1973, p. 133.

[169] R. M. Fristrom, Proc. 9th Symp. (Interat.) Combust., 1963, p. 560; A. A. Westenberg and R. M. Fristrom, *J. Phys. Chem.*, 1961, **65**, 591; C. P. Fenimore and G. W. Jones, *ibid.*, 1961, **65**, 2200.

normally difficult under these conditions. To accommodate the high temperature values as well as those at lower temperatures Ernst et al.[15] modified Zellner's previous expression for k_{17} to $k_{17} = 10^{3.19} T^{2.13} \exp(-1233 K/T)$ dm³ mol⁻¹ s⁻¹. The work to date on this reaction provides excellent examples of the difficulty of characterizing the curvature of Arrhenius plots unless data over a very wide range of temperature are available, and also the difficulty, if curvature is suspected, of extrapolating from lower temperature results into the region >1000 K. Also, as pointed out by Cvetanovic et al.,[170] the expression derived for k_{17} is only one of a number of possible 'best fits' to the data depending upon the criteria adopted for fitting.

There are far fewer data for reaction (37) than for the methane case: as for

$$OH + C_2H_6 \rightarrow C_2H_5 + OH \qquad (37)$$

methane, the low temperature data are in reasonable agreement. Recent data in the range 300—500 K, obtained by the methods cited for methane[80, 161] and also from pulse radiolysis,[11] are in accord with the earlier work of Greiner[163] who derived the expression $k_{37} = 1.12 \times 10^{10} \exp(-1232 K/T)$ dm³ mol⁻¹ s⁻¹ for the range 300—500 K. Between 500 and 1000 K there are only two recent relative rate constant measurements. In one of them Baldwin and his co-workers determined the ratio k_{37}/k_2 at 773 K by addition of C_2H_6 to H_2-O_2 mixtures. In their most recent paper on this reaction[8] corrections have been made for self-heating and effects of minor side-reactions to give a value of 5.7. Using the value of k_2 at 773 K obtained from the expression[6] $k_2 = 1.28 \times 10^5 T^{3/2} \exp(-1480 K/T)$ dm³ mol⁻¹ s⁻¹ gives $k_{37} = 2.3 \times 10^9$ dm³ mol⁻¹ s⁻¹ at that temperature, in very close agreement with the value from the expression of Greiner. In the other relative rate measurement Hucknall et al.[171] used the decomposition of H_2O_2 as an OH source which, in the presence of $CH_4-C_2H_6$ mixtures, yielded products from the analysis of which k_{37}/k_{17} could be obtained. Taken with Ernst et al.'s[15] expression for k_{17} we get $k_{37} = 2.3 \times 10^9$ dm³ mol⁻¹ s⁻¹ at 653 K, also in reasonable agreement with the Greiner expression (1.7×10^9 dm³ mol⁻¹ s⁻¹) at this temperature. These two results suggest that Greiner's expression may be used with some confidence up to a temperature of at least 800 K.

At higher temperatures (>1000 K), as for reaction (36), the older data of Fenimore and Jones[172] give too high a value, probably because of their doubtful assumption that [OH] in their flames could be calculated on the basis of equilibrium being established in the hot boundary. The scatter on the remaining data[56,173,174] is such that it is hardly more helpful in characterizing the rate constant at these temperatures. The results, such as they are, are in order of magnitude agreement with Greiner's expression and with each other, so that on the available evidence it is not possible to decide whether there is curvature in the Arrhenius plot analogous to that in the methane case.

For C ⩾ 3 alkanes the only attempt to study systematically their reactions with

[170] R. J. Cvetanovic, D. L. Singleton, and G. Paraskevopoulos, *J. Phys. Chem.*, 1979, **83**, 50.
[171] D. J. Hucknall, D. Booth, and R. J. Sampson, *Int. J. Chem. Kinet., Symp. 1*, 1975, 301.
[172] C. P. Fenimore and G. W. Jones, Proc. 9th Symp. (Internat.) Combust., 1963, p. 597.
[173] A. A. Westenberg, and R. M. Fristrom, Proc. 10th Symp. (Internat.) Combust. 1965, p. 473.
[174] B. Smets and J. Peeters, Proc. 2nd European Symp. Combust., 1975, p. 38.

OH [reaction (38)] over a temperature range is that of Greiner.[163] Unlike those of

$$OH + RH \rightarrow H_2O + R \tag{38}$$

methane and ethane, the reactions of the higher alkanes have an added complexity due to the possibility of abstracting primary, secondary, or tertiary hydrogen atoms. Attempts have been made to account for the total rate of OH attack in terms of additivity of rates of abstraction for these three types of hydrogen in the molecule. The most recent of these are those of Darnall *et al.*[32] and of Baldwin and Walker.[8] Both groups used their own values of rate constants for OH attack in combination with the earlier results of Greiner. Darnall *et al.*[32] have obtained their rate constant values for a range of alkanes by photolytic production of OH in a 'smog' chamber, as described earlier. These results can be expressed, at 300 K, as

$$k_{38}/\text{dm}^3 \text{ mol}^{-1} \text{ s}^{-1} = 3.91 \times 10^7 \, n_p + 3.45 \times 10^8 \, n_s + 1.26 \times 10^9 \, n_t$$

where n_p, n_s, and n_t are the number of primary, secondary, and tertiary hydrogen atoms in RH. With the exception of cyclobutane the formula fits all the data (17 alkanes) to $\pm 20\%$, including that for other cyclic alkanes. To extend the formulae for higher temperatures they suggest use of the values for the activation energies for primary and secondary hydrogen abstraction given by Greiner, and that tertiary hydrogen abstraction be assigned a zero activation energy, hence yielding

$$k_{38}/\text{dm}^3 \text{ mol}^{-1} \text{ s}^{-1} = 6.08 \times 10^8 \, n_p \exp(-822 \, K/T) +$$
$$1.45 \times 10^9 \, n_p \exp(-428 \, K/T) + 1.26 \times 10^9 \, n_t \exp(0 \, K/T)$$

In deriving a similar formula Baldwin and Walker[8] have adopted a slightly different approach. They have used their own data at the temperatures 753—773 K together with Greiner's data for 300—500 K. Their own experiments give k_{38}/k_2 and they have expressed Greiner's results in similar form by using his absolute values for k_{38}[163] and k_2,[175] combining the data to give

$$k_{38}/k_2 = 0.214 \, n_p \exp(1070 \, K/T) + 0.173 \, n_s \exp(1820) \, K/T +$$
$$0.273 \, n_t \exp(2060 \, K/T)$$

To obtain values of k_{38} the expression for k_2 given earlier is used. At 300 K this leads to

$$k_{38}/\text{dm}^3 \text{ mol}^{-1} \text{ s}^{-1} = 3.63 \times 10^7 \, n_p + 3.57 \times 10^8 \, n_s + 1.26 \times 10^9 \, n_t$$

Comparison with the expression due to Darnall *et al.* shows that they are essentially identical, any differences being well within the error on the experimental data. At higher temperatures it is difficult to judge one approach better than the other since the only basis for comparison is the one set of results at *ca.* 750 K from Baldwin and Walker which was used in deriving the expression. A recent study on 2,2,3,3-tetramethylbutane suggests that results from such formulae should not be accepted uncritically. Baldwin *et al.*[176] found that the ratio of rate constants for OH attack on this compound and on neopentane were 0.8 ± 0.14 at 773 K, whereas both formulae would predict a value of 1.5 at all temperatures.

[175] N. R. Greiner, *J. Chem. Phys.*, 1969, **51**, 5049.
[176] R. R. Baldwin, R. W. Walker, and R. W. Walker, *J. Chem. Soc., Faraday Trans. 1*, 1979, **75**, 1447.

Halogenoalkanes.—Before 1975 there were virtually no data for reactions of OH with halogenoalkanes. Since then some 90 rate constant measurements have been reported, impressive evidence of the topicality of the field. The interest derives from the potentially harmful effects of halogenocarbons on the ozone layer in which reactions with OH would play an important role, and for this reason all but a few investigations have been carried out at, or below, 300 K.

A number of fluorine-, chlorine-, and bromine-substituted alkanes have been studied. For these the reaction proceeds by hydrogen abstraction since halogen abstraction is endothermic. As a consequence the fully halogenated species are relatively inert and for most of them it has not been possible to measure the rate constants accurately but only to establish upper limits ($< 10^5$—10^6 dm^3 mol^{-1} s^{-1}).[177-179] The one reported instance of halogen abstraction is by Garraway and Donovan,[180] who used flash photolysis–resonance absorption and found OH to react with CF_3I with a rate constant of $(7.2 \pm 1.2) \times 10^7$ dm^3 mol^{-1} s^{-1}, presumably to yield HOI. 'Rapid' reaction of OH with other alkyl iodides (CH_3I, C_2F_5I, and C_3F_7I) was also reported.

The available data[4] are sufficient to show that the rate constant at 300 K is increased, and energy of activation decreased, by bromine and chlorine substitution, while increased fluorine substitution has the reverse effect. These effects correlate with the C—H bond-strengthening effect of fluorination and the weakening effects of the other two halogens. Howard and Evenson[177,181] have shown for several halogenomethanes that there is a good correlation between $\log(k/n^{\frac{1}{2}})$ and D(C—H), where k is the rate constant for reaction with OH at 298 K, n is the number of C—H bonds in the molecule, and D(C—H) is the C—H bond dissociation energy; more recently Nip *et al.*[182] have shown that a linear correlation exists for both $\log(k/n^{\frac{1}{2}})$ and $\log(k)$ with ν_{C-H}, the C—H stretching frequencies. It is interesting to note that these correlations would both suggest a value for the rate constant for the reaction of OH with CHF_3 considerably higher than any of the available experimental values. But of all the halocarbons studied the experimental rate constant values for CHF_3 show the greatest scatter, with three values[177,182,183] in the range $(1.2$—$7.8)$ $\times 10^5$ dm^3 mol^{-1} s^{-1}; it may be that the correct value is higher still, closer to the *ca.* 2×10^6 dm^3 mol^{-1} s^{-1} suggested by correlation with bond strength.

Rate data at ambient temperatures are available for some 36 halogen-substituted methanes and ethanes. Reference to the tabulated values of the rate constants[4] shows that generally there is good agreement between measurements by different groups and by differing techniques. Exceptions are for CHF_3, already discussed, and some of the results of Clyne and Holt[183] which appear to yield significantly higher activation energies and A factors. The reasons for this are not clear, but Clyne and Holt have pointed out first that the scatter of the individual rate constant measurements in any one body of work is considerably greater than the reported

[177] C. J. Howard and K. M. Evenson, *J. Chem. Phys.*, 1976, **64**, 197.
[178] R. A. Cox, R. G. Derwent, A. E. J. Eggleton, and J. E. Lovelock, *Atmos. Environ.*, 1976, **10**, 305.
[179] R. T. Watson, G. Machado, B. Conaway, S. Wagner, and D. D. Davis, *J. Phys. Chem.*, 1977, **81**, 256.
[180] J. Garraway and R. J. Donovan, *J. Chem. Soc., Chem. Commun.*, 1979, 1108.
[181] C. J. Howard and K. M. Evenson, *J. Chem. Phys.*, 1976, **64**, 4303.
[182] W. S. Nip, D. L. Singleton, R. Overend, and G. Paraskevopoulos, *J. Phys. Chem.*, 1979, **83**, 2440.
[183] M. A. A. Clyne and P. M. Holt, *J. Chem. Soc., Faraday Trans. 2*, 1979, **75**, 582.

Arrhenius parameters might suggest, and secondly, that since the reactions being studied are not particularly fast ($k = 10^6$—10^8 dm^3 mol^{-1} s^{-1}), the measurements are sensitive to small traces of highly reactive impurities. Experiments in which the halogenoalkanes had been stringently purified to remove such likely contaminants as halogenoalkenes would be extremely valuable.

The only high temperature studies of these reactions are two shock tube measurements at *ca.* 1300 K both on CHF$_3$. In one of them Bradley *et al.*[56] measured concentration–time profiles of OH radicals, generated from decomposition of H$_2$O$_2$, in the presence of a variety of additives, one of which was CHF$_3$. Computer modelling of the system allowed relative rates of attack of OH on the various additives to be deduced. The value of the rate constant for reaction (39)

$$OH + CHF_3 \rightarrow H_2O + CF_3 \tag{39}$$

relative to that for the corresponding reaction (17) with methane was found to be 0.19 at 1300 K. Using the value of k_{17} (1.6 × 10^9) from the same study gives $k_{39} = 3.1 \times 10^8$ dm^3 mol^{-1} s^{-1}.

Ernst *et al.*[15] used flash photolysis to generate OH from water vapour followed by shock heating of the reaction mixture in the range 1255—1445 K, and the decay of [OH] was followed by absorptiometry. They also studied the reaction with CH$_4$ and at 1300 K the ratio k_{39}/k_{17} is 0.13, in reasonable agreement with the value of Bradley *et al.*[56] The small change in k_{39} over the temperature range covered makes it difficult to derive accurate Arrhenius parameters and therefore the authors combine their own value of k_{39} [(4.0 ± 1.0) × 10^8] at 1350 K with that (1.2 × 10^5) of Howard and Evenson[177] at 298 K to obtain $k_{39} = 4.0 \times 10^9$ exp (-3100 K/T) dm^3 mol^{-1} s^{-1}. While the activation energy so obtained (25.8 kJ mol^{-1}) is higher than for other hydrogen abstractions from halogenated alkanes this is not out of line with the relatively high C—H bond energy in CHF$_3$, but it was noted earlier that the Howard and Evenson value of k_{39} at 298 K may be low, and hence this value of the activation energy should perhaps be taken as an upper limit.

Clyne and Holt[183] studied reaction (39) at 296 and 430 K. Their value of k_{39} at 296 K is a factor of 7 higher than that of Howard and Evenson[177] and shows no detectable increase at 430 K, whereas the Arrhenius expression would predict a 26-fold increase over this range. It is unlikely that a discrepancy of this magnitude could be explained entirely by curvature of the Arrhenius plot or by an incorrectly low activation energy.

Alkenes and Halogenoalkenes.—The reaction of OH radicals with olefins may proceed by addition or H atom abstraction or both. Formation of an OH–olefin adduct is exothermic to the extent of some 150 kJ mol^{-1}, which is sufficient to cause rapid decomposition of the adduct unless it is collisionally deactivated. Consequently the reaction rate is expected to be pressure dependent and this is found to be so. In the case of ethene the only thermochemically favourable decomposition route for the adduct is back to OH and the olefin but for larger olefins fragmentation into other species becomes possible, *e.g.*, by elimination of a CH$_3$ group.

Some of the earliest studies of OH–olefin reaction revealed the existence of adducts for ethene and propene and were able to demonstrate[58] their stabilization with increasing pressure in the range 0.1–0.5 kPa. However, evidence has been produced for the abstraction route and there has been some controversy over the relative importance of the two.

Slagle *et al.*[64] studied the reactions of OH radicals with propene, *trans*-but-2-ene, and propadiene using crossed molecular beams of reactants and photoionization mass spectrometry to detect products. In the case of propadiene the adduct itself was observed, but not for the other compounds. In every case ions were found which were interpreted as arising from radicals produced by hydrogen abstraction. However more recently Hoyermann and Sievert[184] have also studied by mass spectrometry the primary products of the OH + propene reaction, but at slightly higher pressures (< 30—240 Pa). Formation of the adduct is observed to be pressure dependent and its isomerization and decomposition leads to the following overall reaction paths

$$OH + C_3H_6 \rightarrow C_2H_5CHO + H \qquad (a)$$

$$\rightarrow HCHO + C_2H_5 \qquad (b)$$

$$\rightarrow CH_3COCH_3 + H \qquad (c)$$

$$\rightarrow CH_3CHO + CH_3 \qquad (d)$$

with branching ratios $(b)/(a) = 4 (\pm 1.5)$ and $(d)/(c) = 3.5 (\pm 1.5)$. It was concluded that abstraction is < 5 % under these conditions. It must also be appreciated that the experiments of Slagle *et al.* were performed under virtually collision-free conditions, and even if their observations are indicative of abstraction processes, such reaction routes may not play a significant role at higher pressures.

There have been a number of experiments in which OH radicals have been produced photolytically in the presence of olefins at pressures in the approximate range 20—100 kPa, and the products analysed. Meagher and Heicklen[185] used H_2O_2–C_2H_4–N_2 mixtures and found as major photolysis products formic acid, ethanol, formaldehyde, and water. The quantum yields of these products were studied as a function of C_2H_4 pressure, light intensity, added oxygen, and total pressure. The formic acid was attributed to oxidation of C_2H_3 radicals by O_2 and its yield was therefore a measure of the extent of H atom abstraction by OH. However, this interpretation leads to a value of *ca.* 26% for the abstraction pathway which seems unacceptably large compared with other results. This system is complex since O_2 and HO_2 as well as OH contribute to the overall chemistry, and also the fate of the C_2H_4–OH adduct is bound to be more sensitive to pressures, in the range studied, than are the adducts of the higher olefins.

Henri and Carr[186] studied the reactions of OH radicals with propene using N_2O–H_2, NO_2–H_2, and N_2O–H_2O as the photolytic OH sources. Different products in differing yields were found in the three systems. The formation of hexa-1,5-diene was taken as evidence for production of allyl radicals by H abstraction in the primary reaction, its yield suggesting that the abstraction pathway occurs to at least a few percent. However, Cvetanovic,[187] in a very similar

[184] K. Hoyermann and R. Sievert, *Ber. Bunsenges. Phys. Chem.,* 1979, **83**, 933.
[185] J. F. Meagher and J. Heicklen, *J. Phys. Chem.,* 1976, **80**, 1645.
[186] J. P. Henri and R. W. Carr, jun., *J. Photochem.,* 1976, **5**, 69.
[187] R. J. Cvetanovic, 12th Internat. Symp. Free Radicals, Laguna Beach, Calif., 1976, reported in ref. 4.

study, concluded that the products could be explained by the almost exclusive occurrence of addition rather than abstraction, and that addition was predominantly at the terminal carbon (65%). The best evidence from the study of overall reaction products for C_2H_4 and C_3H_6 appears then to support the contention that abstraction makes a relatively minor contribution ($< 5\%$) to the overall reaction at ambient temperatures.

Kinetics measurements provide strong support for this conclusion but also indicate that there may be exceptions to it among the higher olefins. Howard[188] has shown that the second-order rate constant for reaction of OH with ethene falls with decreasing pressure, in the range 0.1—0.9 kPa, extrapolating to approximately zero at zero pressure. Furthermore the temperature dependence of OH–olefin reactions is small and negative. Both these findings are indicative of addition rather than abstraction processes. The OH radical is strongly electrophilic and both in mechanisms and rates its reactions are analogous to those of ground state atomic oxygen [O(^3P)]. Thus when the logarithm of the rate constant for OH reaction with a particular olefin is plotted against that for O(^3P) attack on the same olefin, a good straight-line correlation is found except for but-1-ene and 3-methylbut-1-ene, both of which appear to react faster with OH than expected.[189] Since O(^3P) is known to react by addition with these two olefins it is suggested that the apparent extra reactivity of OH is due to the occurrence of abstraction as well as addition. Both of the exceptional olefins have weakly bound allylic hydrogens which would imply that abstraction would also be expected for other higher olefins having this structure. For but-1-ene and 3-methylbut-1-ene this correlation indicates that hydrogen abstraction contributes *ca.* 30% to the total reaction rate in the temperature range 300—425 K.

There are no results to indicate the relative importance of abstraction at higher temperatures. Howard[188] has estimated an abstraction rate constant for ethene of *ca.* 10^5 dm^3 mol^{-1} s^{-1} at 300 K. Even if the abstraction has a relatively large activation energy (25—30 kJ mol^{-1}), addition should still predominate even at flame temperatures. However, as the room temperature results demonstrate, this may not be the case for higher olefins, particularly those with branched side chains.

At ambient temperatures the fate of the thermalized OH–olefin adduct is to react with other species in the environment in which it is formed. To date only adducts of ethene, propene, and *trans*-but-2-ene have been studied in any detail, mainly under conditions where other radicals and the parent olefin are the likely co-reactants; however, O_2 has also been added to simulate the fate of the adducts in the atmosphere, where they act as important intermediates in air pollution.[4]

In the case of the halogenoalkenes there appear to have been no direct studies of the mechanisms and reaction products. The reaction might be expected to go by addition, which for vinyl halides is exothermic[190] by *ca.* 150 kJ mol^{-1}. There is then the possibility of elimination of a halogen atom as an alternative to stabilization of the adduct. For vinyl chloride and bromide the elimination is exothermic, but for the fluoride it is not. Despite this Howard[188] has presented kinetic evidence to suggest that for vinyl chloride at least, elimination of chlorine is not the major pathway.

[188] C. J. Howard, *J. Chem. Phys.*, 1976, **65**, 4771.
[189] R. Atkinson, R. A. Perry, and J. N. Pitts, jun., *J. Chem. Phys.*, 1977, **67**, 3170.
[190] D. D. Davies, S. Fischer, R. Schiff, R. T. Watson, and W. Bollinger, *J. Chem. Phys.*, 1975, **63**, 1707.

There are rate data available for a wide range of olefins at ambient temperatures[4] but few at higher temperatures. For ethene and propene the second-order rate constants are pressure dependent in the pressure range commonly used in kinetics studies. For ethene the fall-off region between second- and third-order kinetics occurs at pressures below *ca.* 40 kPa of He[190] and 30 kPa of Ar[191] at 300 K. Recent results by a variety of techniques[26, 191−193] suggest a limiting high pressure value of $(7 \pm 2) \times 10^9$ dm^3 mol^{-1} s^{-1} at 300 K.

For propene the second-order rate constant[194] appears to be pressure independent down to *ca.* 2.7 kPa of He, but the behaviour at lower pressures is not clear. Two of the measurements of the rate constant by discharge flow methods have given much lower values (*ca.* 3×10^9 dm^3 mol^{-1} s^{-1})[195, 196] than those from other techniques. The discrepancy seems too large to be attributed entirely to the fact that the flow discharge measurements were made at pressures of *ca.* 1 kPa and it is worth noting that the results of Pastrana and Carr[196] for but-1-ene, where no pressure effects apply, are similarly low. Recent relative rate measurements and absolute values from flash photolysis are extremely concordant, giving rate constant values of $(1.5 \pm 0.2) \times 10^{10}$ dm^3 mol^{-1} s^{-1},[26,29,194,197−199] but there is also a set of slightly earlier results[58,200] grouped around *ca.* 8×10^9 dm^3 mol^{-1} s^{-1}. The more recent values are to be preferred.

The rate constant for the reaction of OH with propadiene might also be expected to be pressure dependent at pressures $<ca.$ 3 kPa of He but this has not been established experimentally. The reactions of the higher alkenes will be in their second-order pressure-independent region at all pressures used in kinetics studies. The vast majority of the measurements on the higher alkenes come from two laboratories, that of Pitts and that of Niki. Pitts *et al.* have used two techniques: flash photolysis–resonance fluorescence, yielding absolute values of rate constants, and photolysis of NO$_x$-hydrocarbon mixtures, giving relative values. The results from the two techniques are in excellent agreement.[4] In some instances Niki's earlier studies[58, 59] using discharge flow–mass spectrometry appear to give slightly high values but the more recent work from this laboratory, using photolysis of NO$_x$–hydrocarbon mixtures, yields relative values generally agreeing well with those of Pitt's group. The flash photolysis–resonance fluorescence measurements of Ravishankara *et al.*[194] agree with those of Pitts and Niki for propene and but-1-ene but appear low for *cis*-but-2-ene and 2,3-dimethylbut-2-ene. It has been suggested[4] that this might be due to adsorption effects since a static system was used, whereas the results of Pitts were obtained using a flowing system.

The only new data to appear since the review by Atkinson *et al.*[4] are a flash

[191] R. Atkinson, R. A. Perry, and J. N. Pitts, jun., *J. Chem. Phys.*, 1977, **66**, 1197.

[192] N. R. Greiner, *J. Chem. Phys.*, 1970, **53**, 1284.

[193] R. Overend and G. Paraskevopoulos, *J. Chem. Phys.*, 1977, **67**, 674.

[194] A. R. Ravishankara, S. Wagner, H. Fischer, G. Smith R. Schiff, R. T. Watson, G. Tesi, and D. D. Davies, *Int. J. Chem. Kinet.*, 1978, **10**, 783.

[195] J. N. Bradley, W. Hack, K. Hoyerman, and H. Gg. Wagner, *J. Chem. Soc., Faraday Trans. 1*, 1973, **69**, 1889.

[196] A. V. Pastrana and R. W. Carr, jun., *J. Phys. Chem.*, 1975, **79**, 765.

[197] R. Atkinson and J. N. Pitts, jun., *J. Chem. Phys.*, 1975, **63**, 3591.

[198] A. M. Winer, A. C. Lloyd, K. R. Darnall, R. Atkinson, and J. N. Pitts, jun., *Chem. Phys. Lett.*, 1977, **51**, 221.

[199] W. S. Nip and G. Paraskevopoulos, *J. Chem. Phys.*, 1973, **71**, 2170.

[200] F. Stuhl, *Ber. Bunsenges. Phys. Chem.*, 1973, **77**, 674.

photolysis–resonance absorption study by Nip and Paraskevopoulos[199] at 297 K. Their values for propene of $(1.48 \pm 0.17) \times 10^{10}$ and for but-1-ene of $(2.0 \pm 0.15) \times 10^{10}$ dm^3 mol^{-1} s^{-1} are in good agreement with those from other laboratories. They report a value for pent-1-ene of $(2.39 \pm 0.23) \times 10^{10}$ dm^3 mol^{-1} s^{-1}.

The data at *ca.* 300 K are sufficiently well defined for the relationship between the structure of the alkene and the rate of OH attack to be seen. The basic pattern is similar to that for reaction of O(^3P) with alkenes, with the rate constants increasing in value with the number of substituents on the double bond, as discussed by Atkinson *et al.*[4]

The temperature dependence of the rate constant has been measured for a number of these reactions but only over a small temperature range and in every case a small negative temperature coefficient has been found ($|E_a| \leqslant -4.5$ kJ mol^{-1}). Two possible reasons for the sign and magnitude of the temperature coefficient have been suggested, namely that the reaction has a near-zero energy barrier and a pre-exponential factor with a negative temperature coefficient,[191] and secondly that the initially formed OH adduct is so weakly bound that it preferentially decomposes back to reactants and only occasionally goes forward to products, *i.e.*, the following mechanism pertains:

$$\text{Alkene} + \text{OH} \underset{k_{-1}}{\overset{k_1}{\rightleftharpoons}} \text{Adduct} \overset{k_2}{\longrightarrow} \text{Product}$$

Thus the measured rate constant, k_m, is given by $k_m = k_1 k_2 / (k_{-1} + k_2)$ and the conditions for a negative temperature coefficient are $k_{-1} \gg k_2$ and $E_{-1} > E_1 + E_2$. The negative temperature coefficient for O(^3P) reactions with olefins has been interpreted in these terms.[201]

Ethene is the only alkene for which there are data at high temperatures. Bradley *et al.*[56] in a shock tube study at 1300 K have obtained rate constant ratios for reactions of OH with CO, H$_2$, CH$_4$, CHF$_3$, C$_2$H$_4$, and C$_2$H$_6$. Substituting values for the rate constants for the other reactions give values in the range (5×10^9)–(1×10^{10}) dm^3 mol^{-1} s^{-1} for the rate constants of the following reaction.

$$\text{OH} + \text{C}_2\text{H}_4 \rightarrow \text{Products}$$

Other data are a value of *ca.* 10^{10} at 1250—1400 K from a flame study,[173] and values of 10^{10} and 3×10^9 dm^3 mol^{-1} s^{-1} at *ca.* 800 K.[202,203] Although the accuracy of these results is questionable, particularly the older measurements, they suggest that the temperature coefficient of the rate constant is indeed small, although perhaps not negative over the whole temperature range.

There have been far fewer rate constant measurements for reactions of OH with halogenoalkenes. Howard[188] has shown that at 296 K the second-order reaction rate constants for reactions with a number of halogenoethenes are pressure dependent in the range 0.1—1 kPa of He, and, as for ethene itself, the rate constants have small negative temperature coefficients. A notable exception is C$_2$Cl$_4$ which reacts relatively slowly with OH and for which reaction there is a positive activation

[201] D. L. Singleton and R. J. Cvetanovic, *J. Am. Chem. Soc.*, 1976, **98**, 6812.
[202] R. R. Baldwin, R. F. Simmons, and R. W. Walker, *Trans. Faraday Soc.*, 1966, **62**, 2476 and 2486.
[203] D. E. Hoare and M. Patel, *Trans. Faraday Soc.*, 1969, **65**, 1325.

energy of *ca.* 12 kJ mol^{-1}. This unexpected and so far unexplained result has been obtained in two studies using quite different experimental techniques.[204,205]

Alkynes.—Only the reactions of acetylene, methylacetylene, and phenylacetylene have been studied. In a molecular beam experiment using photoionization mass spectrometry to detect products, Kanofsky *et al.*[63] identified some of the possible reaction pathways for acetylene and methylacetylene.

For acetylene the only important channel was

$$OH + C_2H_2 \rightarrow C_2HOH + H \tag{40}$$

and the expelled H atom was shown to originate from the acetylene and not the hydroxyl radical. For methyl acetylene a number of channels were found [equations (41)—(44)]. Channels (41) and (42) are the major pathways, (43) and (44) being

$$OH + C_3H_4 \rightarrow C_2H_2O + CH_3 \tag{41}$$

$$\rightarrow C_3H_3 + H_2O \tag{42}$$

$$\rightarrow C_3H_4O + H \tag{43}$$

$$\rightarrow C_2H_3 + HCHO \tag{44}$$

much less important; in (42) it is the propargyl rather than the methyl hydrogens which are abstracted. Although (44) is only a minor channel it is of interest because deuterium labelling experiments indicate that it proceeds through formation of a relatively long-lived complex which exists sufficiently long for the hydrogen atoms to migrate internally before decomposition of the adduct. In the case of phenylacetylene[62] observed reaction channels were (45)—(47). Channel (45)

$$OH + C_6H_5C_2H \rightarrow C_8H_7O \tag{45}$$

$$\rightarrow C_6H_5C_2OH + H \tag{46}$$

$$\rightarrow C_6H_4(OH)C_2H \tag{47}$$

represents adduct formation, but the exact position of addition of the OH could not be determined. Reactions (46) and (47) are evidence of attack at both the alkynyl carbon and the ring carbons and no strong preference was found for the one over the other.

These results were obtained under essentially collision-free conditions and it is not possible to assess from them the relative importance of the various channels at higher pressures, nor to assume the absence of other channels, *e.g.*, adduct formation in the case of acetylene. No experiments to identify products at higher pressures seem to have been attempted.

Pitts *et al.*[106] found that the second-order rate constant for reaction of OH with acetylene was pressure dependent below *ca.* 27 kPa of Ar suggesting that the mechanism is, at least in part, one of addition. However, experiments by Davis *et*

[204] J. S. Chang and F. Kaufman, *J. Chem. Phys.*, 1977, **66**, 4989.
[205] D. D. Davis, U. Machado, G. Smith, S. Wagner, and R. T. Watson, unpublished results, cited in ref. 4 and elsewhere.

al.[190] using the same technique (flash photolysis–resonance fluorescence) and covering the pressure range 2.7—67 kPa of He showed no pressure effect and a flash photolysis–resonance absorption study at pressures of 1.3—2.7 kPa gave values of the rate constant which were independent of the diluent gas (He or H_2–N_2O).[21] Furthermore, unlike other additon reactions (ethene, propene), the rate constant of the OH reaction with acetylene appears to have a positive temperature coefficient (E_a *ca.* 2.5 kJ mol^{-1}).[21, 106] Despite these results in view of the high C—H bond energy in acetylene and the predominance of addition over abstraction in reactions of OH with alkenes, it seems unlikely that the reaction with acetylene could proceed by any route other than addition, but obviously more studies are desirable. The rate constant values also show considerable scatter, some of which may be attributable to the different pressures at which they were obtained.

Acetylene is also the only alkyne for which some high temperature data exists. Vandooren and van Tiggelen[206] have measured concentration profiles for a number of species in low pressure (5.3 kPa) acetylene flames. They observe relatively high concentrations of ketene which they suggest arises from reaction (40) which is exothermic by 109 kJ mol^{-1}. They deduce a rate expression $k_{40}/\mathrm{dm^3\ mol^{-1}\ s^{-1}} = 3.2 \times 10^8 \exp(-100\ K/T)$ over the range 570—850 K. Extrapolation to 298 K gives 2.3×10^8 dm^3 mol^{-1} s^{-1}, apparently favouring the results towards the lower end of the range of 298 K values, but if Atkinson and Pitt's findings on the pressure effect are correct this comparison is invalidated. Vandooren and van Tiggelen[206] also suggested another, minor, pathway, reaction (48) for which they obtain $k_{48}/\mathrm{dm^3\ mol^{-1}\ s^{-1}} = 5.5 \times 10^{10} \exp(-6850\ K/T)$ over the range 650—1100 K.

$$\mathrm{OH + C_2H_2 \rightarrow CH_3 + CO} \qquad (48)$$

For methylacetylene there is only one available rate measurement giving a value for the second-order rate constant of $(5.7 \pm 1.0) \times 10^7$ dm^3 mol^{-1} s^{-1} at 300 K.[195] This result must be accepted with caution because of the discordancy of the results for acetylene, the possibility that at the pressure of experiment (0.3 kPa of He) the reaction may have been in its fall-off region, and the fact that the values of rate constants for other OH reactions (with propene and propadiene) obtained in this study appear to be low.

Aromatic Compounds.—The effects of pressure[194, 207] and temperature[208, 209] changes on the reaction rate constant, and results from product analyses,[210] all indicate that OH radicals may react with aromatic compounds by H abstraction and/or addition to the aromatic ring. Abstraction of H atoms from substituent groups is usually energetically favourable compared with that from the ring, and predominates at room temperatures. Thus abstraction from alkylbenzene is almost exclusively from the alkyl groups, but for compounds such as the cresols where the O—H bond energy is large any H-atom abstraction will be largely from the ring.

[206] J. Vandooran and P. J. van Tiggelen, Proc. 16th Symp. (Internat.) Combust., 1977, p. 1133.
[207] D. D. Davis, W. Bollinger, and S. Fischer, *J. Phys. Chem.*, 975, **79**, 293.
[208] R. A. Perry, R. Atkinson, and J. N. Pitts, jun., *J. Phys., Chem.*, 1977, **81**, 296.
[209] R. A. Perry, R. Atkinson, and J. N. Pitts, jun., *J. Phys. Chem.*, 1977, **81**, 1607.
[210] R. A. Kenley, J. E. Davenport, and D. G. Hendry, *J. Phys. Chem.*, 1978, **82**, 1095.

OH radicals are strongly electrophilic and the point of their addition to an aromatic ring will be influenced by directing effects of substituent groups; this point has received little direct experimental study but is expected to parallel the behaviour of $O(^3P)$ atoms. The addition is exothermic, typically[4] by *ca.* 80 kJ mol^{-1}, and decomposition of the adduct will compete with collisional stabilization. The pressure dependence of the rate constants observed in some cases has been interpreted in terms of such competition. Both the newly formed highly energized radical and also the thermalized radical may decompose either by loss of OH (back to the reactants) or by loss of another group as shown in Scheme 2 for *ortho*-addition to toluene,

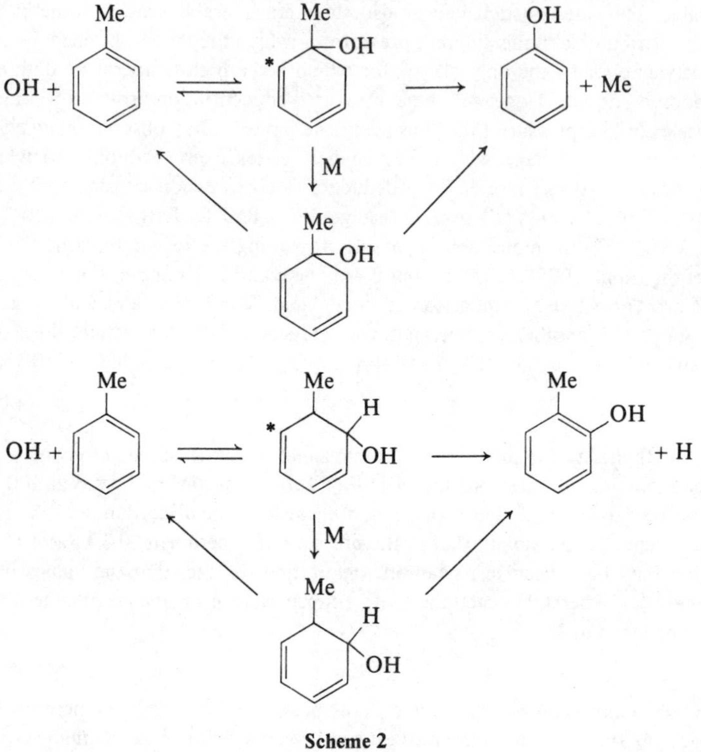

Scheme 2

where the asterisk denotes an energized species; *meta*- and *para*-addition are, of course, also possible.

Only a little information is available regarding the relative importance of addition and abstraction and the various possible elimination routes. Sloane,[61] in an experiment using crossed molecular beams of OH radicals and of benzene, toluene, and 1,3,5-trimethylbenzene, observed mass peaks corresponding to the OH adducts and the products of their decomposition. In every case decomposition by H-atom elimination was found, and for toluene and 1,3,5-trimethylbenzene, methyl group displacement was also observed. Only for toluene was there definite evidence of H-atom abstraction.

Evidence from kinetic measurements clearly demonstrates the importance of addition processes. Thus Davis *et al.*[207] showed that the second-order rate constant for reaction of OH with toluene was approximately 4 times greater than that with benzene and that the toluene rate constant was strongly pressure dependent, increasing by a factor of 2 in going from 0.4—13 kPa of He. They concluded that addition rather than abstraction from the methyl group predominated. In accord with this Ravishankara *et al.*[194] found that changing the length of the alkyl side-chain had little effect on the rate constant but substituting fluorine for hydrogen on the aromatic ring significantly reduced it.

The temperature dependence of the rate constants for OH reactions with a number of aromatic compounds is complex. Perry *et al.*[208,209] find for benzene, toluene, xylenes, trimethylbenzenes, methoxybenzene, and *o*-cresol that between 298 and *ca.* 325 K temperature change has little effect on the rate constants, but in the range *ca.* 325—380 K there is a *decrease* in the rate constants for OH disappearance, their values at 380 K being 4—12 times less than those at 298 K. Beyond 380 K the rate constants rise with temperature up to the highest value used (420—470 K). Under the conditions of the experiments these results could be interpreted in terms of the reactions (49) and (50).

$$\text{(toluene)} + \text{OH} \longrightarrow \text{(benzyl radical)} + H_2O \qquad (49)$$

$$\text{(toluene)} + \text{OH} \rightleftharpoons \text{(OH adduct)} \qquad (50)$$

At the high-temperature end of the range, reaction (49) predominates and the measured rate constant is k_{49}, whereas at the low-temperature end formation of the OH adduct is also significant and the measured rate constant is $k_{49} + k_{50}$. The decrease in the intermediate range is due to the reverse reaction (−50) becoming progressively more important so that by *ca.* 380 K thermal decomposition of the adduct is highly favoured. On this basis Perry *et al.*[207,208] were able to derive values for both $k_{49}/(k_{49} + k_{50})$ and the Arrhenius parameters for reactions (49) and (50). Their interpretation is based on the assumption that the only significant mode of decomposition of the adduct on the time scale of the experiments (1—30 ms) is by loss of an OH group. It has been pointed out[4] that loss of a CH_3 group might be expected to have a lower activation energy than loss of OH, and although phenol has not been found as one of the products of the reaction, the values of $k_{49}/(k_{49} + k_{50})$ should perhaps be treated as upper limits. The activation energies derived in this work can also be used as a basis for calculating the heats of formation of the substituted hexadienyl radicals.[208,209]

Values for $k_{49}/(k_{49} + k_{50})$ have also been obtained by Kenley *et al.*[210] by gas chromatographic analysis of products collected in cold traps and on solid adsorbents from reactions carried out in a discharge flow system at pressures of *ca.* 1 kPa. By attributing the stable products to formation either by abstraction or addition, values for the rate constant ratio could be derived. Agreement with the results of Perry *et al.*[208] is satisfactory, within the fairly large error limits of the kinetic study. However, in making the comparison it should be remembered that the rate constants for some of these reactions are pressure dependent and the two studies were performed at very different pressures. Furthermore the product yields in the flow tube work were expressed as a percentage of the total quantity of gas phase carbon detected. The possibility of condensed products would modify the total yields and possibly the distribution, and in this context lower yields for some products have been reported.[211]

Despite these reservations it seems clear that in alkyl-substituted benzenes both addition and abstraction occur, the latter typically contributing up to 20% to the total rate at 300 K, but becoming relatively more important with increase in temperature and becoming predominant by *ca.* 400 K. Thus for high temperature systems, such as flames, only abstraction need be considered.

The subsequent reactions of the OH–aromatic adduct are of considerable importance in the troposphere and the present rudimentary state of such knowledge has been reviewed in refs. 4 and 159.

Experimentally determined values for the rate constants for OH reactions with aromatic compounds have been tabulated and discussed by Atkinson *et al.*[4] Two techniques have been used in their determination: flash photolysis–resonance fluorescence[194, 207−209, 212] and relative rate constant measurements by photolysis of NO_x–aromatic–alkane–air mixtures.[25,26] Where comparison is possible there is excellent agreement between results from the two methods and from different laboratories. The rate constants for a number of the reactions are pressure dependent. For benzene and toluene at 300 K pressure dependence is found below *ca.* 13 kPa of He,[207] while for argon as the bath gas, pressures of >6.7 kPa were shown to have no effect on the rate constant for reaction with benzene,[212] and for toluene pressure dependence is not observed down to 13 kPa of Ar.[212] For *m-* and *p*-xylene, Ravishankara *et al.*[194] have found that the limiting value is achieved at between 0.4 and 2.7 kPa of He or Ar and for ethylbenzene the limit appears to lie below 0.4 kPa.

As for reactions with other organic compounds, the values of the rate constants demonstrate the electrophilic nature of OH, the trends with position and degree of substitution being analogous to those found for other electrophilic species such as $O(^3P)$ atoms.[4]

Oxygen, Nitrogen-, and Sulphur-containing Compounds.—Oxygen-containing compounds for which rate data are available include aldehydes, ketones, epoxides, esters, ethers, and alcohols. Two sulphur-containing compounds (MeSH and Me_2S) and a number of species containing nitrogen [$MeNH_2$, $EtNH_2$, Me_2NH, Me_3N, Et_2NOH, CH_2NOH, $MeCHNOH$, $MeONO$, $MeNO_2$, $MeC(O)O_2NO_2$] have been

[211] R. J. O'Brien, R. J. Green, and R. M. Doty, 'Chemical Kinetic Data Needs for Modelling the Lower Troposphere', ed. J. T. Herron, R. E. Huie, and J. A. Hodgeson, N. B. S. Special Publication 557, 1979, p. 93.
[212] D. A. Hansen, R. Atkinson, and J. N. Pitts, jun., *J. Phys. Chem.*, 1975, **79**, 1763.

studied and data for ambient temperatures compiled and reviewed by Atkinson *et al.*[4] A recent addition to the list is $MeNHNH_2$ for which a rate constant of $(3.9 \pm 0.8) \times 10^{10}$ dm^3 mol^{-1} s^{-1}, independent of temperature over the range 298—424 K, was obtained[213] by flash photolysis–resonance fluorescence. This value is virtually identical with that[214] for Me_2NH, reflecting similar N–H bond strengths in the two compounds, and provides further evidence for the suggestion that reactions of OH radicals with all the above-mentioned compounds occur by H-atom abstraction rather than addition, unless of course the molecule also contains an unsaturated centre, *e.g.*, $CH_2:CHOMe$,[215] $CH_2:CHCH_2OH$.[11] In general it is found that the rate constants correlate with the strengths of the bonds broken and the number of abstractable H atoms of a particular type.[4] Thus at ambient temperatures, for alcohols,[22,28,216,217] abstraction is presumed to occur from the alkyl side-chain rather than the OH group, whereas for aldehydes[10,58–60,104,141,142] it is the relatively weakly bound aldehydic hydrogen that is removed. In the case of amines the N–H bonds in $MeNH_2$[218] and $EtNH_2$[214] are relatively strong and abstraction of the alkyl H atoms is predominant but in Me_2NH[214] and $MeNHNH_2$[213] the secondary N–H bond is sufficiently weak for abstraction from this group to become competitive. Earlier we stressed the close parallel in reactivity of OH and $O(^3P)$, but in their reactions with oxygen-, sulphur-, and nitrogen-containing organic compounds the two differ, since for $O(^3P)$ the initial attack is by addition.

For some classes of compounds expressions have been suggested relating the rate constant for abstraction to the structure. For ethers containing primary and secondary hydrogens Perry *et al.*[215] have suggested, at 305 K

$$k(\text{ether})/dm^3\ mol^{-1}\ s^{-1} = (3.6 \times 10^8)n_p + (1.02 \times 10^9)n_s$$

where n_p and n_s are the number of primary and secondary hydrogen atoms. However, since only four ethers (dimethyl, diethyl, di-n-propyl, and tetrahydrofuran) have been studied, and for only one of these is there more than one rate constant measurement, it remains to be seen how universal is the formula. There are only slightly more data available for the esters. The trends in rate constants[198,219] with structure of the alkoxy and acyl groups indicate that the OH radical reacts predominantly with the alkoxy end of the ester, the rate constants for which at *ca.* 298 K can be expressed as

$$k(\text{ester})/dm^3\ mol^{-1}\ s^{-1} = (4 \times 10^7)n_p + (5.5 \times 10^8)n_s + (2.1 \times 10^9)n_t$$

Measurements of the temperature coefficients of the rate constants for a number of the reactions of OH with amines[214,218] and sulphides[218,220,221] in the temperature range 299—425 K have yielded small negative values for the activation energies in

[213] G. W. Harris, R. Atkinson, and J. N. Pitts, jun., *J. Phys. Chem.*, 1979, **83**, 2557.
[214] R. Atkinson, R. A. Perry, and J. N. Pitts, jun., *J. Chem. Phys.*, 1978, **68**, 1850.
[215] R. A. Perry, R. Atkinson, and J. N. Pitts, jun., *J. Chem. Phys.*, 1977, **67**, 611.
[216] I. M. Campbell, D. T. McLaughlin, and B. J. Handy, *Chem. Phys. Lett.*, 1976, **38**, 362.
[217] A. R. Ravishankara, S. Wagner, R. Schiff, and D. D. Davis, unpublished results reported in ref. 4.
[218] R. Atkinson, R. A. Perry, and J. N. Pitts, jun., *J. Chem. Phys.*, 1977, **66**, 1578.
[219] I. M. Campbell and P. E. Parkinson, *Chem. Phys. Lett.*, 1978, **53**, 385.
[220] M. J. Kurylo, *Chem. Phys. Lett.*, 1978, **58**, 233.
[221] R. Atkinson, R. A. Perry, and J. N. Pitts, jun., *Chem. Phys. Lett.*, 1978, **54**, 14.

the range 1.1—3.3 kJ mol⁻¹. This could be interpreted as indicating an addition mechanism, as for reactions with unsaturated compounds, but the evidence supporting abstraction seems too substantial to ignore. Atkinson *et al*.[214,218,221] therefore attribute this temperature dependence to a near-zero energy barrier in the reactions together with a T^{-n} term in the pre-exponential factor, where n is in the range 0.5—0.7. A similar negative temperature dependence is found for MeCHO,[142] but activation energies for these reactions are not exclusively negative since for HCHO[141,142] and for Me_2O[215] (the only ether to be studied over a temperature range) the activation energies are positive, with values of *ca.* 730 and *ca.* 3.2 kJ mol⁻¹ respectively.

For the majority of compounds considered here there has been only one determination of rate parameters, often only a rate constant measurement, but where comparisons can be made agreement between results from different techniques is good. However, there are some noteworthy exceptions. For MeSH and Me_2S the relative rate measurements of Cox[104] give very much higher rate constant values than those from flash photolysis–resonance fluorescence.[218,220,221] It has been suggested that in these systems there are species which interfere with the detection method used by Cox,[4] and the absolute results are to be preferred.

Rate constant measurements for reaction (30) also show more than usual

$$OH + HCHO \rightarrow H_2O + CHO \tag{30}$$

scatter.[10,60,141,142] Values of $10^{-8} k_{30}/dm^3 mol^{-1} s^{-1}$ at 298 K are $\geqslant 4.0$, 8.4, 5.6, 8.6, and 3.9. In two of the studies the variation of k_{30} with temperature has been determined: Atkinson and Pitts,[142] using flash photolysis–resonance fluorescence obtain $k_{30}/dm^3 mol^{-1} s^{-1} = 7.5 \times 10^9 \exp[-(88 \pm 150)K/T]$ over the range 299—426 K, whereas Smith[141] using discharge flow–mass spectrometry obtained $4.4 \times 10^{10} \exp[-(717 \pm 239)K/T]$ over the relatively small temperature range 268—334 K. Smith's experiment gave k_{30} relative to the rate constant for the mutual reaction of OH radicals, the energy of activation for which is not well established (see earlier). The experimental technique used by Atkinson and Pitts is the more direct and well tested but Smith's expression agrees better with data for high temperatures.

The higher temperature data consists of an older set of measurements[222-226] in the temperature range 700—900 K which give values of $k_{30}/dm^3 mol^{-1} s^{-1}$ scattered in the range $(1.7—4.0) \times 10^{10}$, and more recent results from methane or acetylene combustion studies giving values of 2.5×10^{10} (1600 K),[168] $(1.4—2.0) \times 10^{11}$ (873—1073 K),[227] 1.3×10^{10} (570 K),[206] and 1.0×10^{10} (485 K).[206] The last two of these were determined by Vandooren and van Tiggelen who combined them with the result of Peeters and Mahnen[168] for 1600 K to derive $k_{30}/dm^3 mol^{-1} s^{-1} = 3.9 \times 10^{10} \exp(-700 K/T)$. The similarity of this expression to that of Smith should not, however, be taken too seriously since the combustion systems studied

[222] D. E. Hoare, *Nature (London)*, 1962, **194**, 283.

[223] R. R. Baldwin and D. W. Cowe, *Trans. Faraday Soc.*, 1962, **58**, 1768.

[224] R. V. Blundell, W. G. A. Cook, D. E. Hoare, and G. S. Milne, Proc. 10th Symp. (Internat.) Combust., 1965, p. 445.

[225] D. E. Hoare, *Proc. R. Soc. London, Ser. A*, 1966, **291**, 73.

[226] D. E. Hoare and G. B. Peacock, *Proc. R. Soc. London, Ser. A*, 1966, **291**, 85.

[227] M. Cathonnet and H. James, *J. Chim. Phys.*, 1975, **72**, 247 and 253.

are complex and the modelling leading to these values requires a knowledge of a number of other rate constants, so that the precision of the final result cannot be high. Atkinson and Pitts[142] have noted the discrepancy between their own results and the high temperature data, and have suggested that it may indicate a curved Arrhenius plot, but it is the reviewers' opinion that the data at all temperatures are not sufficiently reliable for any conclusion on this point to be reached.

The work of Vandooren and van Tiggelen[206] also provides a value for the reaction of OH with ketene which is assumed to take the course shown in reaction (51). The value of k_{51} is 2.5×10^{10} dm^3 mol^{-1} s^{-1}, independent of temperature over

$$OH + CH_2CO \rightarrow HCHO + CHO \qquad (51)$$

the range 480—1000 K. The only other value for this rate constant is a lower limit of $\geqslant 1.0 \times 10^9$ dm^3 mol^{-1} s^{-1} at 295 K from a relative measurement by Faubel *et al.*[228]

In interpreting results at high temperatures there is the added complication that under these conditions further reaction channels may be introduced by the competitive abstraction of more than one type of H atom in the molecule. Due attention to this point has been paid by Baldwin *et al.*,[229] but it remains somewhat academic since the only other compounds for which a few rate constant values at high temperatures are available appear to be EtCHO and MeOH. For the former Baldwin *et al.*[230] obtained from a study of the combustion of EtCHO–O$_2$ mixtures a value for the overall rate constant for OH attack by all reaction channels at 713 K of 3.4×10^9, which may be compared with values of 1.2×10^{10} (see ref. 10) and 1.8×10^{10} dm^3 mol^{-1} s^{-1} (see ref. 59) obtained in other studies at 298 K. These values might suggest a negative temperature dependence for the rate constant similar to that for MeCHO, but this conclusion would be premature in view of the paucity of data.

There have been a number of high temperature studies of methanol ignition and oxidation. The results of these shock tube and turbulent flow reactor studies have been collected by Westbrook and Dryer[231] and used as a basis for constructing a comprehensive mechanism for methanol oxidation over the temperature range 1000—2180 K for fuel–air equivalence ratios of 0.05–3.0 and pressures between 100 and 500 kPa. Although there are some 84 reactions used in the model it is found that just a few reactions dominate the oxidation characteristics in certain ranges of conditions. Thus for fuel-lean mixtures the system is particularly sensitive to reaction (52). Fitting of the model to the results of Aranowitz *et al.*[232] obtained

$$OH + CH_3OH \rightarrow CH_2OH + H_2O \qquad (52)$$

under fuel-lean conditions, with k_{52} as an adjustable parameter, yielded a value of k_{52}/dm^3 mol^{-1} s^{-1} of 1.5×10^9 at 1050 K. An Arrhenius expression, k_{52}/dm^3 mol^{-1}

[228] C. Faubel, H. Gg. Wagner, and W. Hack, *Ber. Bunsenges. Phys. Chem.*, 1977, **81**, 689.
[229] R. R. Baldwin, C. J. Cleugh, J. C. Plaistowe, and R. W. Walker, *J. Chem. Soc., Faraday Trans. 1*, 1979, **75**, 1433.
[230] R. R. Baldwin and R. W. Walker, *Trans. Faraday Soc.*, 1969, **65**, 806.
[231] C. K. Westbrook and F. L. Dryer, *Combust. Sci. Technol.* 1979, **20**, 125.
[232] D. Aronowitz, R. J. Santoro, F. L. Dryer, and I. Glassman, Proc. 17th Symp. (Internat.) Combust., 1979, p. 633.

$s^{-1} = 4.0 \times 10^9 \exp(-1000 \ K/T)$, is also given[231] although it is not made clear over what temperature range it has been obtained. At 298 K it gives $k_{52}/dm^3 \ mol^{-1}$ $s^{-1} = 1.4 \times 10^8$ which may be compared with the three available values[22,216,217] at 298 K which are in the range $(6.0—6.4) \times 10^8$. It is obvious that the Westbrook and Dryer expression is not valid over the whole temperature range, but this does not necessarily invalidate its use, for want of a better, at temperatures close to 1000 K.

Organic Free Radicals.—From measurements on a fuel-rich CH_4–O_2 flame Fenimore[233] has derived a value for the rate constant for the reaction between OH and CH_3 radicals of $(4 \pm 2) \times 10^9 \ dm^3 \ mol^{-1} \ s^{-1}$ at 1970—2190 K, and in a similar study, in which small amounts of CF_3Br were added to the methane flame, Biordi *et al.*[234] obtained $5 \times 10^9 \ dm^3 \ mol^{-1} \ s^{-1}$ at 1090—1375 K for reaction between OH and CF_2. This latter result supersedes a previous determination by the same authors.[235]

[233] C. P. Fenimore, Proc. 12th Symp. (Internat.) Combust., 1969, p. 463.
[234] J. C. Biordi, C. P. Lazzara, and J. F. Papp, *J. Phys. Chem.*, 1978, **82**, 125.
[235] J. C. Biordi, C. P. Lazzara, and J. F. Papp, *J. Phys. Chem.*, 1976, **80**, 1042.

6

Gas-phase Chemistry of the Minor Constituents of the Troposphere

BY R. A. COX AND R. G. DERWENT

1 Introduction

Since the original postulation by Chapman[1] in 1933 of the photochemical theory for stratospheric ozone, it has been widely recognized that gas-phase photochemically initiated reactions of neutral species are important in governing the trace gas composition of the upper atmosphere. The role of homogeneous gas-phase reactions in the lower atmosphere was first investigated in the 1950's, in attempts to understand the phenomenon of photochemical smog and the chemistry involved in its formation.[2] However, in 1962, Junge[3] in his treatise on atmospheric chemistry did not consider that homogeneous gas-phase reactions played a major role in the life cycles of tropospheric trace gases. It was Levy[4] in 1971 who first suggested that free-radical chemistry, driven by photochemical dissociation of ozone and nitrogen oxides, might be important in the background troposphere. He proposed that relatively high concentrations of highly reactive hydroxyl radicals could be maintained in steady state in the natural sunlit troposphere and that this could provide an efficient scavenging mechanism for both natural and man-made trace constituents on a global scale. Since then attempts have been made to unravel this aspect of tropospheric chemistry, and to quantify its role in the trace gas cycles.

Interest in tropospheric chemistry has been further stimulated by the problem of depletion of stratospheric ozone by chlorofluoromethanes and other chlorine-containing species.[5] The extent to which the chlorine compounds injected at the earth's surface reach the stratosphere depends on the efficiency with which they are scavenged in the troposphere. Quantitative determination of the sink strength is required to assess the impact of the various chlorine-containing species, both natural and manmade, on stratospheric ozone.

The scavenging processes acting in the troposphere may be divided into physical removal processes, in which species are absorbed irreversibly at the earth's surface or in precipitation elements (cloud and rain droplets, aerosols), and chemical

[1] S. Chapman, *Mem. R. Soc., London*, 1930, **3**, 103.
[2] P. A. Leighton, 'Photochemistry of Air Pollution', Academic Press, New York, 1961.
[3] C. E. Junge, 'Air Chemistry and Radioactivity', Academic Press, New York, 1963.
[4] H. Levy, *Science*, 1971, **173**, 141.
[5] 'Halocarbons: Effects on Stratospheric Ozone', National Academy of Sciences, Washington D.C., 1977; 'Chlorofluorocarbons and Their Effect on Stratospheric Ozone', Pollution Paper No. 5, Dept. of Environment, H.M.S.O., London, 1976.

removal processes in which species are chemically transformed within the atmosphere.

The physical removal is often referred to as wet and dry deposition (for precipitation and surface removal respectively). Chemical transformation may occur either by homogeneous gas-phase reactions or less generally by hetero-geneous reactions occurring in the liquid phase in cloud and rain droplets. For some compounds, *e.g.*, SO_2, both mechanisms may occur depending on conditions. The chemical transformations usually result in oxidation of the compounds, ultimately to stable oxidation products such as sulphate or nitrate aerosols, CO_2, and H_2O. These products are then removed from the atmosphere by physical processes.

2 Physical and Chemical Properties of the Troposphere

General Features of the Troposphere.—The troposphere describes that portion of the earth's atmosphere which lies between the surface and the stratosphere which overlies it. The importance of the troposphere and its overall behaviour are dominated by those features which form its boundaries.

The temperature of the troposphere is maintained by heating from the earth's surface so that the temperature decreases with altitude above the ground. The atmosphere receives more heat in the tropics than in high latitudes and so temperatures are maximum close to the ground in the tropical regions and fall away both with altitude and latitude (Figure 1). Temperatures decrease with altitude, then level off at some minimum value at *ca.* 10 km, and then start to increase again. The position of this temperature minimum marks the upper boundary of the troposphere and is called the tropopause. The stratosphere constitutes the region of constant, then increasing, temperature.

Movement of air across the tropopause is slow and spasmodic. The main dynamical effect of the stratosphere is to provide a 'lid' for the motions generated in the troposphere, damping out the upward flow of energy from the earth's surface.[6]

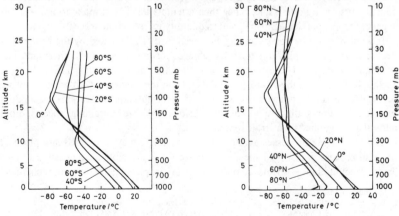

Figure 1 *Temperature profiles in the troposphere and stratosphere for December–February (constructed from data of ref. 11)*

[6] A. Gilchrist, *Rep. Prog. Phys.*, 1979, **42**, 503.

Atmospheric pressure decreases exponentially with altitude through the troposphere as is evident from the axes of Figure 1. The pressure falls to about one half of its surface value of 1000 mb ($\equiv$760 Torr)* by *ca.* 5 km and at the tropopause the pressure is typically *ca.* 100—200 mb ($\equiv$75—150 Torr). The troposphere therefore contains *ca.* 80—90% of the total atmospheric mass which amounts[7] to *ca.* 5.1×10^{21} g.

In addition to containing the bulk of the atmospheric mass, the lowest 10 km of the atmosphere contains the bulk of the minor trace gas burden. The radius of the earth[7] is *ca.* 6370 km, so that the minor trace gases and the motions that transport them are found in a relatively thin layer around the global surface.[6] The concentration distributions of the trace gases tend to be layered parallel to the earth's surface as is exemplified by the water vapour distribution illustrated in Figure 2. This stratification arises from the vertical decrease in pressure and from the atmospheric motions which move approximately horizontally at constant altitudes above the earth's surface. Windspeeds in the horizontal plane are two to three orders of magnitude greater than in the vertical.[6]

In addition to providing the major energy source in the lower atmosphere, the earth's surface acts as a source of minor trace gases. These sources may be natural

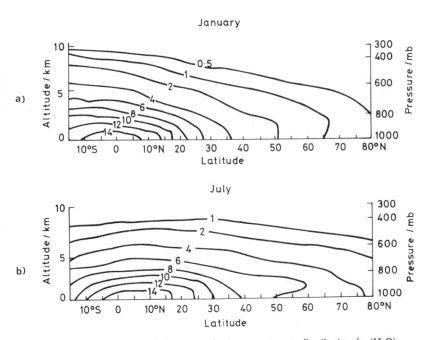

Figure 2 *Cross-sections through the tropospheric water vapour distribution,* [g (H_2O) *per* kg(air)] *during* (a) *December–February, and* (b) *June–August (constructed from data of ref. 11)*

---

* 1 Torr = 133 N m^{-2}.
[7] F. D. Stacey, 'Physics of the Earth', Wiley, New York, 1969.

emissions or releases of material from manmade processes. The latter tend to be concentrated in the large urban and industrial regions of the northern hemisphere. The emissions from the natural biosphere tend to have a more even distribution between the hemispheres though this greater uniformity is somewhat limited by the asymmetry in the land–ocean distribution between the hemispheres. Land areas cover 1.44×10^{18} cm^2 and represent 37% of the northern and 19% of the southern hemisphere.[8] Atmospheric motions will therefore have an important role in removing the discontinuities in the concentration distributions of the minor trace gases generated by the uneven distribution of their sources.

The emissions from the surface comprise a whole range of minor trace gases which are surveyed below. These trace gases are transported throughout the troposphere whilst being acted upon by a variety of sink processes, those of most interest in the present report being the chemical processes driven by sunlight, involving gas-phase photo-oxidation.

Absorption by stratospheric ozone and Rayleigh scattering by air molecules (particularly in the early mornings and evenings) limits the shortest wavelength radiation that reaches the earth's surface to *ca.* 280 nm. This is enough to give a small but important overlap between the solar spectrum in the troposphere and the visible and u.v. absorption spectrum of several photochemically labile species, the most important of which are ozone (O_3), nitrogen dioxide (NO_2) and formaldehyde (HCHO). Light absorption by these three species generates a small steady-state concentration of the highly reactive free-radical species hydroxyl (OH).

It is the hydroxyl radical which executes much of the gas-phase photo-oxidation in the troposphere. Its wide spectrum of reactivity ensures a rapid sink for most of the species which are not readily deposited to land or ocean surfaces or washed out by the rain-forming processes. These include the organic compounds, sulphur compounds, halogenocarbons, and oxides of nitrogen.

The major species which survive tropospheric sinks are transported to the stratosphere include methyl chloride (CH_3Cl), the chlorofluorocarbons (CCl_3F and CCl_2F_2), nitrous oxide (N_2O), carbon dioxide (CO_2), methane, and hydrogen. Stratospheric photo-oxidation of these species leads to the generation of HO_x, NO_x, and ClO_x free radical species which, through their rapid reactions with O and O_3, maintain a steady state with ozone production from molecular oxygen photolysis. It is therefore apparent that the stratospheric ozone layer is controlled largely by tropospherically derived trace gases.

Transport of Minor Trace Gases in the Troposphere.—In the following paragraphs the life history of a pollutant is followed from its emission into the atmosphere to its eventual sink, as it is transported through the troposphere.

Most emissions enter the atmosphere in the boundary layer, which is the region of turbulent mixing next to the ground. The depth of the well mixed boundary is variable. During daytime, mixing is usually confined to the lowest $\frac{1}{2}$ to 2 km by an overhead layer, the free troposphere, in which turbulence is suppressed. At night, particularly when the sky is clear and the wind light, a stable layer forms next to the ground, suppressing turbulence and confining mixing to a much shallower layer,

---

[8] H. V. Sverdup, M. W. Johnson, and R. H. Fleming, 'The Oceans, Their Physics, Chemistry and General Biology', Prentice Hall, New York, 1946.

10—400 m in depth. Convection can extend mixing over much deeper layers under particular circumstances such as during the formation of cumulus clouds and showers.[9]

The atmospheric boundary layer contains a number of sink processes, the most important of which are deposition to the underlying surface and gas-phase photo-oxidation. These act on the trace gases as they are advected away from the source area by the wind field. This wind field usually exhibits substantial changes with height, which when taken with the general uncertainty associated with the mixing depth itself, means that the determination of the effective track taken by the trace gases can be difficult. However, on a timescale of days, the trace gases, or their oxidation products not removed by sink processes in the boundary layer, are incorporated into the atmospheric circulation of the free troposphere.

The exchange between the boundary layer and the free troposphere is largely achieved by the large-scale ascent in depressions, fronts, orographic systems, and large cumulus and cumulonimbus clouds.[10] Many of these systems have precipitation associated with them to remove water-soluble trace gases and degradation products which may have been formed in the boundary layer. Once in the free atmosphere, the trace gases are subject to mean and eddy motions on much larger distance scales than in the boundary layer.

Figure 3 shows the distribution of the east–west wind component averaged around lines of latitude. Values are of the order of 10—30 m s^{-1} in the middle latitudes[11] where the length of a latitude circle is *ca.* 30 000 km. This implies a time of *ca.* 10—30 days to convey an air parcel completely around the globe in an east–west direction. In comparison, the mean wind components along lines of longitude in a north–south direction and in a vertical direction are much smaller. Provided the lifetimes of trace gases are longer than *ca.* 1 month, it can be assumed that east–west gradients in trace gas concentration are much smaller than north–south or vertical gradients. A two-dimensional (altitude–latitude) approach is therefore most suitable in the troposphere and is adopted in this discussion for the long-lived (>1 month) minor trace gases.

Figure 4 illustrates the atmospheric mass fluxes which characterize the two-dimensional behaviour of the mean circulation. This mean circulation is vital to present understanding of the north–south transport of heat, mass, water vapour, and trace gases, together with the exchanges of the latter with the stratosphere,[11] which are central to the present report.

The streamline patterns in December–February and June–August show large circulations dominating the tropical and sub-tropical regions. These are the Hadley cells in which rising air in the tropics descends in the sub-tropical regions. At middle latitudes smaller cells are apparent: these are the Ferrel cells. In September–February, the tropospheric cells extend into the stratosphere and tilt polewards with increasing altitude thus providing an important mechanism for transporting trace gases from the troposphere into the stratosphere.[11] To an extent, the stratosphere is driven dynamically from the troposphere below.

[9] F. Pasquill, 'Atmospheric Diffusion', Ellis Horwood, Chichester, 1974.
[10] F. B. Smith and D. J. Carson, *Boundary Layer Meteorol.*, 1977, **12**, 307.
[11] R. E. Newell, J. W. Kidson, D. G. Vincent, and G. J. Boer, 'The General Calculation of the Tropical Atmosphere and Interactions with Extratropical Latitudes', vol. 1, M.I.T. Press, U.S.A., 1972.

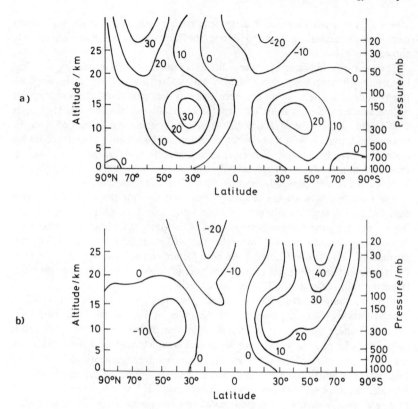

Figure 3 *Contours of the east–west wind component* (m s^{-1}) *during* (a) *December–February, and* (b) *June–August (constructed from data of ref. 11). Positive values are eastward, negative westward*

These streamlines show that transfer from the northern to the southern hemisphere across the equator is only achieved in the upper troposphere and between June and November.[11] The reverse transfer occurs only in the lower troposphere during the same period. As a result interhemispheric mixing occurs on the timescale of a year or so and involves both the upper and lower troposphere.

Apart from transport by the large-scale mean motions of the atmosphere, mixing in the free troposphere is also accomplished by eddies. The distribution of large-scale eddies varies markedly with latitude, altitude, and season.[12] The day-to-day variability of the wind shows a minimum in the lower tropical troposphere which is only matched in the quiescence of the summer stratosphere.[11] In midlatitudes, wind variability may be a factor of two to four larger than in the tropics at the same altitude. It was the stability of the low-level winds which enabled Halley in 1686 to construct mean wind maps, and Hadley in 1735 to propose his atmospheric circulation.

[12] P. J. Crutzen and J. Fishman, *Geophys. Res. Lett.*, 1977, **4**, 321.

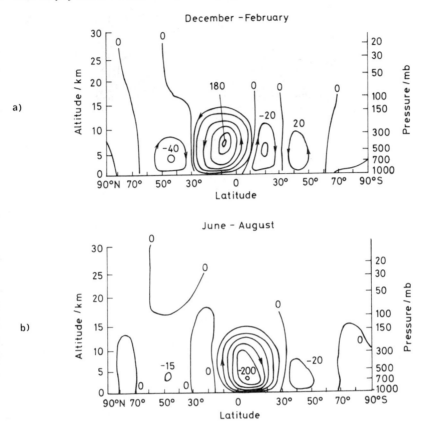

Figure 4 *Contours of the atmospheric mass flux* (10^{12} g s^{-1}) *during* (a) *December–February, and* (b) *June–August* (*constructed from data of ref. 11*). *Signs indicate clockwise or anticlockwise circulation*

Figure 5 shows the horizontal eddy diffusion coefficients which, when taken with the mean circulation and vertical eddy diffusion coefficients, describe the north–south and vertical transport in the troposphere.[13] There is a tendency for the mean motion and eddy transports to overlap so that they tend to reinforce each other in atmospheric regions where the other is least efficient.

Gas-phase Chemistry of the Troposphere.—In the following paragraphs, a description is given of the free radical chain reactions which serve to generate the OH steady state. These reactions constitute a low-temperature combustion system in which methane, hydrogen, hydrocarbons, and carbon monoxide are oxidized to typical combustion products, CO_2, and H_2O.

[13] J. F. Louis, 'Mean Meridianal Circulation: The Natural Stratosphere of 1974', C.I.A.P. Monograph 1, DOT TST 75 51, Washington D.C., U.S.A., 1975.

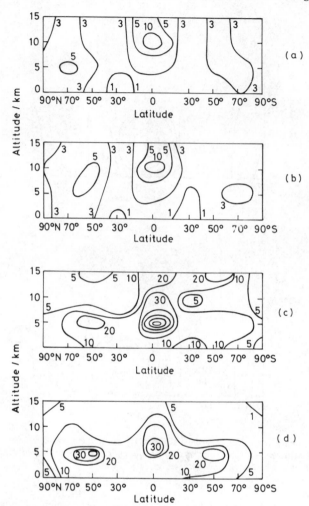

Figure 5 *Contours of the horizontal eddy diffusion coefficient ($K_{yy}/10^9$ cm² s⁻¹) during* (a) *December–February, and* (b) *June–August, and of the vertical eddy diffusion coefficient ($K_{zz}/10^3$ cm² s⁻¹) during* (c) *December–February, and* (d) *June–August (constructed from data of ref. 13)*

Chain Carriers and their Initiation. The two main chain carrying species are OH and HO_2. OH is produced by the u.v. absorption of ozone at wavelengths $\leqslant 310$ nm when electronically excited oxygen atoms in the $O(^1D)$ state are formed, which react with water vapour:

$$O_3 + h\nu \rightarrow O_2 + O(^1D) \tag{1}$$

$$O(^1D) + H_2O \rightarrow OH + OH \tag{2}$$

The alternate and dominant fate of $O(^1D)$ is electronic quenching by O_2 and N_2 to yield ground state atoms which then recombine with O_2 to form ozone:

$$O(^1D) + M \rightarrow O(^3P) + M \tag{3}$$

$$O(^3P) + O_2 + M \rightarrow O_3 + M \tag{4}$$

$$(M = O_2 \text{ or } N_2)$$

The rate of OH production is therefore obtained from equation (5). Substituting the various values of the rate coefficients (k), photolysis rates (J), and species concentrations, noon-time OH production rates of the order of 10^6 molecule $cm^{-3} s^{-1}$ are obtained for the lower troposphere in summer.

$$\frac{d}{dt}[OH] = 2J_1 \frac{k_2[H_2O]}{k_3[M] + k_2[H_2O]} \tag{5}$$

HO_2 is formed by the photolysis of formaldehyde:

$$HCHO + h\nu \rightarrow H + HCO \tag{6}$$

which is rapidly followed by reactions (7) and (8)

$$H + O_2 + M \rightarrow HO_2 + M \tag{7}$$

$$HCO + O_2 \rightarrow HO_2 + CO \tag{8}$$

From the photolysis rate and formaldehyde concentrations, noon-time HO_2 production rates of the order of 10^6 molecule $cm^{-3} s^{-1}$ are expected for the lower troposphere in summer. It is therefore apparent that the two chain carriers OH and HO_2 have similar production rates which at their peak approach 10^6 molecule $cm^{-3} s^{-1}$.

Free-radical Interconversion Reactions. Interconversion of OH and HO_2 occurs by way of fast bimolecular reactions with other trace gases. The main processes which convert OH into HO_2 are:

(*a*) with CO: $\qquad\qquad$ $OH + CO \rightarrow H + CO_2$ $\qquad\qquad$ (9)
$\qquad$ followed by reaction (7)

(*b*) with H_2: $\qquad\qquad$ $OH + H_2 \rightarrow H_2O + H$ $\qquad\qquad$ (10)
$\qquad$ followed by reaction (7)

(*c*) with CH_4: $\qquad\qquad$ $OH + CH_4 \rightarrow H_2O + CH_3$ $\qquad\qquad$ (11)
$\qquad$ followed by: $\qquad$ $CH_3 + O_2 + M \rightarrow CH_3O_2 + M$ $\qquad\qquad$ (12)
$\qquad\qquad\qquad\qquad$ $CH_3O_2 + NO \rightarrow CH_3O + NO_2$ $\qquad\qquad$ (13)
$\qquad\qquad\qquad\qquad$ $CH_3O + O_2 \rightarrow HO_2 + HCHO$ $\qquad\qquad$ (14)

(*d*) with H_2O_2: $\qquad\qquad$ $OH + H_2O_2 \rightarrow H_2O + HO_2$ $\qquad\qquad$ (15)

The main processes which convert HO_2 into OH are:

(*a*) with NO: $\qquad\qquad$ $HO_2 + NO \rightarrow NO_2 + OH$ $\qquad\qquad$ (16)

(*b*) with ozone: $\qquad\qquad$ $HO_2 + O_3 \rightarrow OH + O_2 + O_2$ $\qquad\qquad$ (17)

These free-radical interconversions are able to account for the scavenging of methane, carbon monoxide, and hydrogen. Furthermore, through the light absorption by nitrogen dioxide, they also lead to secondary ozone production which occurs as follows:

$$RO_2 + NO \rightarrow RO + NO_2 \quad \text{for R = Me or H} \tag{18}$$

$$NO_2 + hv \rightarrow NO + O \tag{19}$$

$$O + O_2 + M \rightarrow O_3 + M \tag{4}$$

the net result of which is equation (20):

$$RO_2 + O_2 + hv \rightarrow RO + O_3 \tag{20}$$

This has important consequences for the ozone budget in the troposphere.[13] Hydrogen and carbon monoxide are also produced by the free-radical inter-conversion reactions as a byproduct of CH_4 oxidation through a second channel in formaldehyde photolysis:[4]

$$HCHO + hv \rightarrow H_2 + CO \tag{21}$$

Free-radical Termination Reactions. Hydroxyl radicals may be removed by combination with NO_2 to form nitric acid [equation (22)] which is mainly removed from atmospheric circulation by wet or dry deposition.

$$OH + NO_2 \rightarrow HNO_3 \tag{22}$$

HO_2 radicals are removed by mutual termination in reaction (23) to form hydrogen peroxide, a species which is also removed by precipitation. Additionally, minor termination routes involve the radical–radical reactions (24) and (25). Methyl hydroperoxide (CH_3O_2H) can also be removed by precipitation in a manner analogous to that for H_2O_2.

$$HO_2 + HO_2 \rightarrow H_2O_2 + O_2 \tag{23}$$

$$OH + HO_2 \rightarrow H_2O + O_2 \tag{24}$$

$$HO_2 + CH_3O_2 \rightarrow CH_3O_2H + O_2 \tag{25}$$

The OH Steady State. Applying steady state to the atmospheric photochemical system at noon and neglecting minor termination routes, relations (26) and (27) can

$$2J_6[HCHO] + 2J_1 \frac{k_2[H_2O]}{k_2[H_2O] + k_3[M]} = k_{22}[OH][NO_2] + 2k_{23}[HO_2]^2 \tag{26}$$

$$[OH]\{k_9[CO] + k_{10}[H_2] + k_{11}[CH_4] + k_{15}[H_2O_2]\} = [HO_2]\{k_{16}[NO] + k_{17}[O_3]\} \tag{27}$$

be obtained. By taking $NO_x = NO + NO_2 = 2.5 \times 10^9$ molecule cm^{-3} and putting in the various rate constants we obtain:

$$2 \times 10^6 = 2.3 \times 10^{-2}[OH] + 4.6 \times 10^{-12}[HO_2]^2 \tag{28}$$

$$[OH] = 4.4 \times 10^{-3}[HO_2] \tag{29}$$

with the second ratio dominated by the OH + CO and HO_2 + NO reactions.

From relations (28) and (29) values of $OH = 3 \times 10^6$ cm^{-3} and $HO_2 = 6 \times 10^6$ cm^{-3} can be obtained for the noon concentrations of these radicals in the background atmosphere close to the ground. The kinetic chainlength, k.c.l., defined by equation (30) is *ca.* 3 at noon, showing that in the background atmosphere the free-radical chains are fairly short.

$$\text{k.c.l.} = \frac{\text{Rate of OH–HO}_2 \text{ interconversion}}{\text{Rate of termination}} \qquad (30)$$

These OH concentrations are quite adequate to account for a significant destruction of hydrocarbons, halogenocarbons, and other species in the troposphere.

Sources and Sinks for the Minor Constituents of the Troposphere.—*Sources of the Minor Constituents.* As interest in atmospheric chemistry has expanded over recent years, the number of minor constituents identified in the atmosphere or postulated from source inventories has increased dramatically. It is not possible to list here all the species and examples will only be considered of species which illustrate the nature of tropospheric scavenging processes. The concepts developed can then be applied to the wide range of other species whether natural or manmade, toxic or harmless.

Attention has therefore been directed to atmospheric emissions of organic, and nitrogen-, chlorine-, and sulphur-containing trace gases from a variety of natural or manmade sources. Of these the organic compounds and nitrogen-containing trace gases are the most relevant to this Report since it is their presence together with sunlight that generates hydroxyl radicals throughout the boundary layer and free troposphere.

Table 1 attempts to identify some of the major sources of the particular trace gases of interest in this Report, and indicates the complex range of sources of minor trace gases in the atmosphere, which require a quantitative-understanding before life-cycles can be constructed. Some of these sources can be assessed with an accuracy of $\pm 10\%$ at best, whereas others are no better than guesses. Various budgets and life-cycle studies have been produced for the more well studied species, but even for these, *e.g.*, CO_2, considerable uncertainties remain.[14]

It frequently occurs that the atmospheric degradation of one minor trace gas generates others which survive long enough to have separate existences. For example, the photo-oxidation of methane involves a range of highly reactive intermediates leading to formaldehyde, which has a lifetime of several hours. Formaldehyde is photochemically labile giving amongst its products hydrogen and carbon monoxide. Hydrogen has a lifetime of several years and carbon monoxide several months. A set of atmospheric degradation pathways is thereby set up linking the behaviour of many minor trace gases, and these are being slowly unravelled for both natural and manmade species. It is readily apparent, however, that there is much work yet to be done in this field. Table 1 includes a few of the major degradation pathways.

[14] B. Bolin, E. T. Degens, S. Kempe, and P. Ketnor, 'The Global Carbon Cycle', SCOPE 13, Wiley, Chichester, U.K., 1979.

Table 1 *Natural and manmade sources of the minor trace gases of the troposphere*

Compound	Natural sources	Manmade sources
Carbon-containing trace gases		
Carbon monoxide (CO)	Oxidation of natural methane;[a] oxidation of natural C_5, C_{10} hydrocarbons;[c] forest fires[c]	Oxidation of manmade hydrocarbons; incomplete combustion of wood, oil, gas, and coal, in particular motor vehicles;[d] industrial processes;[d] blast furnaces
Carbon dioxide (CO_2)	Oxidation of natural CO; destruction of forests;[e] respiration by plants[f]	Combustion of oil, gas, coal, and wood;[g] limestone burning[g]
Methane (CH_4)	Enteric fermentation in wild animals;[h] emissions from swamps, bogs, *etc.*,[i] natural wet land areas;[j] oceans[h]	Enteric fermentation in domesticated ruminants;[h] emissions from paddies;[i] natural gas leakage;[h] sewerage gas;[h] colliery gas;[h] combustion sources[h]
Light paraffins, C_2—C_6	Aerobic biological source[d]	Natural gas leakage; motor vehicle evaporative emissions; refinery emissions
Olefins, C_2—C_6		Motor vehicle exhaust;[k] diesel engine exhaust[k]
Aromatic hydrocarbons		Motor vehicle exhaust;[k] evaporative emissions;[k] paints, petrol, solvents
Hemiterpenes, C_5H_8 / Terpenes, $C_{10}H_{16}$ / Diterpenes, $C_{20}H_{32}$	Trees, broadleaves, and conifers; plants[b]	
Nitrogen-containing trace gases		
Nitric oxide (NO)[d]	Forest fires; anaerobic processes in soil; electric storms	Combustion of oil, gas, and coal
Nitrogen dioxide (NO_2)[d]	Forest fires; electric storms	Combustion of oil, gas, and coal; atmospheric transformation of NO
Nitrous oxide (N_2O)	Emissions from denitrifying bacteria in soil;[l] oceans[l]	Combustion of oil and coal[m]
Peroxyacetyl nitrate (PAN)	Degradation of isoprene[n]	Degradation of hydrocarbons[o]
Ammonia (NH_3)[d]	Aerobic biological source in soil / Breakdown of amino acids in organic waste material	Coal and fuel oil combustion; waste treatment

Sulphur-containing trace gases

Compound	Natural sources	Anthropogenic sources
Sulphur dioxide (SO_2)	Oxidation of H_2S;[p] volcanic activity[p]	Combustion of oil and coal;[q] roasting sulphide ores
Hydrogen sulphide (H_2S)	Anaerobic fermentation;[p] volcanoes and fumaroles[p]	Oil refining; animal manure;[r] Kraft paper mills;[l,r] rayon production;[r] coke oven gas[r]
Carbon disulphide (CS_2)	Anaerobic fermentation[s]	Viscose rayon plants;[t] brick making;[t] fish meal processing[t]
Carbonyl sulphide (OCS)	Oxidation of CS_2;[u] slash and burn agriculture;[c] volcanoes and fumaroles	Oxidation of CS_2;[u] brick making;[t] effluent from Kraft mills;[v] blast furnace gas;[v] coke oven gas;[v] shale and natural gas[v]
Sulphur trioxide (SO_3)		Combustion of S-fuel
Methyl mercaptan (CH_3SH)[s,t]	Anaerobic biological sources	Animal rendering; animal manure; pulp and paper mills; brick manufacture; oil refining
Dimethyl sulphide (CH_3SCH_3)[s,t]	Aerobic biological source	Animal rendering; animal manure; pulp and paper mills
Dimethyl disulphide (CH_3SSCH_3)[s,t]		Animal rendering; fishmeal processing
Other organic sulphur compounds:[s,t] C_2–C_4 mercaptans, dialkyl disulphides, dimethyl trisulphide, alkyl thiophenes, benzothiophenes	Anaerobic biological sources	Animal rendering; fishmeal processing; brick making

Chlorine-containing trace gases[w]

Compound	Natural sources	Anthropogenic sources
Hydrogen chloride (HCl)	Volcanoes and fumaroles; degradation of CH_3Cl	Coal combustion; degradation of chlorocarbons
Methyl chloride (CH_3Cl)	Slow combustion of organic matter; marine environment: algae	PVC and tobacco combustion
Methylene dichloride (CH_2Cl_2)		Solvent
Chloroform ($CHCl_3$)		Pharmaceuticals; solvent; combustion of petrol; bleaching of wood pulp; degradation of C_2HCl_3
Carbon tetrachloride (CCl_4)		Solvent; fire extinguishers; degradation of C_2Cl_4
Methyl chloroform (CH_3CCl_3)		Solvent; degreasing agent
Trichloroethylene (C_2HCL_3)		Solvent; dry cleaning agent; degreasing agent
Tetrachloroethylene (C_2Cl_4)		Solvent; dry cleaning agent; degreasing agent
Other chlorofluorocarbons: CCl_3F, CCl_2F_2, $C_2Cl_3F_3$, $C_2Cl_2F_4$, C_2ClF_5		Aerosol propellants; refrigerants; foam blowing agents; solvents

Table 1—*continued*

Compound	Natural sources	Manmade sources
Other minor trace gases		
Hydrogen[x]	Oceans; soils; oxidation of methane, isoprene and terpenes *via* formaldehyde	Motor vehicle exhaust; oxidation of methane *via* formaldehyde
HF	Volcanoes and fumaroles	
Ozone	Stratosphere: natural $NO-NO_2$ conversion	Manmade $NO-NO_2$ conversion
H_2O	Evaporation from oceans	Insignificant
SF_6		Electrical insulator
CF_4[y]		Aluminium industry
CH_3Br[z]	Aerobic biological source	Fumigation of soil and grain
CH_3I[z]	Aerobic biological source	Insignificant

[a] W. Seiler, *Tellus*, 1974, **26**, 116; [b] R. A. Rasmussen and W. Seiler, *Nature (London)*, 1979, **282**, 253; [c] P. J. Crutzen, L. E. Heidt, J. P. Krasnec, W. H. Pollock, and F. W. Went, *Proc. Natl. Acad. Sci.*, 1965, **53**, 215; [d] E. Robinson and R. C. Robbins, SRI Project PR 6755, Stanford Research Institute, California; [e] B. Bolin, *Science*, 1977, **196**, 613; [f] B. Bolin, E. T. Degens, S. Kempe, and P. Ketner, The Global Carbon Cycle SCOPE 13, Wiley, Chichester, U.K., 1976; [g] C. D. Keeling, *Tellus*, 1973, **25**, 174; [h] D. H. Ehhalt, *Tellus*, 1974, **26**, 58; [i] T. Koyama, 'Recent Researches in the Field of Hydrosphere, Atmosphere, and Nuclear Geochemistry', Murucen, Tokyo, 1964, p. 143; [j] A. Baker-Blocker, T. M. Donahue, and K. H. Mancy, *Tellus*, 1977, **29**, 245; [k] 'Vapour Phase Organic Pollutants', Natl. Acad. Sci., Washington, D.C., 1976; [l] J. Hahn and C. E. Junge, *Z. Naturforsch.*, 1977, **329**, 190; [m] R. F. Weiss and H. Craig, *Geophys. Res. Lett.*, 1976, **3**, 751; [n] R. A. Cox, R. G. Derwent, and M. R. Williams, *Environ. Sci. Technol.*, to be published; [o] P. A. Leighton, 'Photochemistry of Air Pollution', Academic Press, New York, 1961; [p] L. Granat, H. Rodhe. and R. O. Hallberg. 'The Global Sulphur Cycle', SCOPE 7, *Ecol. Bull. (Stockholm)* 1976, **22**, 89; [q] J. P. Friend, 'Chemistry of the Lower Atmosphere', Plenum Press, New York, 1973, p. 177; [r] A. C. Stern, 'Air Pollution', vol. 1, Academic Press, London, 1976; [s] 'Report of Working Party on Suppression of Odours from Offensive and Selected Other Trade', Dept. of Environment, H.M.S.O., London, 1975; [u] M. J. Kurylo, *Chem. Phys. Lett.*, 1978, **58**, 238; [v] P. L. Hanst, L. L. Spiller, D. M. Watts, J. W. Spence, and M. F. Miller, *J. Air Pollut. Control. Assoc.*, 1975, **25**, 1220; [w] IMOS Report, 'Fluorocarbons and the Environment', USGPO, Washington, Dept. of Environment, Pollution Paper No. 15, H.M.S.O., London, 1979; Nat. Acad. Sci. 'Chloroform, Carbon Tetrachloride, and other Halomethanes: An Environmental Assessment'; [x] U. Schmidt, *Tellus*, 1974, **26**, 78; [y] R. J. Cicerone, *Science*, 1979, **206**, 59; [z] J. E. Lovelock, *Nature (London)*, 1975, **256**, 193; J. E. Lovelock, *ibid*, 1974, **252**, 292; H. B. Singh, L. J. Salas, H. Shigeishi, and E. Scribner, *Science*, 1979, **203**, 899.

Sinks of the Minor Constituents. The main sink processes which act in the lower atmosphere are summarized in Table 2, together with the species for which each sink process is thought to be significant. These sink processes may be highly variable with time of day, season, and geographical distribution.

Table 2 *Sinks for the minor trace gases*

Sink process	Species for which this is, or may be, significant removal process
Deposition to water surfaces	SO_2, O_3, CO_2
Deposition to plants and vegetation canopies	SO_2, CO_2, SO_3
Deposition to soil	O_3, SO_2, CO, H_2
Wet deposition in rain or by rain-making process	HCl, HF, HNO_3, (NH_3, SO_2, NO_x, after prior conversion into aerosol species)
Direct photolysis in the troposphere	CH_3I, NO_2, HCHO, O_3
OH photo-oxidation	Hydrocarbons, CO; sulphur compounds except (OCS); nitrogen compounds (except N_2O); chlorine compounds (except CFC's)
Aerosol scavenging	Nitric acid, ammonia, iodine
Stratospheric photolysis	CFC's, N_2O, OCS

(*a*) *Deposition to the underlying surface.* Deposition to water surfaces, plants, vegetation, and soil surfaces, *viz* dry deposition, requires both the transport of trace gases within the atmospheric boundary layer and their subsequent reaction or adsorption at the surface or on the surface elements.[15] Dry deposition therefore only acts efficiently on trace gases close to the surface where a specific chemical or biological interaction is available.

The flux of a trace gas to the surface, F, and the concentration, c, of the trace gas both measured at the same height in the boundary layer, usually 1 m, are related [equation (31)] by the deposition velocity, v. For trace gases which are efficiently adsorbed or removed chemically at the surface, values of v may approach 10 mm s^{-1}. For example, v for SO_2 in north-west Europe[16] has a mean value of *ca.* 8 mm s^{-1}, and ozone to most vegetated surfaces has *ca.* 6 mm s^{-1}.[17]

$$F = cv \qquad (31)$$

The effect of scale becomes evident when comparing the cases of manmade SO_2, which is deposited from a 1 km well mixed layer, with that of background ozone which is well mixed up to the tropopause at a height of *ca.* 10 km. The timescale for deposition, t, is related to the scale height of the trace gas, h, and the deposition velocity by equation (32). The deposition timescales for manmade SO_2 and background ozone are accordingly 35 h and 14 days, respectively, illustrating the influence of trace-gas scale height.

$$t = \frac{h}{v} \qquad (32)$$

[15] A. C. Chamberlain, *Int. J. Air Water Pollution*, 1960, **3**, 63.
[16] J. A. Garland, *Proc. R. Soc. London, Ser. A*, 1977, **354**, 245.
[17] J. A. Garland, W.M.O. Symposium No. 538, 1979, pp. 95, World Meteorological Office, Geneva, Switzerland.

The dry deposition of aerosol particles is highly dependent on their particle size distribution. Very large particles deposit chiefly by sedimentation, but particles in the range 1—100 μm are also transported towards the surface by turbulence where sedimentation is supplemented by impaction. Aitken particles (<0.1 μm) diffuse readily to the surface by Brownian motion, so that again deposition is efficient. As a result, the deposition velocity of airborne particulate material shows a minimum between 0.1 and 1 μm where none of these processes is efficient.[18]

Much of the mass of the sulphate aerosol found in the atmosphere falls in the above size range, and a mean value of v of 1 mm s^{-1} has been estimated.[18] Removal of the important submicron atmospheric aerosol by dry deposition would therefore require several weeks. Other mechanisms, *viz* wet deposition, are more important for these aerosol particles.

(b) *Removal by precipitation.* The removal of trace gases by precipitation, referred to as wet deposition, results from the incorporation of minor constituents into falling precipitation (wash-out) and by incorporation into cloud droplets (rain-out). These sink processes are necessarily only significant for those species which are readily water soluble.

The exchange of water-soluble gases in the atmosphere such as SO_2,[19] and HTO,[20] with falling rain drops depends on both rainfall rate and drop size. The removal rate of the trace gas increases from 10^{-4} s^{-1} at a rainfall rate of 1 mm h^{-1} to *ca.* 3.5×10^{-4} s^{-1} at 10 mm h^{-1}.[19] Removal times for a soluble gas such as sulphur dioxide are thus expected to be several hours in moderate rain.

Wash-out is, however, unable to account for the presence of much of the sulphate found in precipitation. The most important route whereby sulphate is incorporated into precipitation appears to involve the rain-out of cloud condensation nuclei.[18] Within the cloud, water vapour condenses on the hygroscopic sulphate aerosol resulting in the formation of cloud droplets. Precipitation elements are formed by coagulation and, below freezing point, by distillation onto ice crystals.[21] It therefore appears that the oxidation of SO_2 either homogeneously by OH radicals or heterogeneously in droplets, combined with rain-out of the aerosol, provides an efficient combined sink process.

The combination of oxidation and rain-out appears to be a general atmospheric sink process which competes with wash-out of the parent trace gas. In addition to SO_2, this sink may operate efficiently for other sulphur- and chlorine-containing minor constituents.

(c) *Aerosol scavenging.* Heterogeneous scavenging by aerosol particles can also serve as an important sink process for such gaseous species as nitric acid, hydrogen peroxide, sulphuric acid, nitrogen pentoxide, and ammonia. Aerosol scavenging followed by the subsequent rain-out of the aerosol may well compete against wash-out for these gaseous minor constituents.

[18] J. A. Garland, *Atmos. Environ.* 1978, **12**, 349.
[19] A. C. Chamberlain, A.E.R.E. Report HP/R, 1261, 1953.
[20] A. C. Chamberlain and A. E. J. Eggleton, *Int. J. Air Water Pollution*, 1964, **8**, 135.
[21] B. J. Mason, 'Physics of Clouds', Clarendon Press, Oxford, 1971.

Aerosol scavenging processes may be separated into: condensation and nucleation processes which mainly involve water soluble species and are usually irreversible, and adsorption, which may involve any gas-phase species and may be reversible or irreversible.

Sulphuric acid vapour will undergo heteromolecular nucleation with water under atmospheric conditions but this has been shown by theoretical considerations to be unlikely for nitric acid[22] and other more volatile acidic vapours.

The rate constants for the uptake of minor constituents onto aerosols have been estimated for each particle size range.[23] From the typical values of accommodation coefficients for adsorption,[24] aerosol particle number densities and size distributions, life-times of 3–6 days have been estimated for scavenging of gases by aerosols in the background troposphere.[25]

Aerosol scavenging, wet and dry deposition may thus be important for certain specific minor constituents restricting their atmospheric lifetimes to *ca.* 1—10 days even if they are not chemically transformed. For those constituents not readily removed by these physical processes, atmospheric photolysis and photo-oxidation may provide alternative sink processes. A review of the basic photochemical and chemical kinetic data required for an investigation of the chemical removal processes is given in Section 3.

Modelling the Behaviour of Minor Constituents in the Troposphere.—An understanding of the behaviour of the minor constituents in the lower atmosphere requires, in principle, a description in three dimensions of atmospheric dynamics, atmospheric chemistry, and sources and sinks. The total problem is extremely complex and the computer requirements for any comprehensive scheme are enormous. In view of the requirement to make progress against this background with present limited understanding, simplified models have been employed to obtain useful results.

The simplifications which are usually employed involve:[26]

(*i*) reducing the number of species,
(*ii*) reducing the dimensionality,
(*iii*) removing the requirement of using meteorological equations,
(*iv*) limiting the spatial resolution, and
(*v*) averaging over time intervals.

Reducing the number of species required to describe the behaviour of the minor constituents entails examining the relative contribution of the different chemical processes and retaining only the most important. Section 3 reviews the present knowledge and understanding of the chemical processes with a view to identifying the most important processes and the gaps in present understanding.

[22] C. S. Kiang and D. J. Stauffer, *Faraday Symposia Chem. Soc.*, 1973, **7**, 26.
[23] A. C. Chamberlain, A. E. J. Eggleton, W. J. Megaw, and J. B. Morris, *Discuss., Faraday Soc.*, 1960, **30**, 162.
[24] J. A. Garland, *J. Nucl. Energy*, 1967, **21**, 677.
[25] R. G. Derwent and A. R. Curtis, A.E.R.E. Report R8853, H.M.S.O., London, 1977.
[26] 'Chlorofluorocarbons and Their Effect on Stratospheric Ozone', Pollution Paper No. 15, Dept. of Environment, H.M.S.O., London, 1979.

Instead of using the full three-dimensional (altitude–latitude–longitude) co-ordinate system, it is possible to consider, after averaging around latitude circles, only the two-dimensional latitude height cross-sections or to consider only the average global profile in a one-dimensional model. Since the lifetimes of the minor constituents of major interest here are longer than the transport time around latitude circles, a two-dimensional treatment should be adequate. However, since the OH radical shows marked latitudinal, seasonal, diurnal, and vertical gradients, one-dimensional treatments of troposphere chemistry are unlikely to be of value.

The wind fields in two-dimensions can be calculated by use of a general circulation model using the various meteorological equations. These require calculations of the radiation budget and hydrological cycle, together with the equations of motion and of state. The problem is more tractable in two-dimensions but nevertheless it is not certain how adequate the numerical representations are and whether adequate resolution is available to represent the major scales of atmospheric motion. The meteorological equations can be replaced by climato-logical values for the wind fields, pressures, temperatures, and atmospheric densities, producing a model that is *kinematic* rather than *dynamical*. This involves the use of parameterized values for transports by the large scale eddy motions in the form of diffusion coefficients.

A two-dimensional model of a minor constituent, i, could therefore be represented by equation (33):

$$\frac{\partial c_i}{\partial t} + \mathrm{div}(c_i u) = S_i + \mathrm{div}[KN \, \mathrm{grad}(c_i/N)] \qquad (33)$$

where c_i is concentration of the species i, u represents the wind field, s_i the net source term, K is an eddy diffusion tensor, and N is the atmospheric density. The net source term comprises: the chemical and photochemical production and destruction of the constituent, emission from the surface, and deposition or other physical removal. In principle all these various parameters and processes have been described so that equation (33) can be solved with useful resolution for a number of practical problems.[25]

3 Photochemical and Kinetic Data

The photochemical and kinetic data required as input for any calculations in tropospheric chemistry comprises rate coefficients for all the processes, both thermal and photochemical, which are identified as important in the description of the behaviour of a particular trace gas. In this Section we review the available data for reactions involved in the basic tropospheric trace gas cycles.

Thermal Reactions.—Most of the reactions we consider are elementary, and the rate expression can be derived from the reaction format (34). For unimolecular and

$$A + B \rightarrow \text{products} \qquad (34)$$

i.e. $-\mathrm{d}[A]/\mathrm{d}t = -\mathrm{d}[B]/\mathrm{d}t = \mathrm{d}[\text{Product(s)}]/\mathrm{d}t = k[A][B]$

bimolecular reactions the rate coefficients are normally expressed in Arrhenius form, giving the temperature-dependence, *i.e.*, $k = p \exp(-Q/T)$; $Q = E/R$, where E is the activation energy. Since mean temperature varies greatly with altitude and latitude in the troposphere, knowledge of the temperature dependence, particularly at temperatures <300 K, is important.

The rates of combination and the reverse dissociation reactions between atoms, radicals, and small molecules may exhibit pressure-dependences in the relevant pressure range, equation (35). The combination reactions can be described by a

$$A + B + M \rightleftharpoons AB + M \quad \text{([M] proportional to total pressure)} \quad (35)$$

pseudo-second-order rate law in which the rate coefficient depends on [M], as in (36). The low-pressure third-order limit is characterized by k_0 {lim([M] $\rightarrow$ 0)

$$d[AB]/dt = k[A][B], \quad k = f([M]) \quad (36)$$

$= k[M]$} which is directly proportional to [M]. The high-pressure second-order limit is characterized by k_∞ {lim([M] $\rightarrow \infty) = k$} and is independent of [M]. The dependence of k on [M] in the transition region between the third- and the second-order regime is generally complex and depends on the detailed nature of the reactants. A simplified approximate procedure for determining k in the transition region has been developed by Troe,[27] which has been employed in recent evaluations of rate constants for atmospheric chemistry.[28, 29] In this treatment k is given in terms of k_0, k_∞, and F, a 'broadening factor' describing the shape of the fall-off curve, equation (37). The quantity F is a function of [M] and is given by equation (38) where F_c is the broadening factor at the pressure corresponding to the

$$k = \frac{k_0 k_\infty}{k_0 + k_\infty} F \quad (37)$$

$$\log F = \frac{\log F_c}{1 + [\log([M]/[M_c])]^2} \quad (38)$$

centre of the fall-off curve, [M_c]. Here $k_0 = k_\infty$, *i.e.*, the extrapolated third-order rate constant is equal to k_∞. The temperature-dependence of third-order reactions is expressed in the T^n-form since this gives a better fit to data over a wider range of temperature for this type of reaction. The thermal dissociation reactions can be treated analogously to the association reactions using pseudo-first-order rate constants, $k[M]$.

A few of the reactions considered as elementary may in fact occur by a complex mechanism, *e.g.*, disproportionation of HO_2 and reaction of OH with CO, and consequently may exhibit pressure- and temperature-dependences not consistent with the foregoing format. In this case rate coefficients for a pressure of 760 Torr are given in the Arrhenius form.

[27] J. Troe, *J. Phys. Chem.*, 1979, **83**, 114.
[28] NASA Ref. Publ. 1010, 'Chlorofluoromethanes and the Stratosphere, ed. R. D. Hudson, N.A.S.A. Ref. Publn. 1010, 1977.
[29] 'Evaluated Kinetic and Photochemical Data for Atmospheric Chemistry', CODATA Bull. 33, I.C.S.U. CODATA Paris, 1979; also *J. Phys. Chem. Ref. Data*, 1980, **9**, 295.

Photochemical Reactions.—Photodissociation of a trace gas species can be represented as a simple first-order process (39) and the rate is defined as

$$A + hv \rightarrow \text{products} \tag{39}$$

$-d[A]/dt = J[A]$, where J is the photodissociation rate coefficient with units of time^{-1}; J is a function of the spectral composition and intensity of sunlight, the absorption cross-section σ, and the overall quantum yield for photodissociation Φ through all identified channels, the last-named quantities being dependent on wavelength, λ. For the trace gases present in the troposphere it is assumed that the approximate form of the Beer–Lambert law for weak absorption applies. Assembling the variables we obtain equation (40) where the summation is carried out over all relevant λ.

$$J = \sum_{\lambda} J_{\lambda} = \sum_{\lambda} \sigma_{\lambda} I_{\lambda} \Phi_{\lambda} \tag{40}$$

Information on σ and Φ is generally obtained from laboratory measurements and has been reviewed and compiled.[28-30]. More problematical is the specification of the solar spectral irradiance, I_{λ}, in the atmosphere as a function of location and time. Observational data on the solar flux is insufficient for the requirements of all but the simplest calculations, and models of the troposphere rely on a radiative transfer calculation to determine the spectral irradiance. However, radiative transfer in the u.v. and visible regions of the spectrum is a particularly complex problem in the troposphere, involving multiple molecular scattering, aerosol scattering, surface albedo, and cloud reflection, and can only be solved approximately. A comprehensive discussion of the problem is given by Leighton,[2] and Isaksen *et al.*[31] have described a simplified scheme for calculating photochemical rate coefficients in the lower atmosphere as a function of solar zenith angle, surface albedo, and taking into account multiple scattering. The J values given in the following sections were calculated using the latter scheme for a latitude of 50°N at the summer solstice, *i.e.*, zenith angle $z = 27°$, unless indicated otherwise.

Ozone Photochemistry and O Atom Reactions.—Kinetic and photochemical data for reactions of ozone and O atoms are given in Table 3.

The calculation of the primary production rate of OH radicals in the lower atmosphere requires accurate rate data for the processes already shown in equations (1)—(3). The photodissociation process (1) has a threshold wavelength

$$O_3 + hv \rightarrow O(^1D) + O_2(^1\Delta_g) \tag{1}$$

$$O(^1D) + H_2O \rightarrow 2OH \tag{2}$$

$$O(^1D) + M \rightarrow O(^3P) + M \quad (M = O_2 \text{ or } N_2) \tag{3}$$

of 310 nm (based on ΔH_0^0) and the evidence strongly suggests that $O(^1D)$ atoms are not produced at longer wavelengths *via* the spin-forbidden process (41) yielding

$$O_3 + hv \rightarrow O(^1D) + O_2(^3\Sigma_g)(\lambda \leqslant 411 \text{ nm}) \tag{41}$$

[30] R. F. Hampson, *J. Phys. Chem. Ref. Data*, 1973, **2**, 267.
[31] I.S.A. Isaksen, K. H. Midtbo, J. Sunde, and P. J. Crutzen, *Geophys. Norveg.*, 1977, **31**, 11.

Table 3 *Photolysis of ozone and O atom reactions*

Ozone photolysis	J/s^{-1}	Notes
$O_3 + hv \rightarrow O(^1D) + O_2$	5×10^{-6}	$\Phi = 1, \lambda < 300$ nm
		$0 < \Phi < 1, \lambda\ 318\text{—}300$ nm

$O(^1D)$ reactions	k/cm^3 molecule^{-1} s^{-1}	
$O(^1D) + O_2 \rightarrow O(^3P) + O_2$	$3.7 \times 10^{-11} \exp(67/T)$	O_2 in $^1\Sigma$ or $^3\Sigma$ state[a]
$O)^1D) + N_2 \rightarrow O(^3P) + N_2$	$3.2 \times 10^{-11} \exp(107/T)$	[a]
$O(^1D) + H_2O \rightarrow 2OH$	2.8×10^{-10}	[a]

$O(^3P)$ reactions		
$O(^3P) + O_2 + M \rightarrow O_3 + M$	$5.6 \times 10^{-34} (T/300)^{-2.23}.$ M $M = N_2$[a]	

[a] CODATA value: see ref. 29.

ground state O_2. For example, Jones and Wayne[32] found a value near zero for the quantum yield of $O(^1D)$ formation, Φ, at 334 nm. It is generally accepted that at wavelengths below 300 nm, $\Phi = 1$, although the accuracy of previous work cannot exclude a significantly lower value, as obtained in a recent investigation of the photofragment spectroscopy of O_3, *i.e.*, $\Phi = 0.9$ at 274 nm.[33] The remaining fraction led to $O(^3P)$ formation. This question requires further study.

At wavelengths >300 nm, Φ decreases steadily to near zero at 320 nm, and in recent years a great effort has been made to determine the exact $O(^1D)$ quantum yield curve as a function of wavelength.[34,35] Moreover a temperature dependence of the quantum yield has been observed[36] in which Φ decreases with temperature at a given wavelength in the wavelength-dependent region. The available data on the temperature- and wavelength-dependence of Φ are now in good agreement, and Moortgart and Kudszus[37] have developed a useful empirical expression giving these dependences over the range 230—320 K and 295—320 nm, so that J_1 can be calculated for the considerably variable temperatures and sunlight spectra, as a function of altitude and latitude, in the troposphere.

The important thermal kinetic parameters for calculation of primary OH production are the rates of reactions (2) and reaction (3). Since the quenching and reaction of $O(^1D)$ is very rapid only the relative values of the rate constants are critical. Thus the uncertainty arising from the disparity which exists between the absolute $O(^1D)$ rate constant measurements using the $O(^1D) \rightarrow O(^3P)$ 630 nm emission technique[38] and the atomic resonance absorption technique,[39] does not apply to the relative rates. In a recent CODATA evaluation,[29] relative rates constants of $O(^1D)$ reactions have been assessed relative to the reaction (or quenching) rate with CO_2. The resultant rate constant ratio $k_2/k_1 = 6.3$ (where k_1 is for $M = N_2 + O_2$ in the atmospheric ratio) has an uncertainty of only $\pm 15\%$. It is

[32] I. T. N. Jones and R. P. Wayne, *Proc. R. Soc. London, Ser. A*, 1970, **319**, 273.

[33] C. E. Fairchild, E. J. Stone, and G. M. Lawrence, *J. Chem. Phys.*, 1978, **69**, 3632.

[34] G. K. Moorgat and P. Warneck, *Z. Naturforsch., Teil A*, 1975, **30**, 835.

[35] I. Arnold, F. J. Comes, and G. K. Moorgat, *Chem. Phys.*, 1977, **24**, 211.

[36] S. Kunis, R. Simonaitis, and J. Heicklen, *J. Geophys. Res.*, 1975, **80**, 1328.

[37] G. K. Moorgat and E. Kudszus, *Geophys. Res. Lett.*, 1978, **5**, 191.

[38] J. A. Davidson, C. M. Sadowski, H. I. Schiff, G. E. Streit, C. J. K. Howard, D. A. Jennings, and A. L. Schmeltekopf, *J. Chem. Phys.*, 1976, **64**, 57.

[39] R. F. Heidner, D. Husain, and J. R. Wiesenfeld, *J. Chem. Soc., Faraday Trans. 2*, 1973, **69**, 927.

generally assumed, and the evidence indeed suggests, that the major channel for reaction (2) is that yielding 2OH. A small contribution from the alternative pathways yielding either $H_2 + O_2$ or $O(^3P) + H_2O$ (*i.e.*, quenching) cannot be ruled out on the basis of available information.

Reactions of HO_x and O_x Species.—In the background troposphere removal of the active 'odd hydrogen' radical species OH and HO_2 is mainly by the radical combination reactions (23) and (24). Despite the recognition of their importance,

$$HO_2 + HO_2 \rightarrow H_2O_2 + O_2 \qquad (23)$$

$$OH + HO_2 \rightarrow H_2O + O_2 \qquad (24)$$

the kinetics of neither of these reactions is well understood. There are now three reported measurements[40–42] of k_{24} in low-pressure fast-flow systems with direct observation of OH and HO_2 giving values in the range 2 to 5.1 $\times$ 10^{-11} cm^3 molecule^{-1} s^{-1}. These values disagree markedly from those obtained from flash photolysis of $H_2O-H_2-O_2$ mixtures[43] and from steady state photolysis experiments involving $O_3-H_2O-O_2$ mixtures conducted by DeMore and co-workers,[44,45] which indicate that k_{24} lies between 1.0 and 2 $\times$ 10^{-10} cm^3 molecule^{-1} s^{-1}. These results can be reconciled if k_{24} is subject to a pressure effect, but this is unexpected for an apparently simple bimolecular reaction between two radical species.

Recent extensive kinetic investigations of the disproportionation of HO_2, reaction (4), have confirmed the earlier measurements of k_4 at 298 K, and pressures near 1 atm.[46,47] However, the rate of reaction (23) has been shown to decrease at pressures below *ca.* 10 Torr, and to possess a surprisingly large negative temperature dependence at pressures near 1 atm.[46] Furthermore k_{23} is apparently increased in the presence of water vapour.[47] The effects have been observed in both direct and relative rate studies of HO_2 kinetics, and have been interpreted in terms of an H_2O_4 intermediate species and the participation of an HO_2-H_2O hydrate. Irrespective of the mechanistic details, the temperature and water vapour effects need to be taken into account in modelling tropospheric chemistry. Unfortunately good data for k_4 as a function of H_2O pressure is only available for 298 K, and since the indications are that the water vapour effect is markedly temperature dependent, more experimental data are needed in this area.

Both OH and HO_2 are interconverted through reaction with ozone, and these reactions [(17) and (41)] play a role in the background troposphere, particularly in regions of very low NO_x concentration. There are now reliable data for k_{41} and k_{17} over a quite large range of temperature and pressure. The values given summarized in Table 4 are mostly taken from a recent CODATA evaluation.[29] The activation

[40] J. P. Burrows, G. W. Harris, and B. A. Thrush, *Nature (London)*, **267**, 233.
[41] W. Hack, A. W. Preuss, and H. Gg. Wagner, *Ber. Bunsenges, Phys. Chem.*, 1978, **82**, 1167.
[42] J. S. Chang and F. Kaufman, *J. Phys. Chem.*, 1978, **82**, 1683.
[43] C. K. Hochanadel, J. A. Ghormley, and P. J. Ogren, *J. Chem. Phys.*, 1972, **56**, 4426.
[44] W. B. DeMore and E. Tschuikow-Roux, *J. Phys. Chem.*, 1974, **78**, 1447.
[45] W. B. DeMore, *J. Phys. Chem.*, 1979, **83**, 1113.
[46] R. A. Cox and J. P. Burrows, *J. Phys. Chem.*, 1979, **83**.
[47] E. J. Hamilton and R. R. Lii, *Int. J. Chem. Kinet.*, 1977, **9**, 875.

Table 4 *Reactions of* HO_x *species*

Reaction	k/cm^3 molecule^{-1} s^{-1}
$OH + HO_2 \rightarrow H_2O + O_2$	3.5×10^{-11a}
$OH + O_3 \rightarrow HO_2 + O_2$	$1.9 \times 10^{-12} \exp(-1000/T)^a$
$OH + H_2O_2 \rightarrow H_2O + HO_2$	$7.6 \times 10^{-12} \exp(-670/T)^a$
$OH + H_2 \rightarrow H_2O + H$	$1.8 \times 10^{-11} \exp(-2330/T)^a$
$HO_2 + HO_2 \rightarrow H_2O_2 + O_2$	$3.8 \times 10^{-14} \exp(1245/T)^b$
$HO_2 + O_3 \rightarrow OH + 2O_2$	$1.4 \times 10^{-14} \exp(-580/T)^a$

Photolysis of H_2O_2	J/s^{-1}
$H_2O_2 \rightarrow 2OH$	6.5×10^{-6c}

[a] Ref. 29; [b] 760 Torr value; R. A. Cox and J. P. Burrows, *J. Phys. Chem.*, 1979, **83**, 2560; [c] $\Phi = 1$, $\lambda < 350$ nm.

energy and the 'A' factor for reaction (17) are remarkably low, indicating an unusual transition state for this reaction.

$$OH + O_3 \rightarrow HO_2 + O_2 \qquad (41)$$

$$HO_2 + O_3 \rightarrow OH + 2O_2 \qquad (17)$$

Reactions Involving H_2O_2.—The only significant source of H_2O_2 in the troposphere is by reaction (23), and it is removed by photodissociation, by reaction with OH, and by absorption both at the earth's surface and in cloud and rain droplets.

There is now excellent agreement between the measured u.v. absorption cross-sections of H_2O_2 in the region 350—190 nm.[48] The weak absorption above 290 nm is of particular importance for the troposphere. The experimental evidence suggests that process (43) is the only channel for dissociation in this wavelength

$$H_2O_2 + h\nu \rightarrow 2OH \qquad (43)$$

region, with a quantum yield of unity. The value for J at ground level, a solar zenith angle of 27°, gives a photodissociation rate of H_2O_2 of *ca.* 2.3% h^{-1}. The alternative gas-phase sink is reaction (15) for which $k_{15} = 8.0 \times 10^{-13}$ cm^3

$$HO + H_2O_2 \rightarrow H_2O + HO_2 \qquad (15)$$

molecule^{-1} s^{-1} at 298 K. This gives a removal rate of the order of 1% h^{-1} in sunlight, and thus photodissociation is the dominant daytime tropospheric sink in the absence of cloud and rain droplets. An interesting aspect of tropospheric chemistry is the behaviour of H_2O_2 dissolved in cloud droplets, where it can oxidize dissolved ionic species such as sulphite (from SO_2) and halide.[49]

Molecular hydrogen H_2 is also removed in the troposphere by reaction (8) with OH. The rate constant and its temperature-dependence are well established.[29]

$$OH + H_2 \rightarrow H_2O + H \qquad (8)$$

[48] L. T. Molina, S. D. Schinke, and M. J. Molina, *Geophys. Res. Lett.*, 1977, **4**, 580.
[49] S. A. Penkett, B. M. R. Jones, K. A. Brice, and A. E. J. Eggleton, *Atmos. Environ.*, 1979, **13**, 123.

Reactions Involving NO_x.—The most important odd nitrogen species in tropospheric chemistry are NO, NO_2, and HNO_3. In addition the higher oxides NO_3 and N_2O_5 and the oxyacids HONO (nitrous) and HO_2NO_2 (peroxynitric) also play a role in the NO_x cycle. The relative concentrations of the 'active' species, NO and NO_2, are controlled by the photochemical equilibrium resulting from reactions (4), (19), and (44). The photochemical and kinetic data needed to define this equilibrium, *i.e.*, J_{19} and k_{44}, are now well known and the equilibrium has been shown to give a good description of the simultaneously observed concentration of NO, NO_2, and O_3, in the atmosphere.[50]

$$O + O_2 + M \rightarrow O_3 + M \tag{4}$$

$$NO_2 + h\nu \rightarrow NO + O \tag{19}$$

$$NO + O_3 \rightarrow NO_2 + O_2 \tag{44}$$

Both NO and NO_2 react significantly rapidly with OH, HO_2, O_3, and NO_3 and rate constants for all these reactions are given in Table 5. The key reactions are

Table 5 *Reactions of* NO_x *species*

Reaction	k/cm³ molecule⁻¹ s⁻¹	Notes
$OH + NO + M \rightarrow HONO + M$	$k_0 = 6.5 \times 10^{-31} (T/300)^{-2.4}.$ [M]	$[M] = N_2$,[a]
	$k_\infty = 1.0 \times 10^{-11}$	
$OH + NO_2 + M \rightarrow HONO_2 + M$	$k_0 = 2.6 \times 10^{-30} (T/300)^{-2.7}.$ [M]	$[M] = N_2$,[a]
	$k_\infty = 1.6 \times 10^{-11}$	
$HO_2 + NO \rightarrow NO_2 + OH$	$4.3 \times 10^{-12} \exp(200/T)$	[a]
$HO_2 + NO_2(+M) \rightarrow HO_2NO_2(+M)$	1×10^{-12}	760 Torr value[b]
$O_3 + NO \rightarrow NO_2 + O_2$	$2.3 \times 10^{-12} \exp(-1450/T)$	[a]
$O + NO_2 \rightarrow NO_3 + O_2$	$1.2 \times 10^{-13} \exp(-2450/T)$	[a]
$NO_3 + NO \rightarrow 2NO_2$	2×10^{-11}	298 K value[c]
$NO_3 + NO_2 + M \rightarrow N_2O_5 + M$	$k_0 = 1.5 \times 10^{-30} (T/300)^{-4.6}.$ [M]	[a]
	$k_\infty = 5 \times 10^{-12}$	
$N_2O_5 + M \rightarrow NO_3 + NO_2 + M$	$k_0 = 8.8 \times 10^{-6}$	[a]
	$\exp(-9700/T).$ N_2/s^{-1}	
	$k_\infty = 5.7 \times 10^{14}$	
	$\exp(-10600/T)/s^{-1}$	
$HO_2NO_2 + M \rightarrow HO_2 + NO_2 + M$	$1.4 \times 10^{14} \exp(-10420/T)/s^{-1}$	760 Torr value[d]
$OH + HNO_3 \rightarrow H_2O + NO_3$	8.5×10^{-14}	Independent of T,[a]

Photolysis of NO_x *species*	J/s^{-1}	
$NO_2 + h\nu \rightarrow NO + O$	7.5×10^{-3}	$\Phi = 1, \lambda < 398$ nm
$HNO_3 + h\nu \rightarrow OH + NO_2$	6.0×10^{-7}	[e]
$HONO + h\nu \rightarrow OH + NO$	4.4×10^{-4}	[f]
$HO_2NO_2 + h\nu \rightarrow$ Products	$1.0 \times 10^{-4} (z = 0°)$	[g]
$N_2O_5 + h\nu \rightarrow$ Products	1.8×10^{-5}	[c]
$NO_3 + h\nu \rightarrow NO + O_2$	3.2×10^{-2}	Based on results of
$\rightarrow NO_2 + O$	8.4×10^{-2}	footnote[c]

[a] Ref. 29; [b] R. A. Cox and K. G. Patrick, *Int. J. Chem. Kinet.*, 1979, **11**, 635; [c] R. A. Graham and H. S. Johnston, *J. Phys. Chem.*, 1978, **28**, 254; [d] R. A. Graham, A. M. Winer, and J. N. Pitts, *J. Chem. Phys.*, 1978, **68**, 4505; [e] H. S. Johnston and R. A. Graham, *J. Phys. Chem.*, 1973, **77**, 62; [f] R. A. Cox and R. G. Derwent, *J. Photochem.*, 1976, **6**, 23; W. R. Stockwell and J. G. Calvert, *ibid.*, 1978, **8**, 193; [g] R. A. Graham, A. M. Winer, and J. N. Pitts, jun., *Geophys. Res. Lett.*, 1978, **5**, 909.

[50] J. G. Calvert, *Environ. Sci. Technol.*, 1976, **10**, 250.

those of HO_2 with NO (16) and OH with NO_2 (22). Reaction (22) provides the major sink for the active NO_x species, since HNO_3 is relatively stable towards gas phase removal either by photolysis or reaction with OH, and is removed mainly through rain or absorption at the surface. The reaction of HO_2 with NO serves to convert the relatively unreactive HO_2 radical into OH. The two reactions have been the subject of intensive study since their significance was realized. The pressure- and temperature-dependence of the rate constant k_{22} has been investigated for the range of conditions relevant to the troposphere[51] and numerous studies at low pressure give a reliable value for the limiting third-order rate constant. The high value of k_{12}, near 8×10^{-12} cm^3 molecule^{-1} s^{-1} at 298 K, first reported in 1977,[52] has been independently confirmed in several laboratories and its temperature-dependence has been investigated.[53] The reaction exhibits a slight negative temperature coefficient which possibly suggests the involvement of an association complex. There is no compelling evidence for a pressure effect on this very important reaction, however.

$$HO_2 + NO \rightarrow NO_2 + OH \qquad (16)$$

$$OH + NO_2 + M \rightarrow HNO_3 + M \qquad (22)$$

The combination reactions (45)—(47) all lead to rather unstable products which do not provide net sinks for NO_2 or NO. Thus HONO is rapidly redissociated by photolysis in the near u.v. region,[54] HO_2NO_2[55] and N_2O_5[56] are both thermally unstable, at least at temperatures prevalent near the earth's surface, redissociating back to their precursor radicals, reverse reactions (46) and (47). However, in the cooler regions of the troposphere, these molecules are stable and may be removed by photolysis. Photolysis also potentially provides an important loss process for NO_3 at all altitudes, since it has a strong absorption in the visible region.[57] Unfortunately data on absorption cross-sections and on the various possible dissociation pathways of NO_3, N_2O_5, and HO_2NO_2, are far from satisfactory at the present time. Table 5 gives references to recently published data for the photodissociation processes.

$$OH + NO + M \rightarrow HONO + M \qquad (45)$$

$$HO_2 + NO_2 + M \rightarrow HO_2NO_2 + M \qquad (46)$$

$$NO_3 + NO_2 + M \rightarrow N_2O_5 + M \qquad (47)$$

Also less well defined from the kinetic standpoint is the chemistry of NO_3 and N_2O_5, particularly reaction (47) and the reaction of NO_3 with NO, owing to the chemical complexity of the system in which they have been studied.[57] There is also uncertainty in the rate of reaction of N_2O_5 with H_2O to give HNO_3 which seems to occur by a heterogeneous reaction.

The pressure- and temperature-dependence of the thermal decomposition of peroxynitric acid has been determined over a wide range of conditions[55] but not the

[51] C. Anastasi and I. W. M. Smith, *J. Chem. Soc., Faraday Trans. 2*, 1976, **72**, 1459.
[52] C. J. Howard, *J. Chem. Phys.*, 1979, **71**, 2352.
[53] M.-T. Leu, *J. Phys. Chem.*, 1979, **70**, 1662.
[54] R. A. Cox and R. G. Derwent, *J. Photochem.*, 1976, **6**, 23.
[55] R. A. Graham, A. M. Winer, and J. N. Pitts, jun., *J. Chem. Phys.*, 1978, **68**, 4505.
[56] P. Connell and H. S. Johnston, *Geophys. Res. Lett.*, 1979, **6**, 553.
[57] R. A. Graham and H. S. Johnston, *J. Phys. Chem.*, 1978, **82**, 254.

temperature dependence of the reverse reaction. Consequently the rate of formation of HO_2NO_2 cannot be defined with confidence over the whole range of conditions in the troposphere.

CH₄ Oxidation.—The reactions which result in the oxidation of atmospheric methane play an important role in governing the steady state of OH and other radicals in the troposphere. CH_4 is removed by OH attack and oxidized to formaldehyde through methyl, methylperoxy, and methoxy radicals[58,59] [reactions (11)—(14)]. Kinetic data for these reactions have been the subject of investigation

$$OH + CH_4 \rightarrow CH_3 + H_2O \tag{11}$$

$$CH_3 + O_2 + M \rightarrow CH_3O_2 + M \tag{12}$$

$$CH_3O_2 + NO \rightarrow CH_3O + NO_2 \tag{13}$$

$$CH_3O + O_2 \rightarrow HCHO + HO_2 \tag{14}$$

over a number of years in view of their importance in low temperature combustion, but there are still quite large uncertainties on some of the key reactions and the alternative pathways. Table 6 surveys the methane cycle reactions.

Table 6 *Reactions involved in* CH_4 *oxidation*

Reaction	k/cm^3 molecule^{-1} s^{-1}	Notes
$OH + CH_4 \rightarrow CH_3 + H_2O$	$2.36 \times 10^{-12} \exp(-1710/T)$	a
$CH_3 + O_2(+M) \rightarrow CH_3O_2(+M)$	2×10^{-12}	k_∞ value^a
$CH_3O + O_2 \rightarrow HCHO + HO_2$	$5 \times 10^{-13} \exp(-2000/T)$	a
$CH_3O_2 + NO \rightarrow CH_3O + NO_2$	7.5×10^{-12}	298 K value^a
$CH_3O_2 + HO_2 \rightarrow CH_3O_2H + O_2$	$7.7 \times 10^{-14} \exp(1296/T)$	760 Torr value^b
$CH_3O_2 + CH_3O_2$		
$\rightarrow CH_3OH + HCHO + O_2$	3.0×10^{-13}	298 K values^c
$\rightarrow 2CH_3O + O_2$	1.6×10^{-13}	
$OH + HCHO \rightarrow H_2O + HCO$	1.3×10^{-11}	a
$HCO + O_2 \rightarrow CO + HO_2$	5.1×10^{-12}	a
$OH + CO \rightarrow CO_2 + H$	1.5×10^{-13}	<100 Torr^a
$OH + CO(+O_2) \rightarrow CO_2 + HO_2$	$1.0 \times 10^{-13} \exp(260/T)$	760 Torr air
Photochemical processes	J/s^{-1}	
$CH_3O_2H + h\nu \rightarrow CH_3O + OH$	6.5×10^{-6}	Same as J (H_2O_2)
$HCHO + h\nu \rightarrow H + HCO$	2.0×10^{-5}	$\Phi = 0$, $\lambda > 338$ nm
$\rightarrow H_2 + CO$	4.5×10^{-5}	Φ pressure and wavelength dependent 760 Torr value given for J

^{a} Ref. 29; ^{b} R. A. Cox and G. S. Tyndall, *Chem. Phys. Lett.*, 1979, **65**, 357; *J. Chem. Soc., Faraday Trans. 2*, 1980, **76**, 153; ^{c} D. A. Parkes, *Int. J. Chem. Kinet.*, 1977, **9**, 451.

[58] H. Levy, *Planet. Space Sci.*, 1973, **21**, 575.
[59] R. A. Cox, R. G. Derwent, P. M. Holt, and J. A. Kerr, *J. Chem. Soc. Faraday Trans. 1*, 1976, **72**, 2044.

The rate of reaction of OH with CH_4 is now well established for the relevant temperature range in the troposphere.[29] Although k_{12} and k_{14} are subject to some uncertainty, they are fast enough to be the dominant pathways for these radicals in the lower atmosphere. Recent measurements[60-62] of k_{13} have revealed that this reaction is fast, the value at 298 K being comparable with that for the reaction of HO_2 with NO, *i.e.*, *ca.* 8×10^{-12} cm^3 molecule^{-1} s^{-1}. Nevertheless, the CH_3O_2 radical may react by other pathways in regions where NO concentrations are very low. Thus the reactions (48)—(50) of CH_3O_2 with other peroxy radicals need to be

$$CH_3O_2 + HO_2 \rightarrow CH_3OOH + O_2 \qquad (48)$$

$$CH_3O_2 + CH_3O_2 \rightarrow CH_3OH + HCHO + O_2 \qquad (49)$$

$$\rightarrow 2CH_3O + O_2 \qquad (50)$$

taken into account. Reaction (44) of CH_3O_2 with NO_2 is also rapid[62] but the product, methyl peroxynitrate, is unstable.[63] Recent data for the analogous compound, isopropyl peroxynitrate, indicate that it redissociates to its precursor radicals ($C_3H_7O_2 + NO_2$) with a lifetime of 0.3 s at 298 K and 7 h at 225 K.[64] Thus it would appear that the alkyl peroxynitrates are not significant chemical species in the atmospheric oxidation of hydrocarbons, occurring in the lower troposphere.

$$CH_3O_2 + NO_2 \rightarrow CH_3O_2NO_2 \qquad (51)$$

The rate constant for reaction (48) has only been measured very recently.[62] The large negative temperature-dependence is similar to that observed in the disproportionation of HO_2 and may imply a complex mechanism and possibly a pressure-dependence. Even at 298 K, the reaction is relatively fast, $k_{48} = 6.5 \times 10^{-11}$ cm^3 molecule^{-1} s^{-1}, and consequently assumes an important role in the absence of NO. On the other hand, the disproportionation of CH_3O_2 is relatively slow at 298 K and probably of minimal importance. The temperature-dependence of this reaction has not been investigated, however, and there is also considerable uncertainty in the branching ratio k_{49}/k_{50} and its temperature-dependence.

The occurrence of reaction (48) leads to the requirement of photochemical and kinetic data for methyl hydroperoxide. In the absence of experimental data, it has been generally assumed that its behaviour with respect to photolysis, gas phase reactions, and involvement in physical removal mechanisms is analogous to that of hydrogen peroxide. A limited amount of data is now available on the u.v. absorption spectrum of CH_3OOH,[62] which suggests that photodissociation in the near-u.v. region may be somewhat faster than for H_2O_2.

The fate of atmospheric formaldehyde is determined by three competing processes, *i.e.*, photodissociation, attack by OH radicals, and absorption in cloud and rain droplets.

[60] I. C. Plumb, K. R. Ryan, J. R. Steven, and M. F. R. Mulcahy, *Chem. Phys. Lett.*, 1979, **63**, 255.
[61] H. Adachi and N. Basco, *Chem. Phys. Lett.*, 1979, **63**, 490.
[62] R. A. Cox and G. S. Tyndall, *Chem. Phys. Lett.*, 1979, **65**, 357.
[63] H. Niki, P. D. Maker, C. M. Savage, and L. P. Breitenbach, *Chem. Phys. Lett.*, 1978, **59**, 78.
[64] E. O. Edney, J. W. Spence, and P. L. Hanst, *J. Air Pollut. Control Assoc.*, 1979, **29**, 741.

Photolysis of formaldehyde proceeds *via* two channels, a molecular channel (6) producing H_2 + CO and dissociation (21) to produce radical fragments. Although

$$HCHO + hv \rightarrow H + HCO \qquad (6)$$

$$HCHO + hv \rightarrow H_2 + CO \qquad (21)$$

the primary processes in the photolysis of HCHO have been the subject of study for nearly 40 years, large disparities have existed regarding both the relative importance of the two channels as a function of wavelength and the energy threshold for channel (6). Recent intense study has largely solved the uncertainties, as will be seen from Figure 6, where the results for the quantum yields Φ_6 and $\Phi_6 + \Phi_{21}$ as a function of wavelength are illustrated. There is good agreement regarding the threshold wavelength, 338 ± 2 nm, above which $\Phi_6 = 0$. The data for $\Phi_6 + \Phi_{21}$ refer to high pressures and it will be seen that the total quantum yield falls off at wavelengths >330 nm. Two recent studies[65, 66] have shown that this effect is pressure-dependent and is an effect on Φ_{21} only, the values of Φ_{21} at $\lambda > 330$ nm increasing to near unity at pressures of a few Torr. The pressure-dependence on Φ_{21} must be taken into account when calculating the rate of production of H_2 from this source in the troposphere.

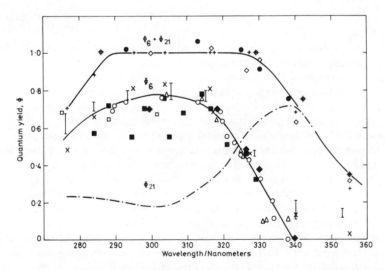

Figure 6 *Wavelength dependence of* HCHO *photolysis quantum yields; taken from data of* ■ *R. S. Lewis, K. Y. Tang, and E. K. C. Lee, J. Chem. Phys., 1976,* **65**, *2910;* ▲ *J. Marling, J. Chem. Phys., 1977,* **66**, *4200;* ◆, ◊ *J. H. Clark, C. B. Moore, and N. S. Nogar, J. Chem. Phys., 1978,* **68**, *1264;* □ *R. S. Lewis and E. K. C. Lee, J. Phys. Chem., 1978,* **82**, *249;* I, ⊕ *G. K. Moortgat, F. Slemr, W. Seiler, and P. Warneck, Chem. Phys. Lett., 1978,* **54**, *444;* ●, ○ *ref. 65; and* ×, + *ref. 66*

[65] A. Horowitz and J. G. Calvert, *Int. J. Chem. Kinet.* 1978, **10**, 805.
[66] G. K. Moortgat and P. Warneck, *J. Chem. Phys.*, 1979, **70**, 3639.

Computation of the rates J_6 and J_{21} also requires data for the absorption cross-section as a function of wavelength. The only complete coverage of the wavelength region of interest seems to be a low resolution (*ca.* 1 nm) measurement of the molar extinction coefficients at 373 K.[67] Higher resolution measurements at lower temperatures are badly needed.

The alternative gas-phase sink for HCHO is reaction (52) for which the rate constant is reasonably well established, although the possibility of a minor pathway producing H + HCOOH cannot be discounted. Both the major route of (52) and photolysis *via* process J_6 lead to formation of the HCO radical and the oxidation of this species has been the subject of some controversy.[68] In particular the question of whether the addition of O_2 to form HCO_3 is competitive with the fast bimolecular reaction (8). Interpretation of the chemistry in systems designed to investigate HCO_3 is difficult and no conclusive evidence for its formation has been presented.

$$OH + HCHO \rightarrow H_2O + HCO \qquad (52)$$

The oxidation of CO by OH is one of the most important reactions in tropospheric gas-phase chemistry and is also one of the most exhaustively studied gas reactions. The rate constant for the bimolecular reaction (9) is very well established at 298 K

$$OH + CO \rightarrow CO_2 + H \qquad (9)$$

and is known to exhibit a markedly non-Arrhenius temperature-dependence, with virtually no activation energy at temperatures <300 K. The overall rate of reaction of OH with CO, however, is pressure-dependent, the rate constant increasing with pressure to a value near 3×10^{-13} cm^3 molecule^{-1} s^{-1} at 760 Torr.[69,70] The effect appears to require the presence of O_2 and it has been suggested that an alternative channel to reaction (29), involving an HOCO intermediate, may be involved, *i.e.*, reactions (53) and (54). Whatever the origin of the pressure effect it is clear that it must be taken into account in modelling tropospheric chemistry. Values of the overall pressure-dependent rate constant over the range 20—760 Torr have been reported.[69] The temperature-dependence of the high-pressure rate constant has not been determined directly. Relative rate studies of Sie *et al.*[70] in which the OH + CO rate constant was measured relative to OH + H_2, give an activation energy of +4 kJ mol^{-1}, which is not consistent with a mechanism involving an energy transfer step to produce vibrationally relaxed HOCO. On the other hand, the high-pressure (700 Torr CO_2) value[71] for the ratio k_9/k_{35} together with the recent consensus value for k_{41}, give $k_9 = 1.0 \times 10^{-13} \exp(260/T)$, *i.e.*, a slight negative temperature-dependence which is more in line for an energy transfer step.

$$OH + CO \rightleftharpoons HOCO^* \rightarrow CO_2 + H \qquad (53)$$

$$HOCO^* \underset{}{\overset{+M}{\rightleftharpoons}} HOCO \xrightarrow{O_2} HO_2 + CO_2 \qquad (54)$$

[67] R. D. McQuigg and J. G. Calvert, *J. Amer. Chem. Soc.*, 1969. **91**, 1590.
[68] 'Chemical Kinetic Data Needs for Modeling the Lower Troposphere', N.B.S. Special Publications No. 557, eds. J. T. Herron, R. E. Huie, and J. A. Hodgeson, Washington, D.C., 1979, 46.
[69] H. W. Biermann, C. Zetzsch, and F. Stuhl, *Ber. Bunsenges. Phys. Chem.*, 1978, **82**, 633.
[70] B. K. T. Sie, R. Simonaitis, and H. Heicklen, *Int. J. Chem. Kinet.*, 1976, **8**, 85.
[71] W. B. DeMore, *Int. J. Chem. Kinet., Symp. 1*, 1975, 273.

Miscellaneous Reactions.—In addition to the reactions involved in the basic tropospheric photochemical–kinetic system, we present reaction kinetic data for reactions involving sulphur compounds, simple organic halogen compounds, a number of other organic compounds including peroxyacetyl nitrate, and some miscellaneous reactions. Most of the reactions involve the attack by OH radicals on various substrate molecules and therefore allow the scavenging rates of these molecules by atmospheric photochemically induced oxidation, to be evaluated. The primary products of these reactions have, in most cases been identified but the detailed chemical mechanism of the subsequent conversion of the initially formed species to stable products are largely unknown, and much more work needs to be done in this area, in order that a full description of the chemical aspects of the processes involved in the life cycle of a particular trace gas can be given.

Reactions of Sulphur Compounds. The reactions of OH with sulphur compounds, particularly SO_2, have received considerable attention in recent years and a survey is given in Table 7. Available data for reaction with H_2S and for SO_2 now seems to be in moderately good agreement; the pressure- and temperature-dependence of $k(OH + SO_2)$ are as expected for an addition reaction to form $HOSO_2$, but there appears to be no satisfactory description of the fate of the $HOSO_2$ radical; however, it seems clear that the oxidized SO_2 ends up as H_2SO_4 aerosol.

There is a considerable discrepancy between the reported rate coefficient values

Table 7 *Reactions of sulphur compounds*[2]

OH *Reactions*	$k/cm^3 \ molecule^{-1} \ s^{-1a}$	*Notes*
$OH + SO_2 + M \rightarrow HSO_3 + M$	$4.0 \times 10^{-14} \exp(960/T)$	Bimolecular rate[b] constant at 1 atmos
$OH + H_2S \rightarrow HS + H_2O$	$\begin{cases} 5.3 \times 10^{-12} \\ 1.4 \times 10^{-11} \exp(-225/T) \end{cases}$	[d]
$OH + CH_3SH \rightarrow$ Products	3.4×10^{-11}	Low pressure, direct[c]
$OH + (CH_3)_2S \rightarrow$ Products	9.0×10^{-11}	[c]
$OH + CS_2 \rightarrow HS + OCS$	$\begin{cases} 1.9 \times 10^{-13} \\ \leqslant 7 \times 10^{-14} \end{cases}$	[e] [f]
$OH + OCS \rightarrow HS + CO_2$	$\begin{cases} 5.66 \times 10^{-14} \\ \leqslant 7 \times 10^{-15} \end{cases}$	[e] [f]
Other reactions		
$HO_2 + SO_2 \rightarrow$ Products	$< 2 \times 10^{-17}$	[g]
$CH_3O_2 + SO_2 \rightarrow$ Products	5×10^{-15}	[h]
$O + CS_2 \rightarrow$ Products	$5.8 \times 10^{-11} \exp(-700/T)$	[d]
Photochemical reactions	J/s^{-1}	
$SO_2 + h\nu \rightarrow SO_2^*$	$\Phi(SO_2) \times 1.7 \times 10^{-4} \ (z = 40)$	$-\Phi(SO_2) \ll 1$[i]
$CS_2 + h\nu \rightarrow CS_2^*$	$\Phi(CS_2) \times 1.4 \times 10^{-4} \ (z = 30)$	$-\Phi(CS_2) < 1$[j]

[a] 300 K unless otherwise stated; [b] G. W. Harris, R. Atkinson, and J. W. Pitts, jun., *J. Chem. Phys.*, 1980, **69**, 378; [c] R. Atkinson, K. R. Darnall, A. C. Lloyd, A. M. Winer, and J. N. Pitts, *Adv. Photochem.*, 1979, **11**, 375; [d] Ref. 29; [e] M. Kurylo, *Chem. Phys. Lett.*, 1978, **58**, 238; [f] R. Atkinson, R. A. Perry, and J. N. Pitts, jun., *J. Chem. Phys. Lett.*, 1978, **54**, 14; [g] R. A. Graham, A. M. Winer, R. Atkinson, and J. N. Pitts, jun., *J. Phys. Chem.*, 1979, **83**, 1563; [h] E. Sanhueza, R. Simonaitis, and J. Heicklen, *Int. J. Chem. Kinet.*, 1979, **11**, 907; C. S. Kan, R. D. McQuigg, M. R. Whitbeck, and J. G. Calvert, *ibid.*, 1979, **11**, 921; [i] R. A. Cox, *Philos. Trans. R. Soc., London, Ser. A*, 1979, **290**, 543; *J. Phys. Chem.*, 1972, **76**, 814; [j] W. P. Wood and J. Heicklen, *J. Phys. Chem.*, 1971, **75**, 854.

for the reaction of OH with OCS and CS_2. Sources of the discrepancies in the two flash photolysis measurements are discussed in detail by Kurylo[72] and are attributed to complications due to photolysis of OCS and CS_2 and to the occurrence of secondary reactions. On this basis the higher values for the two rate coefficients would appear to be correct. However, relative rate measurements suggest that the lower limit estimate for k(OH + OCS) is more realistic.[73] Further work is necessary to resolve this problem.

In the presence of O_2, the HS radical is apparently converted into SO_2 but the mechanism is not known. Any of the reactions (55)—(57) may be involved, but their rate coefficients have not been determined with any certainty. SO is oxidized rapidly to SO_2 *via* reaction with O_3, O_2, or NO_2 but the fate of HSO is unknown.[74] Similarly the reactions of the radical fragments produced by attack of OH on the organic sulphur compounds have not been characterized.

$$HS + O_3 \rightarrow HSO + O_2 \qquad (55)$$

$$HS + O_2 \rightarrow OH + SO \qquad (56)$$

$$HS + NO_2 \rightarrow HSO + NO \qquad (57)$$

Oxidation of SO_2 may also occur *via* reaction with peroxy radicals. The rate coefficient for the reaction of CH_3O_2 with SO_2 has been determined independently[75,76] by the technique of flash photolysis, the values being in good agreement. Note the rate constant given in Table 7 is one-half of the measured CH_3O_2 decay constant in the studies because of secondary removal of CH_3O_2. The relatively high rate of this reaction is surprising in view of the slow rates observed for those of other peroxy radicals with SO_2, HO_2,[77] and CH_3COO_2.[78]

The photodissociation of SO_2 by reaction (58) requires light of wavelength

$$SO_2 + h\nu \rightarrow SO + O(^3P) \qquad (58)$$
$$\Delta H^0 = 547.6 \text{ kJ mol}^{-1}$$

$\lambda \leqslant 210$ nm. Therefore photodissociation of SO_2 is not possible in the troposphere where $\lambda \geqslant 280$ nm. Light absorption by SO_2 does occur at longer wavelengths with the formation [(59), (60)] of electronically excited states of the SO_2 molecule. Extensive experimental studies[79] have shown that the excited states of SO_2 undergo rapid physical quenching by the major atmospheric gases and consequently the quantum yields for oxidation of SO_2 by this route are extremely small.[80] This process appears to be of negligible importance in the atmosphere.[81,82] CS_2 appears

[72] M. J. Kurylo, *Chem. Phys. Lett.*, 1978, **58**, 238.
[73] R. A. Cox and D. H. Sheppard, *Nature (London)*, 1980, **284**, 330.
[74] K. H. Becker, M. Inocencio, and U. Schurath, *Int. J. Chem. Kinet., Symp. 1*, 1975, 205.
[75] E. Sanhueza, R. Simonaitis, and J. Heicklen, *Int. J. Chem. Kinet.*, 1979, **11**, 907.
[76] C. S. Kan, R. D. McQuigg, M. R. Whitbeck, and J. G. Calvert, *Int. J. Chem. Kinet.*, 1979, **11**, 921.
[77] R. A. Graham, A. M. Winer, R. Atkinson, and J. N. Pitts, jun., *J. Phys. Chem.*, 1979, **83**, 1563.
[78] C. T. Pate, R. Atkinson, and J. N. Pitts, jun., *J. Environ. Sci. Health, Environ. Sci. Eng., Ser. A*, 1976, **11**, 19.
[79] H. W. Sidebottom, C. C. Badcock, G. E. Jackson, J. G. Calvert, G. W. Reinhardt, and E. J. Damon, *Environ. Sci. Technol.*, 1973, **6**, 72.
[80] R. A. Cox, *J. Phys. Chem.*, 1972, **76**, 814.
[81] J. G. Calvert, Fu Su, and J. W. Bottenheim, *Atmos. Environ.*, 1978, **12**, 197.
[82] R. A. Cox, *Philos. Trans. R. Soc. London, Ser. A*, 1979, **290**, 543.

to undergo slow photo-oxidation at 313 nm[83] but the quantum yield has not been established for atmospheric conditions.

$$SO_2 + h\nu \,(400—340\ nm) \rightarrow SO_2(^3B_1)\ \text{(Triplet state)} \qquad (59)$$

$$SO_2 + h\nu \,(340—260\ nm) \rightarrow SO_2(^1B_1)\ \text{(Probably other singlets)} \qquad (60)$$

Reactions of Halogenated Hydrocarbons. The rate constants for the reactions of OH with hydrogen-containing chlorocarbons and chlorofluorocarbons have been carefully measured in recent years, because of their significance in the fluorocarbon–ozone problem. The data are generally in excellent agreement and consensus values are presented in Table 8.[84,85] There is a rather greater uncertainty

Table 8 *Rate coefficients for the reactions of some halogenated hydrocarbons with OH radicals*

Reactant	$k/cm^3\ molecule^{-1}\ s^{-1a}$
CH_3F	1.6×10^{-14}
CH_3Cl	$2.2 \times 10^{-12}\ exp(-1140/T)$
CH_3Br	$7.6 \times 10^{-13}\ exp(-890/T)$
CH_2Cl_2	$4.3 \times 10^{-12}\ exp(-1088/T)$
$CHCl_3$	$4.7 \times 10^{-12}\ exp(-1128/T)$
CHF_2Cl (CFC22)	$1.3 \times 10^{-12}\ exp(-1670/T)$
$CHFCl_2$ (CFC21)	$1.5 \times 10^{-12}\ exp(-1180/T)$
CH_3CCl_3	$5.4 \times 10^{-12}\ exp(-1820/T)^b$
C_2Cl_4	$1.1 \times 10^{-11}\ exp(-1246/T)$
C_2HCl_3	$5.3 \times 10^{-13}\ exp(+884/T)$

[a] Values taken from CODATA evaluation (ref. 29) and ref. 85; [b] This value is an update of that given in ref. 29, and is based on data from ref. 84.

of the rate constant for the reaction of OH with methyl chloroform and recent measurements[84] give a 298 K value a factor of 0.5 lower than the previous consensus.[29] In view of the current interest in the atmospheric chemistry of this compound, this measurement problem requires attention. The fully halogenated methanes and ethanes do not apparently react with OH radicals.

Reactions of Some Non-methane Hydrocarbon Species. An exhaustive list of kinetic data for the atmospheric reactions of non-methane hydrocarbons is beyond the scope of this Report but Table 9 lists important reactions involving C_2 species, propene, and isoprene. Most of the rate constants are taken from a recent review by Atkinson *et al.*[85] It should be noted that the addition reactions of OH with C_2H_4 and C_2H_2 are pressure-dependent and only values appropriate to 1 atm pressure are given.

[83] W. P. Wood and J. Heicklen, *J. Phys. Chem.*, **1971**, **75**, 854.
[84] K. M. Jeong and F. Kaufman, *Geophys. Res. Lett.*, 1979, **6**, 757; M. J. Kurylo, P. C. Anderson, and O. Klais, *ibid.*, 1979, **6**, 760.
[85] R. Atkinson, K. R. Darnall, A. C. Lloyd, A. M. Winer, and J. N. Pitts, jun., *Adv. Photochem.*, 1979, **11**, 1975.

Table 9 *Reactions of organic species*

OH reactions	k/cm^3 molecule^{-1} s^{-1}
OH + $C_2H_6 \rightarrow H_2O + C_2H_5$	$1.86 \times 10^{-11} \exp(-1223/T)^a$
OH + $C_2H_4 \rightarrow (C_2H_4OH)$	$2.18 \times 10^{-12} \exp(+385/T)^{a,b}$
OH + $C_2H_2 \rightarrow (C_2H_2OH)$	$1.9 \times 10^{-12} \exp(-620/T)^{a,c}$
OH + $C_3H_6 \rightarrow (C_3H_6OH)$	$4.1 \times 10^{-12} \exp(+540/T)^{a,b}$
OH + $C_5H_8 \rightarrow (C_5H_8OH)$	6.2×10^{-11d}
OH + $CH_3CHO \rightarrow H_2O + CH_3CO$	$6.9 \times 10^{-12} \exp(+250/T)^a$
OH + $CH_3COC_2H_5 \rightarrow$ Products	1.4×10^{-11d}
OH + $CH_3COO_2NO_2 \rightarrow$ Products	$\leqslant 1.7 \times 10^{-13a}$

Ozone reactions	
$O_3 + C_2H_4 \rightarrow$ Products	$9 \times 10^{-15} \exp(-2560/T)$
$O_3 + C_3H_6 \rightarrow$ Products	$6.1 \times 10^{-15} \exp(-1900/T)^e$

Acetyl peroxy radical reactions	
$CH_3CO_3 + NO \rightarrow CH_3 + CO_2 + NO_2$	2.6×10^{-12f}
$CH_3CO_3 + NO_2 \rightarrow CH_3COO_2NO_2$	1.4×10^{-12f}
$CH_3COO_2NO_2 \rightarrow CH_3CO_3 + NO_2$	$4.0 \times 10^{15} \exp(-13040/T)^{f,g}$

Photodissociation	J/s^{-1}
$CH_3CHO + h\nu \rightarrow CH_3 + HCO$	4.1×10^{-5f}

[a] Ref. 85; [b] 1 atm pressure; [c] High pressure; [d] R. A. Cox, R. G. Derwent, and M. R. Williams, *Environ. Sci. Technol.*, 1980, **14**, 547; [e] J. T. Herron and R. E. Hull, *J. Phys. Chem.*, 1974, **78**, 2085; [f] R. A. Cox and M. Roffey, *Environ. Sci. Technol.*, 1977, **11**, 900; [g] D. G. Hendry and R. A. Kenley, *J. Am. Chem. Soc.*, 1977, **99**, 3198; [h] K. Demerjian, J. A. Kerr, and J. G. Calvert, *Adv. Environ. Sci. Technol.*, 1974, **4**, 1; solar zenith angle 40°.

From measurements of the rate constants for the reactions of OH with a series of alkanes, Greiner[86] derived an expression fitting the overall rate constants by summing up the contributions from the individual rate constants for attack on the primary, secondary, and tertiary C—H bonds. The original expression has been modified slightly by Atkinson *et al.*[85] in the light of more recent literature room-temperature data to give:

$$k = 1.01 \times 10^{-12} N_1 \exp(-822/T) + 2.41 \times 10^{-12} N_2 \exp(-427/T)$$
$$+ 2.10 \times 10^{-12} N_3 \text{ cm}^3 \text{ molecule}^{-1} \text{ s}^{-1} \qquad (61)$$

where N_1, N_2, and N_3 are the numbers of primary, secondary, and tertiary C—H bonds respectively. This equation can be used to give overall values of k within acceptable limits ($\simeq \pm 20\%$) for $\geqslant C_3$ alkanes, apart from cyclobutane and cyclopropane.

Kinetic data required for description of the formation and removal of peroxyacetyl nitrate (PAN), which is now believed to play an important role in the behaviour of NO_x in the troposphere, are also given in Table 9. The rate data are based mainly on studies of the thermal decomposition of PAN,[87,88] which like other peroxynitrates can exist in thermal equilibrium with its precursor radicals,

[86] N. Greiner, *J. Chem. Phys.*, 1970, **53**, 1070.
[87] R. A. Cox and M. Roffey, *Environ. Sci. Technol.*, 1976, **11**, 900.
[88] D. G. Hendry and R. A. Kenley, *J. Amer. Chem. Soc.*, 1977, **99**, 3198.

acetyl-peroxy and NO_2. The absolute values for the reactions of CH_3COO_2 with NO and NO_2 are only known to within $\pm$ a factor of 10, since they are derived from estimates of the equilibrium constant for PAN formation, which is rather uncertain. The relative rates of these reactions are more well defined. It should be noted that the association reaction to form PAN and its decomposition may be pressure-dependent, and the values given here only apply to 760 Torr. The other peroxyacyl nitrates probably behave in a similar manner to PAN, *i.e.*, they are significantly more stable than the alkyl peroxy nitrates[64] or HO_2NO_2.

Photodissociation is an important facet of the atmospheric oxidation of non-methane hydrocarbons, since the carbonyl compounds are formed as oxidation products. The rates of photolysis of the aldehydes and ketones are generally assumed to be similar and a J value for CH_3CHO is given in Table 9. However, owing to the complex nature of the photolysis of carbonyl compounds,[2] the quantum yields of the various identified pathways which may be operative under atmospheric conditions is still an area of considerable uncertainty.[89]

Reactions of Inorganic Hydrides. The reaction of hydrogen halides and ammonia with OH radicals may play a minor role in the troposphere chemistry of these species. These rate constants are reasonably well known (see Table 10).

Table 10 *Reactions of inorganic hydrides with OH radicals*

Reaction	k/cm^3 molecule^{-1} s^{-1}
$OH + HCl \rightarrow H_2O + Cl$	$3.0 \times 10^{-12} \exp(-425/T)^a$
$OH + HBr \rightarrow H_2O + Br$	8.5×10^{-12} at 300 K ^{a}
$OH + HI \rightarrow H_2O + I$	1.3×10^{-11} at 300 K^b
$OH + NH_3 \rightarrow H_2O + NH_2$	$2.3 \times 10^{-12} \exp(-800/T)^c$

^{a} Ref. 29; ^{b} G. A. Takacs and G. P. Glass, *J. Phys. Chem.*, 1973, **71**, 1948; ^{c} I. W. M. Smith and R. Zellner, *Int. J. Chem. Kinet. Symp. 1*, 1975, 341.

4 The Importance of Photo-oxidation in the Life Cycles of Minor Tropospheric Constituents

Evaluation of the Tropospheric OH Distribution.—A quantitative evaluation of the tropospheric OH distribution requires some form of modelling approach that adequately describes the detailed aspects of tropospheric behaviour which have already been reviewed. A minimum treatment would be two-dimensional so that important variations of temperature, pressure, and solar intensity, and mixing with latitude, altitude, and season can be included.

By use of a two-dimensional model, it is possible to calculate a distribution of OH which may or may not reflect the actual distribution in the atmosphere. Validation of the calculated distribution against measurement is not possible as yet since techniques for measurement of OH concentration are not advanced enough to give an adequate coverage in both space and time. There are, however, a number of minor constituents which serve as useful tracers for tropospheric OH with which to validate model calculations.

[89] A. C. Lloyd, in ref. 68, p. 27.

There are three major sources of uncertainty in the calculation of the OH distribution. First, there are uncertainties in the chemical kinetic data base, both in the presently available rate coefficients and photochemical constants, and also from the involvement of potentially important species which are not yet adequately included, *e.g.*, H_2O_4. Secondly, there are considerable uncertainties in the source strengths of species such as NO_x, CO, and CH_4, which control the OH concentration. Thirdly, there are errors arising from the necessary simplifying approximations made in the modelling technique, *e.g.*, in the calculation of solar intensity and the formulation of atmospheric mixing.

The following discussion is based on calculations performed using the two-dimensional model reported by Derwent and Curtis.[25] The model was set up using available observations[90,91] of tropospheric methane and carbon monoxide (^{12}CO) and the NO_x injection estimated by Galbally.[92] The OH distribution which was obtained showed maximum values close to the ground in low latitudes during summer conditions. Although the hemispheric distributions had slightly different shapes, the hemispheric maximum OH concentrations were almost identical.

The carbon monoxide injection which was required to balance the observed latitudinal CO distribution and the calculated OH distribution was considerably greater than the injection rate from manmade sources. This difference can possibly be accounted for by the natural biospheric sources due to biomass burning[93] or emissions of isoprene and terpenes.[94] There are, however, no currently available quantitative estimates of these source strengths and because of uncertainties in model input, the calculations do not allow further resolution of the two unknowns, *viz* the OH concentration and the biospheric CO source strength. If the OH global concentration increased, then the biospheric source strength would need to increase to maintain the observed latitudinal CO distribution. However, the two-dimensional model does allow the formulation of the relationship between the OH distribution and the biospheric CO source strength.

Present information on the major tropospheric constituents CO, CH_4, and NO_x, alone is not adequate to define precisely the OH concentration distribution. However, by turning attention to the other isotopic varieties of these trace constituents a further relationship between OH and the biospheric CO source strength can be obtained allowing the evaluation of both. In particular, a consideration of ^{14}CO produced continuously in the upper troposphere and lower stratosphere by cosmic-ray bombardment of atmospheric nitrogen is most relevant.[95] Natural biospheric processes liberate a small but significant amount of ^{14}CO whereas manmade processes do not since they consume fossil carbon.

Figure 7 shows a comparison of the observed and calculated ^{14}CO distribution at $51° N$ for various times of the year, using the known cosmic-ray source[96] and a small biospheric injection.[95] The two-dimensional model is able to reproduce the

[90] D. H. Ehhalt, *Tellus*, 1974, **26**, 58.

[91] W. Seiler, *Tellus*, 1974, **26**, 116.

[92] I. E. Galbally, *Tellus*, 1975, **27**, 67.

[93] P. J. Crutzen, L. E. Heidt, J. P. Krasnec, W. H. Pollock, and W. Seiler, *Nature (London)*, 1979, **282**, 253.

[94] P. R. Zimmerman, R. B. Chatfield, J. Fishman, P. J. Crutzen, and P. L. Hanst, *Geophys. Res. Lett.*, 1978, **5**, 679.

[95] A. Volz, D. H. Ehhalt, R. G. Derwent, and A. Khedim, *Ber. K.F.A. Julich*, No. 1604, 1979.

[96] R. E. Lingenfelter, *Rev. Geophys.*, 1963, **1**, 35.

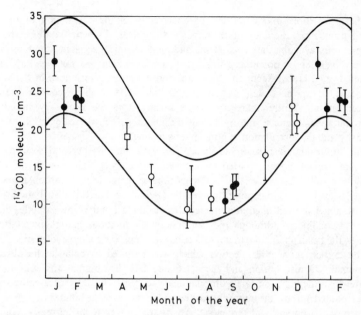

Figure 7 *Annual variations in the observed (taken from data of ref. 95) and calculated concentrations of* ^{14}CO *at 51° N (molecule cm^{-3}). The two calculated lines were obtained using the standard OH distribution calculated using the AERE Harwell two-dimensional (upper line) and doubling the OH concentrations (lower line)*

observed seasonal variations with the correct phase. However, the standard OH distribution gives ^{14}CO concentrations which are consistently higher than those from observations. Doubling the OH distribution at each point in latitude, altitude, and time of year produces ^{14}CO concentrations which are consistently below those from observations. Clearly the OH distribution lies between these two limits and a simple optimization procedure can be employed to obtain the appropriate multiplication factor which minimizes the deviations between model and observation.

This optimization procedure can then be used to obtain the OH distribution for various natural biospheric injections of ^{14}CO. Since the ^{14}C to ^{12}C isotopic ratio is known[97] for the biospheric source, this gives the required relationship between OH and biospheric source strength. Figure 8 shows the OH distribution which simultaneously balances both the ^{14}CO and ^{12}CO cycles. The tropospheric mean OH concentration amounts to $(6.5 \pm 2.5) \times 10^5$ cm^{-3}. The quoted 1σ confidence limits include uncertainties due to transport, chemistry, and tropospheric concentrations.

Life Cycles of Methane, Carbon Monoxide, and Hydrogen.—Table 11 shows the resulting budgets for CO, CH_4, and H_2 which have been obtained using the aforementioned OH distribution. The life cycle of CH_4 is dominated by the biogenic

---

[97] J. Levin, Diplomarbeit Universitat Heidelberg, W. Germany, 1978.

Table 11 *Budgets of atmospheric methane, carbon monoxide, and hydrogen*

	Source strength/g year^{-1}		*Sink strength*/g year^{-1}	
Methane	Swamps and marshes*	450×10^{12}	OH Oxidation	500×10^{12}
	Enteric fermentation[a]	100×10^{12}	Stratospheric removal[a]	100×10^{12}
	Manmade sources[a]	50×10^{12}		
^{12}CO	Terpenes and isoprene*	1150×10^{12}	OH Oxidation	2450×10^{12}
	Ocean[b]	100×10^{12}	Surface removal[d]	320×10^{12}
	Methane oxidation	880×10^{12}		
	Manmade sources[b,c]	640×10^{12}		
^{14}CO	Terpenes and isoprene	2.0×10^3	OH Oxidation*	14.4×10^3
	Methane oxidation	1.5×10^3	Surface removal[e]	1.0×10^3
	Cosmic ray bombardment[e]	11.9×10^3		
Hydrogen	Terpenes and isoprene	25×10^{12}	OH oxidation	33×10^{12}
	Methane oxidation	31×10^{12}	Surface removal*	42×10^{12}
	Manmade sources	15×10^{12}	Stratospheric removal[f]	1×10^{12}
	Biogenic sources	5×10^{12}		

* Calculated by mass balance. [a] Ref. 90; [b] Ref. 91; [c] J. Logan, personal communication; [d] K. H. Liebl and W. Seiler, Proc. Microbial Production and Utilisation of Gases, Goltze Druck, Gottingen, 1976; [e] Ref. 95; [f] U. Schmidt, Proc. Ecole Européenne d'Etat d'Environnnement, Peyresq, September, 1978.

source and the photochemically driven sink. CH_4 oxidation produces a sizable CO source but manmade sources of ^{12}CO are comparable. Both these sources are smaller than the additional CO source, from vegetation either through biomass burning or terpene emissions. Reaction with OH is by far the most important sink for CO. H_2 is produced from methane and terpene oxidation, through photolysis of the intermediate product of photo-oxidation, HCHO. This source is comparable to manmade sources both of which exceed biogenic emissions. To balance the H_2 budget, a surface removal process comparable in magnitude to the photo-chemically driven OH-sink is required.

In drawing up these budgets, it is assumed that the cycles are in steady state. There is at present insufficient observational data to support or reject this assumption for CO, CH_4, or H_2. For CO_2, however, its life cycle is clearly not in balance, since there has been a steady, year-by-year increase in atmospheric CO_2 concentration.[98] Most of the atmospheric CO_2 increase can be attributed to direct emission from fossil-fuel combustion; the contribution of atmospheric CO oxidation to the total $CO_2(C)$ budget is only a few percent.

From estimates of the atmospheric masses of the minor constituents and the total source (or sink) strengths it has been possible in Table 12 to estimate atmospheric lifetimes or turn-over times for these constituents. It is apparent that hydrogen and methane have lifetimes which are of the order of several years whereas that of carbon monoxide is several months.

It is also apparent from Table 12 that the isotopes of carbon monoxide have markedly different atmospheric lifetimes although identical OH rate coefficients for each species were assumed for the model. The two-dimensional nature of the OH

[98] B. Bolin, E. T. Degens, S. Kempe, and P. Ketner, 'The Global Carbon Cycle', SCOPE 13, Wiley, Chichester, U.K., 1979.

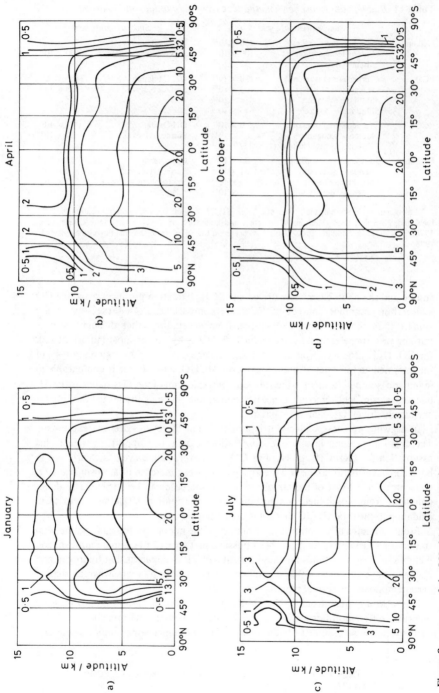

Figure 8 contours of the OH distribution (10^5 molecule cm^{-3}) during (a) January, (b) April, (c) July, and (d) October (constructed from data of

Table 12 *Source strengths, atmospheric masses, and atmospheric life-times for methane, carbon monoxide, and hydrogen*

Minor constituent	Total source strength/g year^{-1}	Total atmospheric mass/g	Lifetime/years
Methane	600×10^{12}	4.4×10^{15a}	7.3
^{12}CO	2770×10^{12}	480×10^{12b}	0.17
^{14}CO	15.4×10^{3}	6×10^{3c}	0.4
Hydrogen	76×10^{12}	171×10^{12d}	2.3

^{a} Based on tropospheric measurements of D. H. Ehhalt, W.M.O. Symposium, No. 511, World Meteorological Office, Geneva, Switzerland, 1978, p. 71; ^{b} A. Volz, Proc. Ecole Européenne d'Etat d'Environnement, Peyresq, September, 1978; ^{c} Ref. 95; ^{d} U. Schmidt, in footnote (*b*).

distribution and of the isotopic CO species ensures that lifetimes due to OH reaction, τ, cannot be adequately represented by a simple equation of the type

$$\text{Lifetime due to OH reaction } \tau = \frac{1}{k(\text{OH})} \qquad (62)$$

where k is the rate coefficient for OH attack on the reactant. The importance of the covariance of the mutually reacting species in causing the difference in behaviour of the two CO species is clearly apparent from the factor of two difference in their lifetimes. ^{14}CO shows highest concentrations where OH is lowest in the upper polar stratosphere, whereas ^{12}CO is highest where OH is highest, in low latitudes close to the gound. The isotopic CO varieties sense entirely different aspects of the tropospheric OH distribution. Halocarbons with markedly different k values see different OH distributions because of similar covariance problems. This problem is discussed again later.

Atmospheric Chemistry of Sulphur Compounds.—The atmospheric chemistry of sulphur has been the subject of intensive study for a number of years because of concern about the impact of the vast amounts of gaseous SO_2, which is produced in the combustion of fossil fuels. There have been a number of assessments of the global sulphur cycle all of which indicate that the manmade sources of SO_2 are comparable with natural sources on a global scale, and in many regions are the dominant source of sulphur.[99, 100] SO_2 is also an interesting atmospheric constituent on account of its diverse sinks. Physical removal, both wet and dry deposition, and chemical conversion, both homogeneous and in solution in water droplets, play a part in its atmospheric life cycle. In the present Report we are concerned mainly with the homogeneous gas aspects of SO_2 chemistry.

Since the direct photo-oxidation following light absorption by SO_2 occurs with a very low quantum yield, the radical + SO_2 reactions provide the main if not exclusive homogeneous pathways for SO_2 oxidation. In urban air, the reaction of SO_2 with an intermediate produced in ozone–alkene reactions may also be significant. On the basis of the atmospheric concentrations of OH, HO_2, and CH_3O_2 generated in the two-dimensional model calculations it can be readily shown

[99] J. P. Friend, in 'Chemistry of the Lower Atmosphere', ed. S. I. Rasool, Plenum Press, New York, 1973, p. 177.
[100] L. Granat, H. Rodhe, and R. O. Hallberg, *Ecol. Bull.* (*Stockholm*), 1976, **22**, 89.

that reaction (63) with OH is the dominant route for homogeneous SO_2 oxidation

$$OH + SO_2 (+ M) \rightarrow HSO_3 (+ M) \quad ---\rightarrow \quad H_2SO_4 \qquad (63)$$

under most conditions. In the boundary layer reactions with CH_3O_2 may also be significant and the typical mean OH and CH_3O_2 radical concentrations give a lifetime of *ca.* 10 days for SO_2, which is rather long compared with that due to dry deposition (2 days) or wet deposition (1–3 days).[101] Thus only a small fraction of the SO_2 released at the ground can be converted into H_2SO_4 and sulphate aerosol *via* gas-phase reaction. In the background troposphere the lifetime of SO_2 with respect to oxidation by OH is even longer. However, since physical removal processes are also inefficient in this region, conversion of SO_2, into sulphate is probably the dominant fate of 'background' SO_2, and this provides a global source of sulphate aerosol particles. Recent observations[102] of a rather constant mixing ratio of SO_2 near 1 part in 10^{10} in the background troposphere supports this picture.

Sulphur is also emitted into the atmosphere in the gaseous form as H_2S, CS_2, OCS, and organic sulphides. The latter react very rapidly with OH and a discussion of their chemistry will be confined to that of the boundary layer where they are converted at least partly to SO_2.[73] The detailed mechanism for this conversion has yet to be elucidated. H_2S is also converted into SO_2 following reaction (64) with OH, when HS radicals are produced initially. The mechanism for HS oxidation is also unclear but SO_2 seems to be the final product.

$$OH + H_2S \rightarrow HS + H_2O \qquad (64)$$

The lifetime of H_2S in the boundary layer, based on reaction with OH and the OH distribution, is 3 days and so this molecule is unlikely to be a significant carrier of sulphur in the background troposphere.

The role of CS_2 and OCS in the sulphur cycle has been the subject of several recent papers.[103–105] An important role for OCS as a carrier of sulphur from the troposphere to the stratosphere has been suggested and this is supported by recent measurements which show that it is widely distributed in the atmosphere with a mixing ratio of 4 parts in 10^{10}.[106,107] Photolysis of OCS in the stratosphere ($\lambda \leq 210$ nm) releases S atoms which are oxidized to SO_2 and subsequently converted into H_2SO_4 aerosol. This aerosol contributes to the total aerosol burden of the stratosphere, which plays a significant role in the atmospheric radiation budget. It has also been suggested that CS_2 may be a significant source of atmospheric OCS, though reaction (65) with OH radicals,[72] or possibly through

$$OH + CS_2 \rightarrow HS + OCS \qquad (65)$$

reactions of excited CS_2 molecules produced by photoabsorption by CS_2 in the 280—350 nm region. There is, however, considerable uncertainty in the life cycle and budget of OCS and CS_2 arising from a lack of knowledge of the magnitude of

[101] R. A. Cox and S. A. Penkett, *J. Chem. Soc., Faraday Trans. 1*, 1972, **68**, 1735.
[102] E. Meszaros, *Atmos. Environ.*, 1978, **12**, 699.
[103] P. J. Crutzen, *Geophys. Res. Lett.*, 1976, **3**, 73.
[104] N. D. Sze and M. K. W. Ko, *Nature*, 1979, **280**, 308.
[105] J. A. Logan, M. B. McElry, S. C. Wofsy, and M. J. Prather, *Nature (London)*, 1979, **281**, 185.
[106] F. J. Sandalls and S. A. Penkett, *Atmos. Environ.*, 1977, **11**, 197.
[107] P. J. Maroulis, A. L. Torres, and A. R. Bandy, *Geophys. Res. Lett.*, 1977, **4**, 510.

the various possible sources and also from the differences in the reported rate constants for the reactions of OCS and CS_2 with OH radicals which provide a potentially large sink for these molecules in the troposphere. If the highest reported values for these OH reactions (see Table 7) apply, the calculated lifetimes for CS_2 and OCS are 0.2 and 1.2 years, respectively, based on the above OH distribution. The corresponding global flux of sulphur from these compounds can be estimated to be 10×10^{12} g (S) year^{-1} based on available tropospheric concentrations.[106] This would constitute a small but significant contribution to the total atmospheric S budget which, including manmade SO_2 emissions, amounts to *ca.* 146×10^{12} g (S) year^{-1}.[100]

Oxidation of Hydrocarbons.—There has been great interest in the chemistry of hydrocarbon photo-oxidation in the atmosphere following the discovery in the 1950's that it led to the formation of photochemical smog. The early treatise by Leighton[2] outlined large areas of photochemistry and chemical kinetics where the then current knowledge was inadequate for a detailed understanding of the smog phenomenon. Since that time there has been great progress in this field.

Most volatile hydrocarbon species emitted into the atmosphere are reactive toward OH radicals and additionally the alkenes undergo reaction with 'ozone'. Hence the atmospheric lifetimes of these species are relatively short and chemical conversion occurs mainly in the boundary layer. The oxidation mechanism is, in general, analogous to that involved in the oxidation of methane, *i.e.*, a chain reaction involving peroxy radicals which are formed by OH attack followed by O_2 addition to the initial radical product. The peroxy radicals undergo O-transfer reactions either with NO or another radical to form alkoxy radicals which react further to produce a carbonyl compound as the first stable oxidation product together with a smaller organic peroxy radical or HO_2. The overall scheme may be written:

$$OH + R^1H \rightarrow R^1 \xrightarrow{\;\;+O_2\;\;} R^1O_2 \tag{66}$$

$$R^1O_2 + NO \rightarrow R^1O + NO_2 \tag{67}$$

$$R^1O + O_2 \rightarrow R^2CO + \text{peroxy radical } (R^3O_2 \text{ or } HO_2) \tag{68}$$

The conversion of NO into NO_2 in this process leads to the formation of ozone since the NO_2 product undergoes photolysis to NO and $O(^3P)$:

$$NO_2 + h\nu \rightarrow NO + O \tag{19}$$

$$O + O_2 + M \rightarrow O_3 + M \tag{4}$$

This is the origin of ozone formation in photochemical smog. The rate of the overall process and the detailed mechanism of oxidation depends on the nature of the hydrocarbon species. Recognition of this fact has led to the concept of 'hydrocarbon reactivity,' a classification which describes the ozone formation potential of manmade organic species present in polluted air. The key factors determining reactivity are those which determine the rate of RO_2 radical production; the OH reaction rate and the rates and nature of the other steps in the mechanism are clearly important.

There is now a good body of kinetic data for OH reactions for a large number of hydrocarbons and other organic species (see Section 3 and refs 85 and 108). The detailed mechanisms are only established for the simpler species. Some of the key features of the mechanism are exemplified in the scheme for oxidation of n-butane outlined in Figure 9. The initial attack of OH is 85% on the secondary H atoms and 15% on the primary H atoms, in accordance with the known reactivity of OH with these types of bonds in the homologous series of alkanes.[85,86] After addition of O_2 to the C_4H_9 radical and NO to NO_2 conversion, two types of butoxy radical are formed, primary and secondary. Three competitive reaction patterns for alkoxy radicals have been identified for the C_4H_9O radical:[109,110]

(a) reaction with O_2 to form a carbonyl compound and an HO_2 radical,

(b) thermal decomposition to form an aldehyde or ketone with fewer carbon atoms, together with an alkyl radical, and

(c) isomerization by internal H abstraction from the 4- or 5-carbon atom, relative to the C—O position.

Process (a) only occurs in primary and secondary alkoxy radicals since the tertiary radical has no α-H atom. For secondary radicals (a) and (b) can compete and for tertiary radicals decomposition predominates. The isomerization reactions require a structure with a straight chain of at least four carbon atoms. The process seems to be dominant for n-alkoxy radicals derived from n-C_5 and n-C_6 hydrocarbons and this agrees with the predictions of thermochemical kinetic calculations. From the examples given in Figure 9 it is clear that the rate of the alkoxy radical decides the nature of the initial carbonyl products during photo-oxidation. The effect of chemical structure on the reactivity of alkoxy radicals is a key area of research required to provide a detailed picture of the mechanism of photo-oxidation of organic compounds in the atmosphere.

The carbonyl compounds formed during the breakdown of hydrocarbons are further oxidized, either by direct photodissociation or by attack by OH radicals, to give smaller carbon fragments. These eventually degrade, mainly to formaldehyde and subsequently to CO. Some of the carbon C ends up as acyl radicals, $R\dot{C}O$ (*e.g.*, from attack of OH on CH_3CHO), which is of particular interest from the point of view of atmospheric chemistry, since they can then form acylperoxy radicals by addition of O_2. These radicals form relatively stable peroxynitrates on reaction with NO_2, *e.g.*, the well known first member of the series, peroxyacetyl nitrate, $CH_3COO_2NO_2$. Furthermore, O-atom transfer by reaction with NO or another peroxy radical leads to the unstable acetate radical RCO_2, which decomposes rapidly to $R + CO_2$. Thus the carbonyl carbon atom is oxidized directly to CO_2 without intermediate formation of CO.

The formation of stable peroxynitrates has important consequences in atmospheric pollution because they are powerful phytotoxicants and are also apparently associated with eye irritation in smog.[111] Their formation is also of interest because it provides a mechanism for the transport of active nitrogen oxides, NO and NO_2,

[108] I. Campbell and D. Baulch, ch. 5 of this volume, p. 137.

[109] A. C. Baldwin, J. R. Barker, D. M. Golden, and D. G. Hendry, *J. Phys. Chem.*, 1977, **81**, 2483.

[110] W. P. L. Carter, K. R. Darnall, A. C. Lloyd, A. M. Winer, and J. N. Pitts, jun., *Chem. Phys. Lett.*, 1976, **42**, 22.

[111] E. R. Stephens, E. G. Darley, O. C. Taylor, and W. E. Scott, *Int. J. Air Water Pollut.*, 1961, **4**, 79.

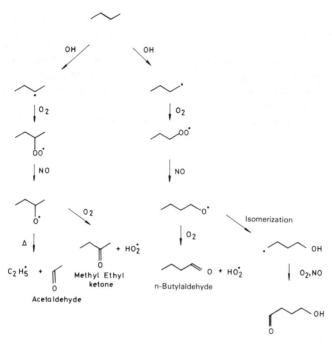

Figure 9 *Degradation pathways in the atmospheric photo-oxidation of n-butane*

from their continental and mainly anthropogenic source regions into the general background circulation of the troposphere. The importance of NO and NO_2 in the chain-carrying reactions in the photo-oxidation of CH_4 in the background troposphere has already been stressed. However, there are apparently no maritime sources of NO_x and since NO_2 is rapidly converted into HNO_3 by its reaction with OH, it is not a suitable molecule for transport over long distances in a tropospheric air mass. On the other hand, HNO_3 is relatively stable towards attack by OH or photodissociation and cannot release 'active' NO_x easily before it is removed from the atmosphere by rain or by absorptions at the ground. In contrast PAN is stable towards OH attack and photodissociation but it can readily release NO_2 by thermal decomposition at temperatures prevailing near the earth's surface ($t_{\frac{1}{2}}$ at 300 K = 20 min). PAN will be relatively more stable at high altitudes where the prevailing temperature is lower and when incorporated into the circulation of the background troposphere it can be transported over long distances. The role of PAN in the tropospheric NO_x budget has been fully discussed in a recent review by Crutzen.[112]

Oxidation of carbonyl compounds directly to CO_2 without intermediate formation of CO is of interest for the tropospheric carbon budget. Model calculations of the life cycles of CH_4 and CO have revealed that a rather sizable source of CO from natural sources other than CH_4 oxidation is required to balance the CO budget, as determined from the observed distributions of $C^{12}O$ and $C^{14}O$

[112] P. J. Crutzen, *Ann. Rev. Earth Planet. Sci.*, 1979, **7**, 443.

and the OH sink strength.[95] The oxidation of non-methane hydrocarbons, mainly isoprene C_5H_8 and the C_{10} terpenes emitted from vegetation, has been suggested as a possible source of the additional CO required.[94] The measured rate constants for reaction of isoprene and terpenes with OH and O_3 indicate that these compounds will be very rapidly oxidized after release into the atmosphere. Some reports[113,114] have indicated that most of the volatile carbon released in this form is converted into aerosol, presumably formed by condensation of low volatility oxygenated products. The fate of these aerosols in the atmosphere has not been investigated. However, recent work[115] has shown that, at least in the presence of NO_x, the reaction of OH with isoprene yields mainly gaseous carbonyl products, HCHO, $CH_3COCH:CH_2$, and probably $CH_2:C(CH_3)CHO$, which themselves are readily oxidized to give CO. Peroxyacetyl nitrate is produced in the oxidation of isoprene and terpenes[114,115] in the presence of NO_x, indicating the production of CH_3COO_2. Thus a fraction of the carbon content of terpenes may be oxidized directly to CO_2, reducing the magnitude of the CO source.

Because of the high reactivity of hydrocarbons and related organic compounds toward OH radicals, it is clear that photochemically initiated oxidation provides a powerful and widespread sink for these molecules in the troposphere, and consequently they have a relatively short atmospheric lifetime. The operation of this sink process in polluted air when mixing is poor and gives rise to photochemical smog. Because of the involvement of a larger variety of hydrocarbons in urban smog, a more complex matrix of chemical reactions is required to describe their oxidation and the production of ozone and other secondary pollutants in the air mass, compared to the simple CH_4 chemistry used to model the background troposphere. Several review articles have covered some of the detailed chemical aspects of photochemical smog modelling.[116-118]

Oxidation of Halocarbons.—There is much current interest in the possible effects of the continued release of chlorofluorocarbons (CFC's), widely used as refrigerants and aerosol propellants, on the integrity of the stratospheric ozone shield. Although the original problem was posed in terms of CFC's 11 and 12, the presence of other chlorine-containing gases has been revealed in the atmosphere. The evaluation of the impact of these other chlorine-containing minor constituents requires information on their sources whether manmade or natural and on the nature and magnitude of their sink processes.

In a first attempt to study the role of the simple chlorocarbons, Cox *et al.*[119] showed that reaction with OH radicals appeared to be a significant sink for those halocarbons which contain H atoms and C=C bonds, *e.g.*, tetrachloroethylene. Using rate coefficient data for reactions of type (69) lifetimes, τ, were evaluated

[113] F. W. Went, *Nature (London)*, 1960, **187**, 671.
[114] B. W. Gay and R. R. Arnts, Proc. Int. Conf. Photochemical Oxidant Pollution and its Control, vol. 11, U.S. Environ. Protection Agency, Res. Triangle Park, N.C. 1977, 745.
[115] R. A. Cox, R. G. Derwent, and M. R. Williams, *Environ. Sci. Tech.*, 1979, in press.
[116] K. Demerijian, J. A. Kerr, and J. G. Calvert, *Adv. Environ. Sci. Technol.*, 1974, **4**, 1.
[117] B. J. Findlayson-Pitts and J. N. Pitts, jun., *Adv. Environ. Sci. Technol.*, 1977, **7**, 75.
[118] W. P. L. Carter, A. C. Lloyd, J. L. Sprung, and J. N. Pitts, jun., *Int. J. Chem. Kinet.*, 1979, **11**, 45.
[119] R. A. Cox, R. G. Derwent, A. E. J. Eggleton, and J. E. Lovelock, *Atmos. Environ.*, 1976, **10**, 305.

using the simple expression (62) where OH is some form of globally averaged concentration of OH radicals.

$$\text{OH} + \text{Halocarbon} \xrightarrow{k} \text{Sink} \qquad (69)$$

Two-dimensional model studies have shown that because of the covariance of k, OH, and the halocarbons with which it reacts, equation (62) must be replaced by a more complicated relationship between rate coefficient and lifetime. This is because the more reactive halocarbons are not distributed uniformly throughout the troposphere and the sink strength reflects only the OH distribution close to the source, normally at the surface. The more reactive the halocarbon the higher the average OH concentration required in equation (69), to calculate a meaningful lifetime.

Table 13 shows the lifetimes of certain halocarbons calculated using the two-dimensional model and the OH distribution for which the ^{12}CO and ^{14}CO cycles are simultaneously balanced. For comparison the lifetimes calculated with equation (62) and the global mean OH concentration of $6 \cdot 5 \times 10^5$ mol cm^{-3} are also shown. The importance of covariance errors are clearly evident. It should be added that lifetimes calculated with one-dimensional modelling techniques would also be in error by about similar magnitudes as the globally averaged model implied by equation (69).

The lifetimes shown in Table 13 confirm that for many of the halocarbons, OH oxidation is an efficient sink. By this means much of the organic halogen budget to the atmosphere is oxidized within the troposphere to inorganic halogen compounds such as HCl and HBr. These products are then removed in precipitation.

Table 13 *Halocarbon lifetimes estimated by use of two-dimensional and globally averaged models*

Halogenocarbon	Atmospheric lifetime/years	
	Two-dimensional model[a]	Global model[b]
CH$_3$Cl	1.4	2
CH$_3$Br	1.7	2
CH$_2$Cl$_2$	0.6	0.9
CHCl$_3$	0.6	0.9
CHF$_2$Cl (CFC 22)	8	30
CHFCl$_2$ (CFC 21)	1.8	3.7
CH$_3$CCl$_3$	5	13
C$_2$Cl$_4$	0.2	0.7

[a] Calculated using OH distribution which fits ^{14}CO and ^{12}CO budgets and the relationship between lifetime and k given by R. G. Derwent and A. E. J. Eggleton, *Atmos. Environ.*, 1978, **12**, 1261; [b] Calculated using mean temperature of 250 K data for Table 8, and mean global OH of 6.5×10^5 cm^{-3}.

However, for the halocarbons with low OH reactivity, the mean lifetimes may extend to several years. Under these conditions, interhemispheric exchange will give a much more uniform distribution between the two hemispheres. As the halocarbon lifetime increases with decreasing OH reactivity then the fraction of the surface injection of the halocarbon which can reach the stratosphere will also increase. For a halocarbon lifetime of *ca.* 10 years, *ca.* 20% of the surface injection could be transported into the stratosphere. It is therefore apparent from Table 13 that methyl chloroform and methyl chloride could be small but significant manmade and natural chlorine carriers to the stratosphere respectively. Using this type of analysis, ozone depletion estimates from all the chlorocarbons in Table 13 have been estimated from their current respective atmospheric release rates.[26]

Acknowledgements. We thank A. Chamberlain, A. Curtis, and J. Garland of A.E.R.E. Harwell, D. Ehhalt, U. Schmidt and A. Volz of K.E.A. Julich, and R. Atkinson of Environmental Research and Technology Inc. for useful discussions.

Author Index

Abauf, N., 2
Ackerhalt, J. R., 76, 77
Adachi, H., 215
Addison, M. C., 119
Agostini, P., 134
Akins, D. L., 112
Alben, K. T., 35
Aldridge, J. P., 6, 77, 109
Aleksakhin, I. S., 134
Alexander, M. H., 54, 68, 69
Alimpiev, S. S., 78
Alkemade, C. T., 123
Allen, L., 123
Allison, J., 13
Amar, D., 105
Ambartzumian, R. V., 73, 76, 80, 83, 97, 100, 109, 110
Anastasi, C., 141, 169, 213
Anderson, J. B., 2
Anderson, J. G., 144, 151, 157, 162, 171
Anderson, P. C., 220
Anderson, R. J. M., 121
Anderson, S. P., 104
Anderson, T. G., 77
Andres, R. P., 2
Andresen, P., 31
Antonov, V. S., 135
Aoiz, F. J., 41
Archie, W. C., jun., 103
Arkhipkin, V. G., 123
Arnold, G. S., 9, 44
Arnold, I., 209
Arnts, R. R., 232
Aronowitz, D., 187
Arthurs, A. M., 51, 62
Ashfold, M. N. R., 84, 94, 95, 112, 113
Asprey, L. B., 110
Asscher, M., 124
Atkinson, R., 138, 143, 150, 151, 156, 161, 163, 164, 165, 168, 177, 178, 181, 184, 185, 218, 219, 220
Atri, G. M., 161
Atvars, T. D. Z., 106
Auerbach, A., 35
Auerbach, D. J., 24, 51
Ausschnitt, C. P., 134
Avouris, P., 133

Babrowicz, F. W., 44

Badcock, C. C., 219
Bado, P., 82
Bagratashvili, V. N., 110
Baker-Blocker, A., 202
Bakos, J. S., 133
Baldwin, A. C., 230
Baldwin, R. R., 140, 161, 171, 173, 179, 186, 187
Balint-Kurti, G. G., 11, 49, 52
Bandy, A. R., 228
Barker, D. L., 119
Barker, J. R., 80, 89, 116, 230
Baronavski, A. P., 119, 128, 131
Barrat, J. P., 55
Bartoszek, F. E., 23
Basco, N., 215
Bauer, S. H., 98
Bauer, W., 24
Baulch, D. L., 2, 153, 171, 230
Baybutt, P., 44
Beaty, E. C., 122
Beauchamp, J. L., 81
Beck, D., 24
Beck, S. M., 10
Becker, K. H., 145, 219
Beerlage, M. J. M., 134
Behrens, R., 14, 15, 16
Benmair, R. M. J., 102, 110
Bemand, P. P., 143
Bennewitz, H. G., 48, 49
Benson, S. W., 110
Berg, J. O., 133
Bergmann, K., 8
Berman, M. R., 81, 124
Berman, P. R., 67
Bernheim, R. A., 123
Bernstein, R. B., 5, 10, 11, 12, 16, 32, 41, 44, 119, 133
Berry, M. J., 92
Bialkowski, S. E., 96, 112, 113
Bickes, R. W., 5, 32
Biermann, H. W., 154, 217
Billman, K. W., 80, 110
Biordi, J. C., 188
Birely, J. H., 107, 109
Bischel, W. K., 108, 122, 123, 124
Bittenson, S., 110
Bjorklund, G. C., 134
Black, C., 167
Black, J. G., 74, 80

Blackwell, B. A., 11, 26
Blais, N. C., 8, 24, 44
Bloembergen, N., 74, 76, 80
Blundell, R. V., 186
Bly, S. H. P., 21
Boer, G. J., 193
Boesl, U., 133
Bogan, D. J., 97
Bohm, H. D. V., 134
Bokor, J., 108
Bolin, B., 199, 202, 225
Bollinger, W., 177, 181
Bomse, D. S., 81
Booth, D., 141, 172
Bormann, J., 65
Bottenheim, J. W., 219
Bottner, R., 94
Bradley, J. N., 146, 148, 178
Brandt, D., 20, 21, 26
Brauman, J. I., 81, 89, 97, 103
Braun, W., 82, 91, 99
Bray, R. G., 124
Breckenridge, W. H., 131
Breen, J. E., 144
Breitenbach, L. P., 140, 151, 215
Brenner, D. M., 79, 89
Brewer, R. G., 66
Brezinsky, K., 79
Brice, K. A., 211
Brinkmann, U., 15
Brooks, P. R., 7, 46
Brophy, J. H., 9, 10, 14, 119, 143
Browett, R. J., 29
Brown, G. R., 31
Brown, L. C., 34
Broyer, M., 54
Brudzinski, R. J., 147
Brueck, S. R. J., 78
Brumer, P., 44
Brunner, F., 83
Brunner, T. A., 8
Brzychey, A., 18
Buckingham, A. D., 71
Buechele, J. L., 101
Burak, I., 109
Burrows, J. P., 166, 210, 211
Burton, C. S., 150
Buss, R. J., 2, 4, 25
Butler, J. E., 113, 130, 144

Butler, R., 156
Butterfield, K. B., 119

Calvert, J. G., 212, 216, 217, 218, 219, 221, 232
Campani, E., 123
Campbell, I. M., 2, 145, 152, 153, 185, 230
Campbell, J. D., 82, 83, 94, 96, 110, 113
Cantrell, C. D., 76, 77, 109
Capey, W. D., 146
Carr, R. W., jun., 149, 176, 178
Carrington, T., 19, 34, 115
Carson, D. J., 193
Carter, C. F., 45
Carter, W. P. L., 143, 230, 232
Cartwright, D. C., 109
Case, D. A., 35, 36, 40, 54
Castano, F., 113
Castleman, A. W., 169
Cathonnet, M., 186
Caughey, T. A., 65
Celto, J. E., 125
Chalek, C. L., 17
Chaltikyan, V. O., 123
Chamberlain, A. C., 203, 204, 205
Chang, J. S., 116, 144, 163, 180, 210
Chang, S.-G., 141
Chapman, S., 189
Chatfield, R. B., 223
Chekalin, N. V., 100, 109, 113
Cheng, C., 104
Cheung, J. T., 25
Chin, S. L., 135
Chu, S. I., 69
Churney, K., 99
Cicerone, R. J., 202
Clark, M. D., 107
Clark, R., 59
Cleugh, C. J., 187
Clough, P. N., 31, 92
Clyne, M. A. A., 143, 174
Coggiola, M. J., 3, 25, 87
Cohen, R. B., 19
Cole, J. L., 18
Collins, C. B., 134
Colson, S. D., 133, 135
Coltharp, R. N., 143, 160
Colussi, A. J., 110
Combourieu, M., 148
Comes, F. J., 209
Compton, R. N., 133
Conaway, B., 174
Connell, P., 213
Cook, W. G. A., 186
Cool, T. A., 107, 125
Corbin, R. J., 134
Cosandey, M. R., 45
Cotter, T. P., 83
Coveleskie, R. A., 97
Cowe, D. W., 186

Cox, A. P., 66
Cox, D. M., 83, 110
Cox, R. A., 142, 156, 161, 171, 174, 202, 210, 211, 212, 213, 214, 215, 218, 219, 221, 228, 232
Craig, H., 202
Crooks, J. B., 24
Crosley, D., 65
Cross, J. B., 24
Cross, R. J., 35
Crutzen, P. J., 194, 202, 208, 223, 228, 231
Curtis, A. R., 205
Cvetanovic, R. J., 30, 149, 150, 172, 176, 179

Dagdigian, P. J., 13, 15, 54, 68
Dai, H.-L., 81
Dalby, F. W., 133, 134
Dalgarno, A., 51, 52, 62, 69
Dallarosa, J., 108
Damon, E. J., 219
Danen, W. C., 98, 100, 102, 110
Danon, J., 113
Darley, E. G., 230
Darnall, K. R., 138, 143, 178, 218, 220, 230
Datz, S., 16
Davenport, J. E., 116, 181
Davidovits, P., 18
Davidson, E. R., 44
Davidson, J. A., 209
Davis, D. D., 142, 151, 159, 163, 174, 177, 178, 180, 181, 185
Deford, D. D., 110
Degan, G., 123
Degens, E. T., 199, 202, 225
de Haas, N., 138, 148, 163, 168
Dehaven, J., 18
Dehmer, P. M., 19
Del Greco, F. D., 149
Demerjian, K., 221, 232
DeMore, W. B., 166, 210, 217
De Pristo, A. E., 54, 67, 68, 69
Derwent, R. G., 156, 161, 171, 174, 202, 205, 212, 213, 214, 221, 223, 232, 233
Deutsch, T. F., 78, 79
Dever, D. F., 86
De Vries, P. L., 45
Dickson, L. W., 26
Diebold, G. J., 15
Dietz, T. G., 119, 133
Dimoplon, G., 33
Dimpfl, W. L., 9, 143
Din, M., 171
Ding, A. M. G., 26
Discherl, R., 18
Dispert, H. H., 7
Dixon, D. A., 16
Dixon-Lewis, G., 171
Doljikov, V. S., 100, 110, 113

Donahue, T. M., 202
Donnelly, V. M., 113, 119, 131, 132
Donovan, J. T., 158
Donovan, R. J., 117, 119, 128, 131, 174
Doty, R. M., 184
Dove, J. E., 162
Dows, D. A., 83
Doyle, G. J., 143
Driver, R. D., 8
Drouin, M., 109
Dryer, F. L., 137, 187
Drysdale, D. D., 137, 153, 171
Dubois, L. H., 18
Ducloy, M., 67
Duff, J. W., 44
Duley, W. W., 123
Duncan, M. A., 119, 133
Duperrex, R., 82, 93
Durana, J. F., 151
Durant, J. L., jun., 97
Duren, R., 8
Durkin, A., 34

Eberly, J. H., 122
Eckstein, J. N., 123
Edney, E. O., 215
Eggleston, J., 108
Eggleton, A. E. J., 174, 204, 205, 211, 232, 233
Ehhalt, D. H., 202, 223, 227
Eland, J. H. D., 133
El-Sayed, M. A., 127, 133
Ely, D. J., 8
Endo, H., 156
Engelhardt, R., 8
Engelke, F., 7, 8, 14, 15, 24
English, T. C., 51
Ennen, G., 56
Erdweg, M., 26
Erlandson, A. C., 125
Erler, K., 168
Ernst, J., 141
Estler, R. C., 7, 8, 15
Evans, D. K., 82, 110
Evenson, K. M., 148, 159, 166, 171, 174

Fair, R. W., 146, 167
Fairchild, C. E., 209
Faist, M. B., 44
Fano, U., 54
Farrar, J. M., 2, 42
Faubel, C., 187
Faubert, D., 135
Feld, M. S., 67
Feldmann, B. J., 80
Feldmann, D., 113
Feldman, D. L., 10, 133
Fenimore, C. P., 171, 172, 188
Fenn, J. B., 2
Ferguson, A. I., 123
Field, D., 168

Field, R. W., 66
Filip, H., 77
Filippov, E. P., 110
Filseth, S. V., 113, 115
Findlayson-Pitts, B. J., 232
Fischer, H., 178
Fischer, R. A., 80
Fischer, S., 151, 159, 177, 181
Fischer, T. A., 93
Fisher, E. R., 143
Fishman, J., 194, 223
Fisk, G. A., 25
Fitz, D. E., 44, 52
Fleming, R. H., 192
Flicker, H., 77
Flygare, W. H., 66
Flynn, G. W., 66, 87, 92
Fonk, R. J., 134
Fotakis, C., 117, 119, 128, 131
Fournier, J., 115
Fox, K., 77
Franck, J., 58
Frankel, D. S., 67, 79
Freed, K. K., 42
Freedman, A., 14, 15, 16
Freedman, P. A., 124
Freeman, R. R., 134
Freund, S. M., 77, 80, 105, 109, 110
Friend, J. P., 202, 227
Friichtenicht, J. F., 18
Fristrom, R. M., 158, 171, 172
Fullstone, M. A., 112
Furzikov, N. P., 110
Fuss, W., 78, 83

Galbally, I. E., 223
Galbraith, H. W., 76, 77
Gallagher, J. W., 122
Gallagher, T. F., 51
Garcia, D., 104
Gardiner, W. C., 149, 154, 155
Garland, J. A., 203, 204, 205
Garraway, J., 174
Gauthier, M., 109, 110
Gay, B. W., 232
Geddes, J., 31
Gegenbach, R., 48
Gehring, M., 162
Geilhaupt, M., 94
Gentry, W. R., 10, 32
George, T. F., 45
Georges, A. T., 134
Gerber, W. H., 44
Ghormley, J. A., 167, 210
Gies, M. W., 7
Giese, C. F., 10
Gilbert, J. R., 140, 147
Gilbert, R. G., 41
Gilchrist, A., 190
Glass, G. P., 143, 144, 156, 158, 222
Glassman, I., 137, 187
Glatt, I., 102

Göppert-Mayer, M., 120
Golde, M. F., 119
Golden, D. M., 80, 89, 116, 149, 230
Gole, J. L., 17
Gompf, F., 6
González Ureña, A., 41
Goodman, L., 133
Goodman, M. F., 82, 83
Gordon, R. G., 58
Gordon, S., 141
Gorokhov, Yu. A., 73, 76, 83, 110
Gorry, P. A., 4, 27, 28, 29, 44
Goss, L. P., 113, 130
Gottscho, R. A., 66
Gouedard, G., 54
Govini, G., 123
Gower, M. C., 80, 110
Graham, R. A., 212, 213, 218, 219
Graham, R. E., 31, 140, 147
Granat, L., 202, 277
Granneman, E. H. A., 134
Grant, E. R., 77, 87, 91
Gray, D., 141
Green, R. J., 184
Greiner, N. R., 149, 171, 173, 178, 221
Grice, R., 2, 4, 6, 11, 12, 27, 28, 29, 33, 34, 41, 44, 45
Griffiths, P. R., 151
Grigorovich, S. L., 110
Grosser, A. E., 31
Grunwald, E., 86, 104
Guillory, W. A., 99, 113
Gupta, A., 97
Gutman, D., 31, 99, 116, 140, 147

Haas, Y., 97, 124
Haberland, H., 23
Hack, W., 144, 148, 157, 161, 178, 187, 210
Hackett, P. A., 109, 110
Haerten, R., 48, 49
Hahn, J., 202
Hall, J. L., 123
Hall, R. B., 83, 89, 110
Hallberg, R. O., 202, 227
Hallsworth, R. S., 73
Halpern, J. B., 82, 96, 124, 126, 127
Ham, D. O., 78
Hamilton, E. J., 210
Hampson, R. F., 208
Hanazaki, I., 31
Hancock, G., 77, 80, 82, 83, 84, 94, 95, 96, 110, 112 113, 115
Handy, B. J., 145, 152, 185
Hanle, W., 61
Hansch, T. W., 123
Hansen, D. A., 151, 184
Hanst, P. L., 202, 215, 223

Happer, W., 54
Hardaker, M. L., 95
Harker, A. B., 150
Harris, G. W., 166, 167, 185, 210, 218
Harris, R. A., 117
Harrison, R. G., 105
Hartford, A., jun., 100, 102
Hartig, W., 123
Hartmann, J., 78
Hatzenbuhler, D. A., 56
Hayashi, S., 5
Haydon, S. C., 124
Hefter, U., 8
Heicklen, H., 217
Heicklen, J., 141, 142, 156, 161, 176, 209, 218, 219, 220
Heidner, R. F., 209
Heidt, L. E., 202, 223
Helmcke, J., 123
Hendry, D. G., 181, 221, 230
Hennessy, R. J., 32, 77, 80
Henri, J. P., 176
Henry, R. J. W., 52
Henschkel, J., 26
Hepburn, J. W., 3
Hering, P., 8
Herm, R. R., 14, 15, 16
Herman, I. P., 82, 110
Herrero, V. J., 41
Herrmann, A., 10, 134
Herrmann, J. M., 5, 32
Herron, J. T., 30, 91, 99, 147, 170, 217, 221
Herschbach, D. R., 1, 2, 3, 13, 16, 24, 25, 29, 33, 35, 36, 40, 41, 45, 54
Hertel, I. V., 8
Hicks, K. W., 113
Hillier, I. H., 44
Hirschy, V. L., 6
Hoare, D. E., 179, 186
Hobson, J. H., 29
Hochanadel, C. J., 167, 210
Hochstrasser, R. M., 124
Hodgeson, J. A., 170, 217
Hoffbauer, M. A., 10
Hofmann, H. H., 8
Hogan, P. B., 134
Hohla, K., 119
Holland, R. F., 77
Hollis, M. J., 134
Holmlid, L., 42
Holt, P. M., 156, 161, 171, 174, 214
Holtom, G. R., 121
Hopkins, D. E., 171
Hoppe, H. O., 8
Horowitz, A., 216
Horsley, J. A., 83, 110
Houston, P. L., 66, 110
Howard, C. J., 148, 159, 166, 171, 174, 177, 209, 213
Howard, R. E., 45

Hoyermann, K., 148, 161, 162, 176, 178
Hsu, D. S. Y., 106
Hucknall, D. J., 141, 172
Hudgens, J. W., 22, 91, 97, 113, 130, 144
Hudson, J. W., 80, 105
Hui, K. K., 107
Huie, R. E., 30, 91, 170, 217
Hull, R. E., 221
Husain, D., 209
Hwang, R. J., 110

Innes, K. K., 124
Inocencio, M., 219
Isaksen, I. S. A., 208
Isenor, N. R., 73, 134

Jackson, D., 161
Jackson, G. E., 219
Jackson, W. M., 126, 127
Jacobs, M., 68
Jacox, M. E., 154
Jalenak, W. A., 109
James, H., 186
Jang, J. C., 99
Japar, S. M., 143
Jasinski, J. M., 81, 97
Jennings, D. A., 209
Jensen, C. C., 77, 100
Jensen, R. J., 73
Jeong, K. M., 220
Jeyes, S. R., 57, 59, 60
John, P., 105
Johns, J. W. C., 66
Johnson, B. W., 134
Johnson, M. A., 13
Johnson, M. W., 192
Johnson, P. M., 124, 133
Johnson, S. A., 8
Johnson, S. G., 16
Johnston, H. S., 141, 212, 213
Johnston, H. W., 159
Jones, B. M. R., 211
Jones, G. W., 171, 172
Jones, I. T. N., 209
Jones, J. D., 134
Jortner, J., 76, 77, 97
Jourdain, J. L., 148
Junge, C. E., 189, 202

Kahn, L. R., 44
Kaldor, A., 83, 89, 110
Kalelkar, A., 170
Kan, C. S., 218, 219
Kanofsky, J. R., 140, 147
Karlov, N. V., 78
Karny, Z., 7, 97, 131
Kasai, T., 31
Kastler, J. A., 54
Katayama, M., 99, 100
Katô, H., 59, 60
Kaufman, F., 144, 149, 151, 157, 162, 163, 171, 180, 210, 220

Keck, J., 170
Keehn, P. M., 104
Keeling, C. D., 202
Kelley, P. J., 122
Kempe, S., 199, 202, 225
Kendrick, J., 44
Kenley, R. A., 181, 221
Kerr, J. A., 171, 214, 221, 232
Ketley, G. W., 84, 112, 113
Ketner, P., 199, 202, 225
Khedim, A., 223
Khokhlov, E. M., 78
Kiang, C. S., 205
Kidson, J. W., 193
King, D. S., 82, 85, 91, 93, 96, 112
King, J., 77
Kinsey, J. L., 8, 9, 29, 44, 62, 143
Kirby, R. M., 145
Kirsch, L. J., 26
Kittnell, C., 123
Klais, O., 220
Klewer, M., 134
Kligler, D. J., 108, 123, 124
Klimek, D., 3
Kneba, M., 107
Knyazev, I. N., 76, 101, 110, 135
Ko, M. K. W., 228
Kolodner, P., 80
Kompa, K. L., 83, 119
Koren, G., 85, 110
Koster, D. F., 102
Kouri, D. J., 69
Kowalski, J., 123
Koyama, T., 202
Krajnovich, D. J., 77, 79
Kramer, G. M., 83, 110
Kramer, K. H., 48
Krasnec, J. P., 202, 223
Krause, H. F., 16, 34, 116
Kudryavtsev, Y. A., 101, 110
Kudszus, E., 209
Kuebler, N. A., 133
Kung, A. H., 81
Kunis, S., 209
Kuntz, P. J., 44
Kurylo, M. J., 142, 164, 185, 202, 218, 219, 220
Kurzel, R. B., 56, 58
Kuwata, K., 31
Kuz'mina, N. P., 110
Kuz'mina, Y. A., 101
Kwan, L. N. Y., 26
Kwei, G. H., 8, 29, 41
Kwok, H. S., 77, 79

Lambropoulos, P., 117, 134
Langford, A. O., 157, 159
Lantzsch, B., 24
Larsen, D. M., 76, 77
Larsen, R. A., 3
Lawley, K. P., 117

Lawrence, G. M., 209
Lazzara, C. P., 188
Leary, K. M., 110
Le Bras, G., 148
Le Breton, P. R., 2, 24, 25
Lee, H. U., 15, 18
Lee, J. H., 145
Lee, S. A., 123
Lee, S. M., 110
Lee, Y. T., 2, 3, 4, 24, 25, 41, 42, 77, 79, 87
Lehmann, J. C., 54
Lehmann, K. K., 133
Leighton, P. A., 189, 202
Leite, J. R. R., 67
Lengel, R. K., 10, 133
Leo, R., 105
Leroi, G. E., 56
Leseicki, M. L., 99, 113
Lester, W. A., 45
Letokhov, V. S., 73, 76, 80, 83, 100, 101, 109, 110, 113, 135
Leu, M. -T., 159, 213
Leutwyler, S., 10, 134
Levenson, M. D., 134
Levin, J., 224
Levine, R. D., 52
Levy, H., 189, 214
Levy, M. R., 2, 10, 32, 45, 94
Lewis, F. D., 101
Lichtin, D. A., 10, 133
Liebl, K. H., 225
Light, G. C., 144
Lightman, A. J., 10, 134
Lin, C. L., 159
Lin, C. -S., 126
Lin, C. T., 106
Lin, M. C., 113, 130, 144
Lin, S. M., 12
Lin, S. T., 97, 110
Lingenfelter, R. E., 223
Lipmann, H., 145
Lissi, G., 141
Liu, K., 3, 16
Liverman, M. G., 10, 119
Lloyd, A. C., 137, 138, 143, 178, 218, 220, 222, 230, 232
Lobka, V. V., 76
Loesch, H. J., 24
Loewenstein, R. M. J., 105
Loewenstein-Benmair, R. M. J., 100
Logan, J. A., 34, 225, 228
Lokhman, V. N., 100, 113
Lompre, L. A., 134
Loree, T. R., 119
Los, J., 134
Louis, J. F., 195
Love, R. L., 32
Lovelock, J. E., 174, 202, 232
Lu, R. R., 210
Lubman, D. M., 42
Lucas, D., 140, 147
Lübbert, A., 3

Luntz, A. C., 24, 31
Lurie, J. B., 127
Lussier, F. M., 79, 100, 103
Lyman, J. L., 73, 77, 80, 91, 107, 110

McAlpine, R. D., 82, 110
McCaffery, A. J., 55, 57, 59, 60, 65
McClain, W. M., 117, 121
McClelland, G. M., 35, 40, 54
McClusky, F. K., 82, 110
McCormack, J., 60, 65
McDonald, J. D., 2, 22, 24, 25
McDonald, J. R., 113, 119, 128, 131, 132
McDonald, R. G., 26
McDowell, R. S., 77
McElry, M. B., 228
McFarland, M., 155
McGarvey, J. A., 125
McGuire, P., 69
Machado, G., 174
Machado, U., 180
McIntosh, A. I., 124
McKellar, A. R. W., 66
McKenzie, A., 144
McKillop, J. S., 46
McLaughlin, D. T., 185
McLean, A. D., 45
McPhail, S. M., 41
McQuigg, R. D., 217, 218, 219
Madden, P. A., 60
Maeda, Y., 134
Magee, J. L., 33
Maggs, R. J., 202
Mahnen, G., 171
Mainfray, G., 134
Makarov, G. N., 73, 76, 83, 97, 110
Maker, P. D., 140, 151, 215
Malanin, Yu. A., 110
Mallard, W. G., 155
Malpole, J. N., 77
Mancy, K. H., 202
Mangir, M. S., 94, 97, 113, 115
Mani, S. A., 131
Manning, R. G., 82
Manos, D. M., 17, 23
Mantz, A. W., 151
Manuccia, T. J., 79, 106, 107
Manus, C., 134
Marcus, R. A., 41
Margitan, J. J., 144, 162, 171
Mariella, R. P., 24
Marks, T. J., 110
Marling, J. B., 82, 110
Maroulis, P. J., 228
Marsden, D. G. H., 2
Martin, M., 117, 119, 128
Marx, B. R., 123
Mascord, D. J., 44
Mason, B. J., 204
Masui, T., 31

Mathur, B. P., 10, 12, 134
Matsumoto, J. H., 144
Matus, L., 26
Maxson, V. T., 24
Mayer, T. M., 5, 16, 44
Meagher, J. F., 161, 176
Measures, R. M., 134
Megaw, W. J., 205
Meier, K., 113, 115
Meier, U., 94
Menzinger, M., 14, 18, 34
Merchant, V. S., 73, 96
Merritt, J. A., 110
Messing, I., 113, 115
Meszaros, E., 228
Mex, G., 148, 157
Michael, J. V., 145
Michaelis, W., 134
Midtbo, K. H., 208
Mikuni, H., 99
Miller, C. M., 94, 95
Miller, D. R., 4
Miller, M. F., 202
Miller, S. S., 110
Miller, W. B., 13
Miller, W. H., 45
Milligan, D. E., 154
Milne, G. S., 186
Mims, C. A., 14, 34
Minshull-Beech, J. P., 112, 113
Mirza, M. Y., 123
Mochizuki, T., 134
Moerkerken, H., 49
Molin, Yu. N., 77, 107
Molina, L. T., 211
Molina, M. J., 211
Monchick, L., 68, 69
Monkhouse, P. B., 143
Monts, D. L., 10
Moon, P. B., 10
Mooradian, A., 77
Moore, C. B., 76, 81, 107, 109
Moorgat, G. K., 209, 216
Morikawa, M., 134
Morinaga, K., 149, 155
Morley, C., 149
Morris, E. D., jun., 147
Morris, J., 205
Morrison, R. J. S., 91
Moulton, P. F., 77
Movsesyan, M. E., 123
Movshev, V. G., 135
Mrzowski, S., 58
Muckerman, J. T., 44
Müller, G., 48, 49
Mukamel, S., 74, 76
Mulac, W. A., 141
Mulcahy, M. F. R., 144, 215
Munslow, W. D., 98
Murphy, J. T., 10
Murphy, R. E., 160

Naaman, R., 42, 85, 131
Naegeli, D., 137

Nagai, K., 99, 100
Nakane, H., 31
Neoh, S. K., 3
Nereson, N. G., 77
Neumann, R., 123
Neusser, H. J., 133
Newell, R. E., 193
Newton, K. R., 5
Ng, C. Y., 4
Nieman, G. C., 133, 135
Niemann, J., 97
Nienhuis, G., 123
Niki, H., 140, 143, 147, 151, 167, 215
Nilsson, L. C., 110
Ninoyan, Z. O., 123
Nip, W. S., 162, 174, 178
Nitz, D. E., 134
Nitzan, A., 97
Nogar, N. S., 109
Norris, A. C., 171
Nowak, A. V., 110
Nowikow, C. V., 6, 27, 28, 29
Nutt, G. F., 124
Nygaard, K. J., 134

O'Brien, R. J., 184
Ogren, P. J., 167
Oka, T., 66
Okon, M., 110
Oldenborg, R. C., 119
Ollison, W. M., 35
Olschewski, H. A., 149
Olsen, D. B., 154
Olszyna, K. J., 104
O'Neill, G. M., 31
Ono, Y., 32
Oppenheim, U. P., 85, 110
Orel, A. E., 45
Orenstein, A., 113
Orr, B. J., 109
Osif, T. L., 142
Ottinger, C., 56
Overend, R., 142, 149, 167, 174, 178
Owen, P. J., 210
Owens, J. H., 155

Pace, S. A., 11
Pack, R. T., 52, 69
Panfilov, V. N., 77, 107
Pang, H. F., 11
Papp, J. F., 188
Paraskevopoulos, G., 142, 149, 167, 172, 174, 178
Parker, D. H., 133
Parker, G. A., 52
Parkes, D. A., 214
Parkinson, P. E., 145, 185
Parks, E. K., 33
Parks, J. H., 99
Parr, T. P., 14, 15, 16
Parrish, D. D., 16, 29
Parson, J. M., 16, 41

Pasquill, F., 193
Pasternack, L., 15, 113, 132
Pastrana, A. V., 149, 178
Patch, D. F., 4
Pate, C. T., 219
Patel, M., 179
Patrick, K. G., 212
Paul, W., 48
Pauly, H., 8
Payne, W. A., 145
Peacock, G. B., 186
Peeters, J., 171, 172
Penkett, S. A., 211, 228
Penzhorn, R. D., 147
Perry, D. S., 26
Perry, R. A., 156, 161, 163, 164, 168, 177, 178, 181, 185, 218
Person, W. B., 77
Pessine, F. B. T., 106
Peterson, N. C., 82
Petty-Sil, G., 133
Phillips, L. F., 157
Phillips, W. D., 8
Pilon, R., 109, 110
Pippin, H. G., 46
Pitts, J. N., jun., 138, 143, 151, 156, 161, 163, 164, 165, 168, 177, 178, 181, 184, 185, 212, 213, 218, 219, 220, 230, 232
Placzek, G., 56
Plaistowe, J. C., 187
Plumb, I. C., 215
Pobo, L. G., 19
Polacco, E., 123
Polanyi, J. C., 3, 11, 19, 20, 21, 23, 26, 43, 44, 46, 54, 92, 143
Pollack, E., 41
Pollock, W. H., 202, 223
Pope, W. M., 12
Popov, A. K., 123
Poppe, D., 65
Porter, R. A. R., 31
Potter, A. E., 143, 160
Prather, M. J., 228
Prengel, A. T., 18
Preses, J. M., 87
Preston, R. K., 8
Preuss, A. W., 144, 210
Preuss, D. R., 17
Prior, M., 49
Pritchard, D., 123
Pritchard, D. E., 8
Pritchard, D. K., 146
Proch, D., 77, 83
Proctor, A. E., 12
Pruss, F., 140, 147
Pryce, M. H., 133
Pummer, H., 108, 124
Puretzky, A. A., 76, 83, 97, 110

Quack, M., 77, 83
Quelly, T. J., 109

Quick, C. R., jun., 82, 87, 92, 93
Quigley, C. P., 82

Rabinowitz, P., 83, 110
Rabitz, H., 67
Radford, H. E., 166
Radlein, D. St. A. G., 2
Ramsey, N. F., 49
Rasmussen, R. A., 202
Ravishankara, A. R., 142, 157, 159, 163, 178, 185
Rawlins, W. T., 155
Reck, G. P., 10, 12, 134
Redpath, A. E., 34
Reihl, J. L., 62
Reilly, J. P., 119
Reinhardt, G. W., 219
Reiser, C., 77, 100
Reisler, H., 94, 96, 97, 113, 115
Rettner, C. T., 10, 32, 119
Reuss, J., 49, 50, 70
Rhodes, C. K., 108, 122, 123, 124
Rice, S. A., 41
Rice, W. W., 119, 130
Richardson, D. J., 171
Richardson, M. C., 73
Richardson, T. H., 99
Riley, C., 78
Riley, S. J., 40
Ringer, G., 32
Rink, J., 73
Ripley, D. L., 149
Ritter, J. J., 91, 105, 109
Robbins, R. C., 202
Roberts, A. J., 94
Roberts, C. S., 52
Roberts, R. E., 70
Robertson, L. C., 110
Robin, M. B., 133
Robinson, C. P., 73
Robinson, E., 202
Rockwood, S. D., 73, 80, 91, 105, 119
Rodhe, H., 202, 227
Roffey, M., 221
Rohmeld, M., 66
Rohwer, P., 23
Rommel, M., 13
Ronn, A. M., 97, 110
Rosenfeld, R. N., 81, 89, 97
Ross, J., 34, 70
Rossi, M. J., 80, 116
Rost, K. J., 8
Rothe, E. W., 10, 12, 134
Rotzoll, G., 3
Rousseau, D. L., 124
Rowe, M. D., 55, 57, 59, 60
Rusin, L. Y., 24
Ryabov, E. A., 73, 109, 110
Ryan, K. R., 215
Rynefors, K., 42

Sadowski, C. M., 113, 115, 209
Safron, S. A., 13, 25
Salas, L. J., 202
Sam, C. L., 126, 130
Sampson, R. J., 141, 172
Sanchez, A., 67
Sandalls, F. J., 228
Sanhueza, E., 218, 219
Santoro, R. J., 187
Sapondzhyan, S. O., 123
Sarkisian, A. A., 110
Sarkisyan, G. S., 123
Sartakov, B. G., 78
Sathyamurthy, N., 43
Saunders, M., 35
Savage, C. M., 140, 151, 215
Sawyer, R. F., 149
Schatz, G. C., 43
Schearer, L. D., 134
Schek, I., 77
Schenck, P. K., 112
Schiff, H. I., 209
Schiff, R., 151, 177, 178, 185
Schimitschek, E. J., 125
Schinke, S. D., 211
Schlag, E. W., 133
Schmeltekopf, A. L., 209
Schmid, W. E., 78
Schmidt, T., 23
Schmidt, U., 202, 225, 227
Schmidt-Bleek, F. K., 16
Schmiedl, R., 94, 113
Schott, G. L., 149
Schreiber, J. L., 21, 26, 44
Schroder, H., 77
Schügerl, K., 3
Schultz, A., 13, 14
Schultz, M. J., 80
Schulz, P. A., 77, 79, 87
Schumacher, E., 10, 44, 134
Schurath, U., 145, 219
Schwartz, H. L., 50
Scott, W. E., 230
Scribner, E., 202
Seiler, W., 202, 223, 225
Seligson, D., 67
Serri, J. A., 8
Setser, D. W., 98, 99
Shapiro, M., 44
Shaub, W. M., 98
Shaw, M. J., 120
Sheen, S. H., 33
Shen, Y. R., 77, 79, 87
Sheng, S. J., 133
Sheppard, D. H., 219
Shibanov, A. N., 113
Shigeishi, H., 202
Shimazu, M., 134
Shimoni, Y., 69
Shobatake, K., 41
Shoemaker, R. L., 66
Sibener, S. J., 4
Sickin, E. R., 92
Sidebottom, H. W., 219

Sie, B. K. T., 156, 217
Siegel, A., 14
Sievert, R., 176
Silver, J. A., 8, 143
Silvers, S. J., 66
Simmons, R. F., 179
Simonaitis, R., 142, 156, 209, 217, 218, 219
Simons, J. P., 10, 32
Singh, H. B., 202
Singh, T., 149
Singleton, D. L., 172, 174, 179
Siska, P. E., 40, 42
Skrlac, W. J., 19, 21
Slagle, I. R., 116, 140, 147
Slater, R. C., 99
Sloan, J. J., 11, 26, 143
Sloane, T. M., 140, 147
Smalley, R. E., 10, 119, 133
Smets, B., 172
Smith, D. J., 34
Smith, F. B., 193
Smith, G., 142, 163, 178, 180
Smith, G. K., 143, 144
Smith, G. P., 13
Smith, I. W. M., 28, 105, 107, 141, 142, 149, 153, 157, 161, 168, 169, 213, 222
Smith, N., 8
Smith, R. H., 144, 164
Smith, S. J., 134
Smolanek, J., 133
Snelson, A., 156
Solarz, R. W., 8
Solomon, I. J., 156
Spence, J. W., 202, 215
Spencer, J., 163
Spencer, J. E., 143, 144, 156
Spiller, L. L., 202
Sprung, J. L., 143, 232
Stacey, F. D., 191
Starov, V., 105
Stauffer, D. J., 205
Stedman, D. H., 147
Steel, C., 105
Steinert, W., 139, 153, 155
Steinfeld, J. R., 56, 58, 66, 67, 77, 79, 100, 103, 109
Stenholm, S., 66
Stephens, E. R., 230
Stephenson, J. C., 82, 85, 91, 93, 96, 112
Stern, A. C., 202
Steven, J. R., 144, 215
Stewart, G. W., 34
Stief, L. J., 145
Stockwell, W. R., 212
Stöcklin, G., 26
Stoicheff, B. P., 123, 124
Stolte, S., 12, 34, 49, 50, 68
Stone, E. J., 209
Stone, J., 82, 83
Storz, R. H., 134
Streit, G. E., 159, 209

Strunin, V. P., 107
Stuhl, F., 151, 154, 161, 164, 167, 178, 217
Stwalley, W. C., 5, 24
Sudbo, Aa. S., 77, 79, 87
Sukr, H., 123
Sunde, J., 208
Suran, V. V., 134
Sverdup, H. V., 192
Sze, N. D., 228
Sze, R. C., 119
Taguchi, R. T., 92
Tai, C., 133, 134
Takacs, G. A., 158, 222
Takahashi, M., 99
Takeyama, T., 149, 155
Takubo, V., 134
Tal, D., 110
Tang, I. N., 169
Tang, S. P., 18
Taylor, O. C., 230
Telle, H., 15
Tesi, G., 163, 178
Thebault, J., 134
Thrush, B. A., 119, 153, 166, 167, 210
Tiee, J. J., 80, 110, 119, 130, 193
Timofeev, V. P., 123
Toader, E. I., 134
Toennies, J. P., 24, 48
Tomkins, F. S., 134
Torres, A. L., 228
Tracy, D. H., 134
Trainor, D. W., 131, 154, 170
Trias, J. A., 125
Troe, J., 149, 168, 207
Truhlar, D. G., 44
Tsang, W., 91, 99
Tsay, W., 78
Tschuikow-Roux, E., 210
Tsien, T. P., 52
Tsou, L. Y., 51
Tuccio, S. A., 100, 102
Tully, J. C., 25, 44
Tumanov, V. A., 73
Tyndall, G. S., 214, 215

Unland, M. L., 66
Utterback, N. G., 18

Valentini, J. J., 3, 24
van den Bergh, H., 82, 93
van den Ende, D., 34
Van der Wiel, M. J., 134
Vandooran, J., 181
van Tiggelen, P. J., 181
Veltman, I., 34
Verma, K. K., 24
Viard, R., 3
Vidal, C. R., 65
Viers, D. K., 123
Vietzke, E., 26

Vigue, J., 54
Villis, T., 80
Vincent, D. G., 193
Volz, A., 223, 227
Von Dijk, C. A., 123
von Rosenberg, C. W., 154, 170
Wagner, H. Gg., 141, 143, 144, 148, 149, 155, 157, 161, 162, 178, 187, 210
Wagner, S., 174, 178, 180, 185
Walker, B. F., 155
Walker, J. A., 99
Walker, R. W., 138, 140, 161, 171, 173, 179, 187
Wallace, S. C., 3, 124
Walsh, A. D., 24
Wampler, F. B., 119, 130
Warneck, P., 209, 216
Wassell, P. T., 26
Watson, D. G., 26
Watson, R. T., 142, 174, 177, 178, 180
Watts, D. M., 202
Waugh, J. S., 62
Way, K. R., 5, 8, 24
Wayne, R. P., 167, 209
Weil, R., 110
Weiner, J., 35
Weinstein, N. D., 25
Weiss, R. F., 202
Weitkamp, C., 134
Weitz, E., 101, 110
Welge, K. H., 82, 83, 94, 96, 110, 113, 124
Welzbacher, H., 143
Went, F. W., 202, 232
Wessel, J. E., 124
West, G. A., 92
Westbrook, C. K., 187
Westenberg, A. A., 138, 148, 153, 163, 168, 171, 172
Weston, R. E., jun., 87, 92
Wexler, S., 19, 33
Wharton, L., 51
Wheatley, S. E., 134
Whitbeck, M. R., 218, 219
White, J. N., 154
Whitehead, J. C., 2, 7, 12, 13, 15, 34
Whitten, G., 141
Wicke, B. G., 18
Wiesenfeld, J. R., 209
Wilcomb, B. E., 15, 16, 44, 54
Wilhelmi, G., 6
Williams, A., 171
Williams, M. R., 202, 221, 232
Williams, P. F., 124
Williamson, A. D., 133
Willis, C., 109, 110
Wilson, E. B., 66
Wilson, W. E., jun., 137, 153, 158, 171
Wine, P. H., 157, 159

Winer, A. M., 138, 143, 178, 212, 213, 218, 219, 220, 230
Winkler, K., 123
Winterfield, C., 80
Witt, J., 8
Wittig, C., 80, 82, 87, 92, 93, 94, 96, 97, 113, 115
Wöste, L., 10, 134
Wofsy, S. C., 228
Wolfrum, J., 107, 162
Wood, R. W., 58
Wood, W. P., 218, 220
Woodall, K., 54
Woodin, R. L., 81
Worley, S. D., 143, 160
Worry, G., 41
Wren, D. J., 14
Wright, D. C., 134

Wright, J. S., 26
Wu, C. H., 143
Wu, C. Y. R., 24
Wu, K. T., 11

Yablonovitch, E., 74, 76, 79, 80
Yahav, G., 97
Yakozeki, A., 18
Yamada, F., 116
Yamanaka, C., 134
Yang, S. C., 5, 24
Yardley, J. T., 126, 130
Yardley, R. N., 11
Yogev, A., 100, 102, 105, 110
Young, C. E., 19
Yu, M. H., 94, 97, 113
Yuan, J. M., 45

Zacharias, H., 94, 124
Zahniser, M. S., 157, 163
Zakheim, D., 124, 133
Zandee, L., 10, 12, 119, 133
Zapesochnyi, I. P., 134
Zare, R. N., 7, 8, 10, 13, 14, 15, 42, 54, 85, 94, 95, 97, 131, 133
Zeegers, P. J. T., 123
Zeiri, Y., 44
Zellner, R., 139, 141, 142, 143, 153, 154, 155, 157, 161, 168, 222
Zetsch, C., 154, 217
Zimmerman, P. R., 223
Zitter, R. N., 102
Zu Putlitz, G., 123